JOHN FOX BERSHOF, MD

THE FIRST HISTORY OF MAN

Book cover design by Estee Fox Bershof

Editor: Chantel Hamilton
Copy-Editor: Jacklyn Arndt

Acknowledgements: Eva Nix, Janelle, and Family

Library of Congress
John Fox Bershof, MD
The First History of Man / John Fox Bershof

Library of Congress Control Number: 2020924883

ISBN 9798569112470

Independently Published

Printed in the United States of America

To mother and father

CONTENTS

INTRODUCTION

When I was a knucklehead kid, I never really understood the purpose of the introduction to a book, especially if it was a book either I had to read for school, or I knew I was going to read anyway. And the longer the introduction was, the less sense it made to me. I often found myself just skipping it completely. Then it dawned upon me later in life—especially as I undertook writing this book on everything from miasmas and the first microscope to the bark of the cinchona tree and gin & tonics—that perhaps an introduction serves a purpose: to give an overview of the book's structure and topics, and to invite the reader to actually read the thing. I'm a slow learner, but I do learn. Another thing I learned happened in the doctor's lounge one day several decades ago—a moment of inspiration that took me down a long road that, many detours later, eventually brought me to writing this book.

It was a typical crisp fall day in the late 1990s. I was in between surgeries, hanging out in the surgeon's lounge, munching on donuts, avoiding the hospital's version of coffee—a bitter brew, to be sure—and listening to the chatter of the other surgeons with the dim drone of the morning news from the wall-mounted TV humming in the background. The big table at the center of the surgeon's lounge was scattered with the local newspaper and several national dailies, an abandoned half-eaten bagel with cream cheese, an equally abandoned half-drunk Styrofoam cup of coffee, left by a surgeon no doubt summoned to the operating room in a hurry—or maybe just too lazy to clean up after themselves. And, important to the purposes of this introduction, a *Wall Street Journal* was spread out across that table.

I sat there with my donut (okay, maybe two donuts), skimming over articles in the *WSJ*, when one story in particular caught my eye. It was about a young man still in his twenties worth a billion dollars on paper for having started a medical information website. It would be an understatement to say I was taken aback. Here was a kid who hadn't even graduated college, let alone gone to medical school, serving up medical information to the general public—standalone articles he hadn't even written, medical information I wasn't sure he was entitled to publish, collated into one website. His name was Jeff Arnold and his company was called WebMD.

I didn't know it at the time, didn't appreciate it until several years later, that Arnold was pretty smart. He had a vision: a business model much greater than just serving up health information. But that is not what I understood that morning, chomping on my favorite sugar donut, or maybe it was glazed, as I read the article. I said to myself, "If this Arnold character can be worth a billion on paper serving up health articles, maybe I could do something similar and be content with a fraction of that wealth." Because it wasn't really about the money—it was about doing something I thought I could be good at (other than being a surgeon, of course). What I wanted to do was write a textbook of medicine for the general public that tore down the veils of all its (sometimes) complex concepts and broke them down into a flow that was inviting, fleeced of the normal terror-inducing jargon that often accompanies medical and science narratives. I'm not sure how many arrows are in my quiver, but I'd like to think I have at least three: an arrow for surgical skills, an arrow for distilling complex, impenetrable concepts into manageable, sometimes whimsical narratives, and perhaps an arrow for storytelling.

I thought about that *WSJ* article for days and days, weeks and weeks, until one day I finally made the decision that, yes, I would jump into the medical info game. After all, at that time during the latter half of the 1990s the dot-com world was booming. And I mean *booming*. After doing some research, I soon realized that several other internet health information database companies were out there doing the same thing as WebMD—serving up disease-by-disease articles that they, too, had not written. In some cases, competing health information platforms served up the exact same articles. To make matters worse, at least for me, those articles weren't even written by physicians but rather by health care writers. That's not to say a trained health care writer can't write better prose than a physician—they probably can—but, still, it rattled me all the same. Where were all the experts?

Medicine in the waning years of the twentieth century wasn't the only discipline being rocked by the possibilities of the Web. The late 1990s was an incandescent period for the internet. Simply meteoric. Anyone and everyone had a chance to stake their claim in any foothold they could find as the Information Age unfolded. That period is known historically as the "dot-com bubble," when nearly anyone with a halfwit idea was making a bucketload of money—at least on paper—simply by creating an information-based website. But, in the end, many dot-com companies didn't have much value to them. Their entire monetary worth was based on nothing more than the fact they had a million email addresses in their database or a million unique users, which the industry termed "eyeballs." Back then, a website could offer up stuff for free—like, say, digital greeting cards to send over the internet—all while their sole prized possession was the email database they had built. With little in the way of a revenue stream, these internet startups, initially thought to be golden, were doomed from the outset.

One of the preludes to the dot-com bubble was the 1993 release of Mosaic, a precursor Web browser to such behemoths as Google. Mosaic allowed chumps like me to jump on and surf the internet by landing on websites that Mosaic could "read." It and subsequent browsers altered the dynamics of the "digital divide"; that is, with smarter browsers and smarter computers, everyone could get in on the action and enjoy the internet, a degree in computer science no longer required. Between 1995 and 2000, the NASDAQ Composite index rose 400 percent. Investment companies and venture capitalists were throwing money at internet startups, based on the assumption that if even if just one in ten was successful, they'd make a bundle of cash.

"GRANDPA SAYS HE USED TO HAVE A MAILBOX ON THE SIDE OF HIS HOUSE, AND SOMEBODY CALLED A MAIL CARRIER USED TO ACTUALLY PUT MAIL INTO IT."

Perhaps the quintessential example of an overvalued internet company was the pioneering darling of the industry: America Online, a.k.a. AOL. The "You've Got Mail" when you launched your digital AOL mailbox was pretty slick—made you feel important, like, you know, you had mail!—and it spawned a spectacular 1998 movie by the same name starring two perfectly matched performers, the always perfect Tom Hanks as Joe Fox and the adorable Meg Ryan as the adorable Kathleen Kelly. Those two had been paired in a previous Nora Ephron directed romcom *Sleepless in Seattle* 1993, a movie in which you don't realize until the end that for the most part, Tom Hanks' character Sam Baldwin and Meg Ryan's Annie Reed astoundingly don't have a scene together until the very last scene, their chemistry so palpable it transcended the distance of time and space. But as for AOL, the company, not the movie, it was so highly (over)valued that it was able to merge with—or rather, should I say, *consume*—the old-guard media conglomerate Time Warner, an event that happened on January 10, 2000. AOL was then and still is a good company, but buying Time Warner with pocket-change (again, *on paper*) was nevertheless something astounding to behold. AOL was started by two brilliant guys, Steve Case and Ted Leonsis, and their platform was not only terrific, but they really did help to usher into life the entire dot-com world. I still have several AOL email accounts—and, no, I'm not a luddite, tethered to caveman technology. But consider this: although my dates might be murky, Time Warner was, when purchased by AOL, a venerable telecommunications company dating back at least to the early 1970s, while the upstart AOL appeared on the scene only in 1991. But this relative newcomer AOL had a bunch of email addresses, a modest monthly subscription option, and eyeball traffic streaming through its platform, buttressing its value enough that it could buy a stalwart company like Time Warner, with proven assets and a long track record of success.

As the '90s came to a close and the world survived the no-show Y2K or millennium bug—internal computer clocks would go berserk switching from 1999 to 2000—things started to turn south for the dot-com world, bug or no bug. Starting in March 2000, Wall Street began making series of adjustments to its valuations as it became clear that a lot of highly prized dot-com companies were running out of investment capital and that their business models—supported mostly by eyeball traffic, user email lists, and online advertising with some primitive e-commerce thrown in—did not include sustainable revenue streams. Surprise, surprise. Too many dot-com companies had no realistic business plan, if they ever had one at all, to justify their once and future value.

By November 2000, a mere ten months after the Cinderella marriage of AOL and Time Warner and only eight months after the highest NASDAQ peak for dot-com companies, internet stocks had dropped 75 percent, wiping out $1.755 trillion dollars in value—and bursting the dot-com bubble. Those investors who had not been lucky enough to have cashed out a year or two before lost a bundle of dough. Or I should say, lost a bundle of dough *on paper*, because it was cash they never actually had.

But for me, bubble or no bubble, as the new millennium began, I was determined to start my own health information website. I was ready to throw my hat into the ring. Originally a boxing phrase describing how a challenger in the audience could signal he wanted to fight the champ standing in the ring by throwing in his hat, nowadays "throwing one's hat in" mostly refers to running for political office. While I was looking to construct my own medical info website and not run for mayor, I knew I had to do something to stand apart from the crowd, just like an upstart politician has to. Otherwise, what I was offering would just be more of the same.

First of all, I was going to write the entire information database myself rather than cull the same health articles all the other platforms were using. As I mentioned earlier, I fancy myself to be a pretty good distiller of medical information, able to parse out the important material and tips that a patient would want to know about any given topic and dispense with the minutiae that too many doctor and health care writers disappointingly include. To mix a metaphor, if you place too many trees in your forest, the reader won't know which trees they should be listening to. If it is the Ents you're listening to, the talking trees in J. R. R. Tolkien's *Lord of the Rings* 1954-55, then I suppose that last line would not be a mixed metaphor.

Most web health articles were and often still are written by science writers, and while they might be good writers, they're not physicians. Since they've never been in the trenches of the hospital, taking medical hand grenades, they can't easily parse between valuable and not-so-valuable information. The issue isn't just health care writers but physician writers too, whose attempt to write articles for the public often fall short, getting lost in detail or, worse, using jargony words without defining them—an unforgivable assumption. It is one thing to write a medical article for a physician audience, but it's an altogether different challenge to author one for the general reader.

Along with extracting all the essential information on each topic and presenting it for a wide audience, my platform was going to link topics within articles so that a person could be reading about one disease, come across another disease, click on it, and go to that other article. Such links, formally termed hyperlinks, are so ubiquitous today, taken for granted, that now I can scarcely remember a time when they were not embedded into every single webpage. But, back then, they weren't so common. What you experienced as a user of early health information platforms (and still sort of do today) was nothing more than simple, lifeless text pages collected under one umbrella URL. They looked like term papers typed on an IBM Selectric typewriter and then digitized. For you younger readers out there, a "typewriter" is what us "OK Boomers" and "Gen Xers" used before computer word processors were invented.

The third thing my database was going to do—and this was the real gem of my plan—was work on cell phones, not desktop computers. Back in the early 2000s, cell phones weren't "smart" so much as "brick stupid." How stupid you ask? Well, cell phones in 2002 couldn't visit a regular internet website, because those sites then (as now) were written in hypertext markup language, better known as HTML. Early cell phones couldn't read this computer language, let alone format the information it conveyed for their teeny tiny user interface screen. Those cell phones could read only their own language, such as wireless application protocol (WAP) and wireless markup language (WML)—and my database was going to speak those languages, too.

I spent several years writing that textbook of medicine and surgery into a database, with every single hypertext link hand designed. When uploaded on my website server and set "live," all

that information became readable by all the cell phone languages of the day and all of it was accessible over a cellular signal, whereas all the other health information websites—all my competitors—were tethered to an ethernet landline computer along with is HTML limitations. I named my health info website skynetMD simply because, untethered from an ethernet landline, the medical subjects came from the "sky" via a cellular signal.

And, for a few years, I did well. Lots of eyeballs went to skynetMD, many of which were attached to people chuffed with the novelty of reading health information on their phone while they sat at the doctor's office, interminably waiting to be seen. It was certainly a better option than reading the ten-year-old waiting room magazines, at least. We have become spoiled by what modern mobile phones can do. Back then, other than basic voice calling, a few ringtone choices, a handful of pixelated video games, and very limited texting capabilities—256-character short messaging service, or SMS, texts—a cell phone couldn't perform very many tasks. Compared to today's mobiles, those now ancient devices appear closer to caveman technology.

I was pleased with what I had accomplished. I had said I was going to enter the medical information internet game, and I had done it—and done it well. But then something happened that I could not have predicted—largely because I was not very smart nor savvy and, ultimately, I was operating like many of the dot-com companies had before their bubble burst: I was flying by the seat of my pants. That something that happened was the "smartphone"—which came along and made my health information website skynetMD, tailored made for cell phones, obsolete.

The first true smartphone was probably Apple's iPhone, released in 2007. Yes, there were already Blackberries and the Palm OS device, which performed as pretty slick nascent smartphones, but the 2007 iPhone is the poster child of the industry—a truly nifty device if there ever was one. Visionary and brilliant techie Steve Jobs, founder of Apple computers in 1977 along with fellow genius Steve Wozniak, better known as The Woz, introduced the iPhone on January 9, 2007 in San Francisco as only Jobs could and did introduce the sleekest of Apple devices. Steve Jobs died way too young, in 2011 age fifty-six from pancreatic cancer. It was also around this time that the industry dropped the "cell phone" moniker and started calling these newfangled devices "mobile phones." Why? Because smartphones could access the internet anywhere, anytime—they were truly mobile—jumping online using either cellular or Wi-Fi signals. Plus, which

doomed skynetMD, they were able to read HTML webpages *and* they could properly format them for their small user interface; the smartphone screens were bigger than the original cell phone screens. These new, mesmerizing devices could also take pictures with a built-in camera, incorporated a touchscreen interface, and were able to download and run a gazillion applications, from YouTube to Google Maps. With the iPhone and devices of similar ilk, a user could surf to any website on the internet, no longer tethered to a desktop or laptop ethernet port. Unsurprisingly, such capabilities put the kibosh on mobile phones that demanded a language other than HTML, and, with a whimper, skynetMD likewise went kaput.

For the better part of the next five years, I busied myself doing what life chose for me to do: be a surgeon. But always in the back of my mind was this nagging idea—like a song you can't get out of your head, a catchy earworm tune (for me, it's the theme song to *M*A*S*H*). An idea, a thing, that wouldn't go away. And that "thing" was what led me to the beginnings of this book. Although I had enjoyed writing that skynetMD textbook of medicine for the general reader, the parts I really savored and reveled in when building skynetMD were all the behind-the-scenes and side stories I had encountered along the way that don't make it into a medical text. The layers of minutiae behind the complex topics that didn't make it into the distillate I had presented in my database. I'm not just talking about the history of medicine, either—I'm talking about all of science and all of history, for that matter. It all flows together, it's all interconnected. And looking at something so human as the practice of medicine makes this incredibly clear.

Take, for instance, the neurodegenerative disease amyotrophic lateral sclerosis, or ALS. It was first described by Jean-Martin Charcot, who was born in Paris in 1825 and who practiced medicine at the renowned Salpêtrière Hospital in that city. Considered the father of neurology, Charcot described ALS in 1869. Yet many of us know ALS as Lou Gehrig's disease, named after the New York Yankees slugger and first baseman

who, as his ALS progressed, gracefully withdrew from the Yankees' lineup on May 2, 1939, to a stunned crowd. He stood there on home plate that beautiful spring day in Yankee Stadium, the "house that Ruth built," and said to the onlookers:

> For the past two weeks you have been reading about a bad break. Yet today I consider myself the luckiest man on the face of the earth. I have been in ballparks for seventeen years and have never received anything but kindness and encouragement from you fans.

Gehrig died from ALS two years later.

When I was piecing together my skynetMD entry for ALS, I was less interested in the disease itself and more interested in Lou Gehrig the man, including, for instance, his lifetime batting average: .340, sixteenth on the all-time career leaders list, a list atop which sits Ty Cobb with .366. Gehrig's famous speech at Yankee Stadium in turn made me curious about baseball stadiums at large, having gone to many MLB games in my lifetime. Unlike football fields, basketball courts, soccer fields, and ice hockey arenas, which all have standardized dimensions, in baseball, the infields are all the same but that elusive home-run wall is different at each stadium. This unusual discrepancy gives baseball a slight wrinkle—which I like. And what about Abner Doubleday, the Union major general credited with inventing baseball in 1839 in Cooperstown, New York, in some cow pasture—no doubt filled with some cow pies—a cattle grazing field that belonged to some guy named Elihu Phinney? Quite the inauspicious start to America's greatest pastime. But this is why the Baseball Hall of Fame is located in Cooperstown—although presumably not in a cow pasture. Along with being the father of baseball, Doubleday is also credited with firing the first shot at Fort Sumter in South Carolina—the shot that ushered in the American Civil War.

For those too young to know of Lou Gehrig, you're probably familiar with one of the most recognizable people in the world for the last two, maybe three, decades: Stephen Hawking. He lived with ALS for a whopping fifty-five years, diagnosed at age twenty-one in 1963 and sadly passing in 2018 at age seventy-six. How did Hawking do it? How did he live so long with such a debilitating illness when the average lifespan, from diagnosis of ALS to death, is a frighteningly short two to five years? Gehrig was known as the "Iron Horse of Baseball," yet despite his peak physical condition he only lived for two years following his diagnosis. Hawking, on the other hand, while brilliant was no doubt a skinny and probably not so physical adept nerd—his "iron horse" was his nifty motorized/computerized wheelchair—but he lived so much longer after his diagnosis. It's strange, isn't it. We still don't know why, but it's worth pondering. And speaking of pondering, what exactly did Hawking mean by those "strange occurrences" that happen at the "event horizons" of black holes out there in the far reaches of the universe? And can time really run backward?

And what about that famed French neurologist Charcot who first described ALS, peculiarly and improbably, during his lifetime he was more widely known for one special patient he had under his charge than his medical discoveries. Charcot provided for public display—would not fly in today's world of patient privacy and decorum—his female patient Louise Augustine Gleizes, born in 1861 and confined to the mental wing of Salpêtrière Hospital under Charcot's care. She suffered from extraordinary hysteria, and that is putting it mildly. Sadly, her descent into a frenzied madness was partially attributed to a series of unconscionable events that befell her as a child, from brutal beatings at a religious boarding school to molestations and rapes by various men, including her mother's lover. Always on the edge, made more pronounced through hypnosis, Charcot would delve into her unconscious past, all the while onlookers observing, a trance-inducing freakish atmosphere, where Augustine would self-masturbate and simulate acts of copulation. It was like a side show at a traveling circus, but no circus I've ever been to. Among noted observers were

French Impressionist artist Edgar Degas and apparently world-renowned neurologist Sigmund Freud traveled from Vienna to Paris just to witness the sexual delirium of the young, and apparently quite beautiful Augustine. The spectacle all came to an end, fortunately, when in 1880 the 19-year old Augustine dressed like a man escaped the Salpêtrière Hospital mental ward, never to be seen nor heard from again.

All of this is a long way of showing how it occurred to me, several years back, that nearly all of human history can be viewed through the lens of medicine. Sure, this might not be an entirely novel realization. That is, you can take any item or any concept and use it as a jumping off point into the spiraling narratives of history—Mark Kurlansky, for example, did it with table salt in his interesting book *Salt: A World History* (2002), as did the gifted writer Bill Bryson using his Victorian-era rectory-turned-home as the core subject of his splendid book *At Home: A Short History of Private Life* (2010). So it dawned upon me, while having written the numerous medical articles that became skynetMD, that medicine is a perfect cliff from which to take such a dive—or, rather, several such dives.

To this end, the underlying frame of this book is medicine, and in particular human disease. At every turn in the road, it is my desire and hope to impart some understanding of the ailments that plague our species—in as jargon-free way as possible. The *overlying* structure, however, as we travel disease by disease through the pages of this book, are the numerous jumping-off points, from which we will delve into the splendid history, mystery, and serendipity of science, the eddies and currents of human history, in all its splendor and all its tragedy, not to mention dips and dives into all that came before us, when humankind was just a gleam in evolution's eye.

The great and prolific sci-fi writer Stephen King once said, "Kill your darlings, kill your darlings, even when it breaks your egocentric little scribbler's heart, kill your darlings." The translation being: stay on point and mercilessly throttle your little side stories and anecdotes, even if they're cute, because they don't add to, and in fact mostly distract from, the real story. I adore Stephen King, but this book is precisely that, about those darlings. We won't forget about medicine, but you *can* forget that this narrative will stay on a straight and narrow path; we won't be killing these darlings. Instead, we'll journey to the top of Everest and into the center of the Earth, from the beginning of time to the unforgiving time of dogma, collecting every darling—from science, history, art, music, biology, chemistry, physics, math, astronomy, geology, geography, anthropology, archaeology, religion, you name it—we find. We'll cherish all those darlings that make up the human history that remains eternal in all of us.

If the underlying frame of this book is medicine and the overlying structure is the jumping-off points into all our precious darlings, then the overarching roof on top of it all is made up of the personal anecdotes and vignettes threaded throughout its pages. When you're a physician who gets to my age (my precise age necessarily omitted because I can scarcely believe it myself), you've been in the trenches of medicine long enough to have seen a thing or two. These personal stories I hope show how while medical history can reveal the entire world to us, the practice of medicine is ultimately about valuing us each as individuals.

Lewis Carroll's rather famous *Alice in Wonderland* (1865) includes this demand of the White Rabbit, when he is called to the witness stand to report on what he knows about who stole the Queen of Hearts' tarts: "Begin at the beginning and go on till you come to the end; then stop."

Perhaps this is good advice for the White Rabbit and the mystery of the missing tarts, but the thing with medicine is this: it's not really just one simple frame but rather a huge expanse of interlocking frames, all of which have their own innumerable jumping-off points—and so first beginning at the beginning and then going on until the end is far too much to ask of a single book. Such a pursuit needs to be broken down into volumes—which is what I have done. You hold in your hands *The First History of Man*, the initial of a planned eight volumes. And the journey I invite you on in this first volume is this: a foray into the world of infection in general, and bacterial infections in particular, with darling visits with Cleopatra, Oscar Wilde, and a watch-wearing apple out in the emptiness of space along the way, all threaded together with two or three meaningful narratives from my own life. The jumping off adventures awaiting you in the second volume are built on viral, parasitic, fungal and sexually transmitted diseases, with further volumes covering heart and lung diseases, and so on and so forth, through eight volumes, accompanied again by overlying jumping-off points into darlings of science and history and an overarching personal story.

I hope at the turn of every page you are entertained and also learn a thing or two—maybe even a handful of that scary medical jargon stuff—because, quite frankly, before I began taking all the darling high dives that medicine presented to me, I didn't know most of this stuff either.

If I'm the White Rabbit, and if you're Alice, it is high time (or should I say, teatime) to follow me down the rabbit hole.

1 THE PHONE RANG

December 1995—time for the annual school winter break for our children. My wife and I had a routine: every other year, we would stay home and do all the holiday stuff; she'd put on quite a show. Which meant all the other years, we would visit my mother-in-law. Yes, there's a joke in there.

When I was in elementary school a long, long, long, long time ago, it was called Christmas break, not winter break—a direct reminder then that America is largely Christian in origin. "Christmas vacation," in becoming a politically incorrect phrase, went the way of the dinosaurs and was replaced with the religion-neutral "winter break." Christmas is, if you are not Christian and if you look beyond the religious customs, meanings, and relics—that is, if you look the other way—is still a wonderful holiday. As a Jew, I always enjoy Christmas, filled with decorated trees, snow and snowmen, twinkly lights, rather ugly Christmas sweaters, holiday carols (many of which were written by Jewish composers), hot chocolate and eggnog (the latter paired with a splash of rum for the adult version), and classic Christmas films.

I'm the ghost of non-sectarian midwinter public holiday future

At the top of the movie list is the wonderfully perfect *It's a Wonderful Life* (1946), closely followed by *Miracle on 34th Street* (1947), *A Charlie Brown Christmas* (1965), *How the Grinch Stole Christmas* (1966), *National Lampoon's Christmas Vacation* (1989), *Home Alone* (1990), and my personal favorite, the Charles Dickens adaptation *A Christmas Carol*, either the 1951 or 1938 version, starring the zeitgeist transformation of Ebenezer Scrooge.

Then you have all the lit-up shopping districts, whose holiday lights used to go up a week or two before

Christmas, then started creeping up to those frenzied shopping days after Thanksgiving—known as Black Friday—and now, it seems, holiday lights appear even before all the Halloween candy has had a chance to be consumed.

The origin of the term "Black Friday" to describe the shopping day after Thanksgiving is unclear. Some claim it has to do with several calamities that have occurred on Fridays, like the panic of 1869 when financiers Jay Gould and James Frisk attempted to corner the market on gold. Fortunes were made and lost that day. And since the day after Thanksgiving is calamitous in shopping malls across America, and it is a Friday, it made sense. But the more modern interpretation is that many merchants, who once upon a time recorded their losses in red ink in the old ledger book and profits in black ink, would finish the year "in the black," beginning with the day after Thanksgiving—or Black Friday.

In the little part of Denver where I live there's a place called the Cherry Creek Shopping Center—which has long boasted about being the center of the universe, as if it's some celestial event, like a huge black hole at the center of a galaxy—puts up quite the holiday display. Maybe not quite like Saks Fifth Avenue, but still over the top. And one can't forget the private homes putting in a splendid effort to compete with Clark Griswold's house in *Christmas Vacation*.

Then there is, sadly, most predictably, the holiday fruitcake that no one seems to eat and only ever regifts. It is estimated that there are only a handful of such fruitcakes in the entire United States, merely passing along from one family to another—sort of like the human infections upon which this book is partially based. I've truly never seen anyone actually eat a fruitcake. It must be legend.

I do like Christmas. After all, Jesus was a rabbi.

Along with our adorable daughters, Estee, aged seven, and Eva, aged five, my stunningly beautiful wife, Stacey—her age necessarily omitted from this narrative—and I found ourselves that day in December 1995 enjoying the magic of Disney World. No—I had not just won the Super Bowl. On a hard-earned vacation (and when you're a working parent of young children, all vacations are hard earned), what more delightful thing can a father imagine doing to relax than standing in those nerve-racking, fretful, interminable lines at Disney World, with sticky bannisters and crying toddlers. Between wiping churro off your child's face and wondering how much it's costing you to move a mere ten feet in line, waiting for a chance to ride Pirates of the Caribbean or It's a Small World for the third time, Disney is a vacation that definitely needs a vacation. The thing about it is this: while taking your young children to Disney World is a hard vacation, I wouldn't have traded it for anything. Your child's smiles and giggles and wide-eyed moments are all the reward a father needs, a mother, too. A martini at day's end doesn't hurt either.

When we had had our fill of Magic Kingdom rides during the day, and I had had that all-important afternoon beer, nap, and shower, my little bit of family would take the Monorail to Epcot, adjacent to the Magic Kingdom, for a leisurely stroll, for more rides with more lines to wait in, and, if we had no dinner reservations, another interminably long line for a sit-down meal. I actually enjoyed the stroll around Epcot, ambling about as if it were the Champs-Élysées. It was the best part of my day.

Epcot is an acronym for "Experimental Prototype Community of Tomorrow"—Walt Disney's idea for a theme park based on a futuristic city. Although Epcot has a few innovative, future-oriented rides, in reality it is more about showcasing eleven countries through their food-court pavilions—the US, Canada, Mexico, France, the United Kingdom, Germany, Norway, China, Japan, Italy, Morocco—and less about any futuristic city.

After we had had enough of all that Disney magic over the course of too many days—you can listen to the Electrical Water Pageant blaring "It's a Small World" outside your hotel balcony as you're trying to sleep only so many times without pulling your hair out—we left behind all that wizardry of Disney World on New Year's Eve day, December 31, 1995, and made our way from Orlando to the home of my mother-in-law, Barbie, in Tampa.

In Tampa, there were no lines to get something to eat, no noise, no screaming children, no crying toddlers, no rushing—just mosquitos. The mosquito: not only one of the most annoying creatures in existence, but also it is an insect that can carry human infection, acting as what's called a "vector" to spread disease—a recurring theme along the pages of this narrative, as it were. Barbie's quiet home was situated within a gated golf course community, right on a fairway, standing in stark contrast to the Contemporary Resort on the Disney property, our just departed basecamp. Barbie's home was my vacation from a vacation. Not that I played golf—I didn't and don't—but a home along a private community golf course was relaxing. I could relax, I could run, I could sleep, I could read—did I mention relax? I was looking forward to the quietness, catching up on some lost sleep, reading a book here and there, putting distance between myself and the Magic Kingdom, putting distance between myself and the stress of my job back home.

I'm a plastic surgeon in Denver. Sort of sounds like "I had a farm in Africa," the opening salvo of Karen Blixen's memoir *Out of Africa* from 1937—other than the failing coffee plantation bit. Being in the operating room is actually not stressful for me. In fact, it is one of the few times I'm relaxed—that and when I'm off by myself on a long run. I don't even relax when I sleep, apparently doing battle with whatever subconscious demons haunt my dreams, doing battle with what and why, I have no clue.

As a long-suffering insomniac since college, coupled with the added postapocalyptic Disney destruction that had frazzled my brain, my mother-in-law's home was a chance to "not be touched." "Not be touched." Three sacred words. In surgery residency, when you work and are on call nearly all the time, easily working 100 hours or more per week (and no, that's not a typo), we had a phrase: "I cannot be touched." What it meant was that you were truly off duty: you had handed over your pager (in modern parlance, you turned off your mobile) to the next resident on call, who would now be taking the hits, and no one—and I mean no one—could "touch you." It meant you could trudge back to your apartment, grab a pathetic excuse for a meal, unplug the landline phone, crawl into bed, and not be touched.

At the time of my residency, mobile phones did not exist, except for perhaps that Motorola brick phone, a toy only the rich could afford while the rest of us stupidly lamented the absence of a transportable communication device from our lives. Little did we know that decades later, mobile phones would become the ruin of any semblance of tranquility. These days, with mobiles almost as plentiful as people on Earth—there are over seven billion people on our planet, and five billion of them have a cell phone—sometimes I wish, we all wish, they could be uninvented. And if that's too much to ask, then I wish we'd at least have the courage to toss them all into the nearest lake.

I always know when one of my two daughters texts me, in the industry dubbed by the acronym SMS for Short Message Service, not because she has an assigned notification tone but rather because, instead of one long text with whatever she needs me to do—as a father your daughters always need you to do something—I receive a half-dozen single-line texts. She's brilliant, so I know she knows how to construct a paragraph, but for reasons that escape me, she sends her thoughts one line and 256 character bytes at a time.

Way back when, when texting was first introduced in the mid-1990s, text messages were limited in size, to what was known as a 16-bit resource format, which worked out to no more than 256 characters, quite possibly less—meaning, a little over one sentence per text. Texting did not pick up steam until around 2005, when especially youth started using the technology with near unbridled abandon. Oddly the resource format is still the same or similar 256 character limit, but the brilliant nerds of the industry devised—and I love this word—concatenated SMS which basically means in a nutshell they daisy chain short SMS into longer texts, so they appear to you and me as one text. In my day, it was passing a carefully folded note in class; in my daughters' day, it was SMS. But the original size limit of texts is hardly the reason my daughter fires off single-line texts to me like a spray of machine-gun fire.

Since mobile phones did not exist when I was a surgery resident, pagers were how we were "tapped" or "touched." If I have to begrudgingly date myself, then this was sometime during the latter half of the 1980s into the 1990s, those days when I was being touched by a pager. I was such a light sleeper, I could hear the incoming signal before the pager actually went off, a sort of electrical buzz the device made as it was receiving the signal, before that annoying shrill sound pierced the silence, announcing the conversion of the buzz into a phone number to dial. Early pagers provided just that: a phone number. Later pagers offered a simple phrase message. I grew to hate that faint, nearly imperceptible electric buzz almost as much as I hated the ear-splitting beep beep beep that followed. In residency, once you handed off your pager to the next resident on call, you could not be touched.

So there I was in Florida, New Year's Eve 1995, along with my wife and daughters visiting my mother-in-law, four years after completing my residency training, my pager off, my slick Motorola MicroTAC flip phone also turned off, my practice back home turned over to another plastic surgeon—and I, overflowing with joy, could not be touched. Or so I thought. How wrong I was.

That area of Florida where places like Disney World and where my mother-in-law's golf community are built is situated on swampland. It is quite a remarkable feat, really, especially considering that lurking in these fresh and swampy waters are crocodiles with their pointy snouts and alligators with their square snouts—that's how you tell them apart—as well as the fact that the air is chock-full of flying insects. You would like to think that a pointy "A" would be for the "A" in Alligator and a curved "C" for the "C" in Crocodile to make remembering their snout shape easier, but unfortunately it is just the opposite. A swamp might be a fascinating expanse of low-lying ground, seemingly overflowing with water, a wetland, that is generally uncultivatable, where marshes merge seamlessly with trees and shrubs and reeds, where bog gas slowly rolls across murky waters, a veritable paradise for insects—but, to me, swamps are nothing more than home to killer reptiles and annoying insects, some of whom carry infection.

With the sprawl of humanity, the swampland has been tamed, mostly, brought into submission with shopping malls, covenant-controlled golf communities, and parking lots. But I was not fooled. I knew that right outside my mother-in-law's home, lurking in the murky waters along the perfectly manicured fifteenth fairway, those mean-spirited ancient reptiles were wiggling their way around—all uncomfortably too close to my daughters. It gave me an uneasy feeling that I could at any minute be pulled down into the swamp, stuffed—still alive—under a boggy log, and left there for several days until my well-bloated corpse met the alligator's preferred culinary savoriness.

Florida is the only swampland in the world where crocodiles and alligators coexist, although alligators rule the numbers. My mother-in-law's golf course community was near Tampa, and so not exactly built on swamps, but certainly constructed atop boggy wetlands with a high water table. These Florida marsh-like areas, like other swamps just mentioned, are home to not only the alligator and the croc but also a variety of those annoying mosquitos who love to lay their eggs on swampy waters. The northern mockingbird is the state bird of Florida; it should be the mosquito.

There are about eighty varieties of mosquito in Florida, which I have come to understand is more than any other state in the US, with about a dozen or so of those known to carry infectious diseases harmful to humans. The West Nile and Zika viruses are two that come to mind, so is yellow fever. Oddly, or rather fortunately, the genus of mosquito that can transmit malaria—the female *Anopheles*—was eradicated from Florida in the 1940s through successful insecticide campaigns. But the other varieties of mosquito that continue to live in this area, through no fault of their own, are vectors for several other infections, although mostly viral, not parasitic like malaria.

If you're wondering why it is only the female mosquito that carries malaria and other pathogens into humans and not the male mosquito, the answer is quite simple. It is only the female mosquito that sucks the blood of humans, or other mammals, needing that boosted bloody cuisine for her egg-laying performance. The male mosquito must content himself with plant nectar and tiny aphids for his studly duties.

But on that golf course, on that day, December 31, 1995, it simply rattled me to think that the golfing duffers were sharing their fairway with the alligators. Can you imagine hitting a PING 3-wood off the fifteenth tee only to see your ball land in the rough where an alligator is taking in some rays? Me neither. But reptiles are, after all, cold-blooded, in more ways than one, and because of their cold-bloodedness they need to enjoy noontime sun in order to warm up their scaly reptilian bodies, fairway or no fairway.

Cold-blooded animals do not maintain a constant body temperature like us warm-blooded creatures. Instead, their core body heat takes on the temperature of their environment. Which is also why cold-blooded animals can only roam within certain warmer regions of the planet. Examples of cold-blooded life include fish, reptiles, amphibians, and insects. All the rest of us, the mammals and the birds, are warm-blooded, meaning our bodies go to great lengths to keep our core temperatures within a narrowly defined range. People who study such things believe warm-bloodedness developed as a survival advantage allowing those such among us—mammals, birds—to live in a wider range of temperature environments and to respond easier and faster to rapid changes in climate. Why did our ancestor sapiens need to roam north and south "out of Africa?" Looking for food, of course, and for new hospitable climates. That they were warm-bloodied made their journey to the far reaches of the planet, or nearly the far reaches, quite possible.

After my mother-in-law gave us a tour of her home and grounds—me with a watchful eye on my daughters as I nervously scanned for alligators along the fairway—I settled in for a relaxing afternoon with a cold one. Pager off: check. Mobile phone off: check. I can't be touched: check! My daughters scrambled into their bathing suits and jumped into the swimming pool out back, completely screened in to keep the insects out, completely walled in to keep killer alligators out: check—which tells you about all you need to know about that little slice of paradise, the need for screened-in pools, the need for walls. And despite my giggling daughters splashing in that screened-in, high-walled pool, I still nervously scanned for predators. Pager off: check. Mobile phone off: check. Anxiety and stress off: not exactly check. So much for relaxing. So much for not being touched.

While my wife, Stacey, and mother-in-law, Barbie, sat in the living room chitchatting, a few hours before we would head out on the town for a New Year's Eve dinner at a local Tampa restaurant, I kicked back, popped open a can of beer, rested my weary body from all that Magic Kingdom magic, and kept a close eye on the girls in the pool.

Then the phone rang. Not my slick, *Star Trek*–inspired Motorola flip phone, but the house's landline.

That phone rang.

It has been well over two decades, and that ringing phone still haunts me all these years later—a call that forever colored my thoughts about human infection, a call that would unfold into a series of events that drastically wounded my wife and her family.

Keep in mind I was somewhat new to the plastic surgery private practice game, having finished my training just a few years earlier. Despite having narrowed my field of medicine to a highly specialized surgery subfield, medical school was not that many years back, and I still had my eight years of residency training—a period when you see nearly everything medical under the sun—percolating in my brain. So, I had a reasonable command of the breadth of all that is medicine, including being well versed in human infections and infectious diseases. It is only after years and years of being a specialist that you forget mostly everything you learned. Well, not really *forget*, but place it deep into mysterious encoded brain circuits that, with prodding, with enough prodding, can retrieve the memory—maybe in bits and pieces, but retrievable, nonetheless.

I'm not about to bore you with the physiology of memory—that would take a book in itself, and I don't fully understand it anyway. But, in a nutshell, it is believed that how something is received through a collection of neural impulses, whether that thing is a visual, auditory, or some other sensory input, triggers an electric pattern, which is what constructs the memory. The same pattern of impulses that received the original experience fire in the exact same pattern to retrieve the memory. That is, until it is forgotten, or mostly forgotten. A computer circuit board never forgets its patterns, never loses its memory—at least we don't think they do, but they can—but humans do forget, likely as a courtesy to our brains. If we kept too many memories primed and ready to go, our brains would be massively cluttered, like a bunch of annoying flies buzzing about. And with that brief synopsis of memory, in the words of Forrest Gump, that is about all I have to say about that.

After all my training, I had seen my fair share of human infection: abdominal infections that sapped the very life out of patients, brain infections that converted white matter into goo, neck infections that choked away a person's last breath, and the so-called flesh-eating bacteria that ate away limbs and often life. My residency began in the 1980s with the meteoric rise of HIV infection in the human arena, so it is safe to say that between four years of medical school and eight years of residency training, I had seen it all.

But in each and every case, no matter how severe the infection, no matter how many lives lost, no matter how deep and sincere my empathy for patients who had succumbed or were succumbing to overwhelming fulminant infection, there was always that remoteness, that firewall, that safety zone that allowed me to quietly digest the death and dying of my patients by keeping it tucked in a remote, neatly cordoned-off area of my mind. To survive in medicine, assimilating and moving on is a necessary survival tool; otherwise, you just won't last in the trenches, taking hand grenade after hand grenade.

But not this day.

That phone rang.

On the other end of the phone was Alan, the husband of my sister-in-law Deborah, which made Alan oddly not my "brother-in-law" but my "co-brother-in-law." It is all so confusing, the terminology we use to describe family trees, that "second cousin three times removed" fiddle-faddle. Alan was calling from Los Angeles. I was in Florida, 2,500 miles away. Deborah had taken ill, and Alan was describing her symptoms to my mother-in-law, who had answered the phone, outlining a classic respiratory infection, like the cold or flu—or so he thought. Or so he conveyed.

Alan asked to speak to me, believing Deborah's respiratory infection was just your garden-variety cold. And I, having no reason not to believe it was just a cold, took the call.

But what Alan characterized to me over the phone—what was lost in translation when he first spoke to Barbie—was in fact not your run-of-the-mill cold or flu. As I listened to Alan's description of his wife's illness, as the words rolled off his tongue in confusing, rapid succession, what was being described was an entirely different picture altogether. Something was amiss.

Deborah, it appeared to me during that brief interchange with Alan, had fallen off the curve. "Falling off the curve" in medicine means that the patient's disease course is no longer on a normal journey, a normal trajectory, that a crisis point has been reached, a tipping point, and rather than plotting toward recovery, the line has fallen off the graph and into the abyss. This was where Deborah was now, as the words poured out of his mouth.

Alan described Deborah having been ill with some upper respiratory infection a few weeks earlier. She then seemed to have recovered, only to relapse. To the casual observer, that might not seem like such a big deal—get ill, recover, relapse—but to a physician, it raises a red flag, or it should. In human infection, there is a difference between a relapse of the *same* infection, a *chronic* infection that spikes again, and a *secondary* infection with a different pathogen—a method of parsing infections we'll get into later down the road.

Generally speaking, between our immune system and the normal course of most infections, with or without prescribed antimicrobials, when we humans are confronted with an infection, we most often recover. We launch an immune response with memory, it might be a short memory or a long memory. We really shouldn't relapse, as many infections are here and then gone for good. An acute infection that devolves into a chronic infection—like how tuberculosis and malaria work, and how HIV progresses into AIDS—is quite different from an acute infection that clears followed by another acute infection, which is where a different pathogen rears its ugly head. Alan was describing this latter situation. Our immune systems are generally too smart to be fooled by the same bugger twice in as many weeks and, for that matter, even two distinct yet similar buggers one after the other. In can happen—exceptions always exist—but, in general, the principle mostly holds true.

Alan believing, understandably, that his wife's condition was just a cold that had maybe kicked up a notch into a sinus infection had called me in Florida to see if I could arrange some antibiotics for Deborah in California. I, on the other hand, was hearing a different story. I was hearing about a new infection on the heels of an old infection. And it wasn't just that—Alan was describing a woman who was really very sick. It sounded as though Deborah had become terribly ill, unable even to get on the phone to converse with me directly. That was another the red flag. I told Alan, in no uncertain terms, that he must take his wife to the emergency room right away. I explained that she sounded too sick for half-measures and standard oral antibiotics. Something was frightfully wrong with the picture. He said he would take her to the ER. I repeated myself, even more emphatically. He said they would go. I gave the phone back to Barbie, who again reinforced what I had said.

How had Deborah become ill twice in one month? It can happen, but not with the same virus and not the same exact infection. Had she breathed in someone's cough, someone's sneeze, which sends aerosolized viral particles spewing through the air like clouds of flour pluming in a bakery, floating in space, never quite landing on the ground, waiting to infect the next passersby? Had Deborah touched an elevator button or a banister where a tribe of microbes had just landed, then touched her face? I was thinking that what had first manifested several weeks earlier as a standard upper respiratory viral insult—a cold, a flu—had blossomed into a secondary bacterial infection, like pneumonia. Somewhere along the path of recovering from the first infection, things turned south and headed into much more ominous territory.

How is it, or why is it, that humans are plagued with infection in the first place? The way in which simple bacteria and other unicellular microbes came to both haunt and hunt humans is part of the stuff this book is made of—the science, the medicine, the history required and acquired by humans to understand and subsequently beat back, or not beat back, those predators we call "human pathogens." There may be no better place to begin exploring human infection than the Bible, which is where our story now turns.

2 MIASMAS

The Ten Commandments were purportedly given to Moses on Mount Sinai as the Word of God at the outset of the Jews' departure from Egypt and are believed to have been scribed around 1600 BC. Subsequent to the Ten Commandments, the Hebrew Bible soon followed as the divine Word of God also given to Moses while he and his fellow Jews were wandering—likely wondering, too—in the Sinai Peninsula desert during that forty-year exodus from Egypt.

Although the authorship of the Hebrew Bible, that is, the Five Books of Moses, also called the Torah—Genesis, Exodus, Leviticus, Numbers, and Deuteronomy—is sometimes debated, it was likely compiled within the Sinai Peninsula during that exodus around 1400 BC. Although Moses is often credited with having written the entire Hebrew Bible, it is unlikely—unless, of course, Moses was a soothsayer as well, seeing as his own death is described in the final verses of Deuteronomy. More likely, the Torah is a compilation from a few or several or many sheepherders who wandered with Moses from Egypt to the Promised Land.

The oldest known complete Hebrew Bible— termed an "extant Bible," since it was copied from a copy—was apparently written on sheepskin and dates back to around 1155 AD. That is AD, not BC, so the oldest existing complete copy of the Hebrew Bible was transcribed more than 2,500 years after the original. It was discovered recently in a library at the University of Bologna in Italy. For reasons that are unclear to those who concern themselves with these matters—that is, to historians—several millennia of complete Bibles, from the very first Moses edition in 1400 BC or thereabouts, to the Bologna discovery from 1155 AD, have been lost into dust. It makes one wonder: If it was such an important book, a sort of user manual for life, why weren't the first edition and the other early editions better cared for?

I would be remiss, of course, if I did not mention that many religious historians believe the Bible was written over an extended period of time, over 1,000 years to be exact, from 1400 BC through 1200 BC and up until 200 BC. Yet, there are no complete Bibles dating that far back. Even the original Ten Commandments on stone tablet seem to have been misplaced, described in the Book of Genesis as being housed in the Ark of the Covenant, a gold covered wood chest. Supposedly Indy, played by Harrison Ford, ultimately found the Ark as depicted in *Raiders of the Lost Ark* 1981, but it ended up in some abyss of a U.S. government warehouse.

There are other older but incomplete portions of the Hebrew Bible in existence, the most famous of which, and nearly a complete version, is the one discovered as part of the Dead Sea Scrolls. The Dead Sea Scrolls were written sometime around 200 BC onward, discovered inside the Qumran Caves within the Judaean Desert along the north shore of the Dead Sea, and unearthed by a Bedouin shepherd in 1947. The Dead Sea is a salt sea, sort of like a salty lake, where, due to the high salinity—higher than that of oceans and other seas—not much in the way of aquatic life exists within its mysterious depths. Which is probably why our ancestors called it a "dead sea." The Bedouin are nomadic Arabs, mostly Muslim, who occupy areas of the North African desert and the Middle East and continue to this day to live a nomadic existence, smart enough not to be encumbered by city life. That nearly complete Bible found among the Dead Sea Scrolls, scribed over 1,000 years after the Moses Bible, is still a copy of a copy. Given the length of the Bible, one must accept the distinct possibility that a few errors in transcription occurred.

"IT STARTS 'A RABBI AND A PRIEST GO INTO A BAR...' I THINK THIS DEAD SEA SCROLL IS A FAKE."

There is a joke in there, about relying on copies of copies of copies of the Bible. Like the monk who disappears into the monastery's ancient library to read the oldest known copy of the Bible, only to emerge days later in tears. When asked by his fellow monks why he is crying, he mutters, "The word was 'celebrate,' not celibate"—causing all the other monks in unison to promptly break down in tears along with him.

Many if not most historians believe Moses was *not* the sole author of the Five Books and that he had help from other Israelites in compiling the "Word of God." But what other "sources" contributed to the Hebrew Bible during that forty-year exodus out of Egypt?

To take a critical view of the Bible—which, by the way, is the all-time number-one bestseller book, its contents going beyond just the Word of God, if one is inclined to understand its passages in such a way—it is basically a codified doctrine of how humans were to conduct their lives all those centuries ago. It is chock-full of history, philosophy, social customs, judicial precedents, interesting stories, and teachings of life. In other words, the Hebrew Bible contains many passages that God would not have concerned Himself with, passages that arose out of the dos and don'ts that humans learned from living together. Like I said, it's an owner's manual.

God wouldn't tell historical fables and recount adventures—God would make commandments. But wandering sheepherders certainly would share dramas around the campfire. In the end, who exactly wrote the Torah is unknown to history, and likely will remain forever so. Also unknown to history, oddly, is the exact location of Mount Sinai as it is depicted in the Bible. It is commonly taken to be in the Sinai Peninsula in Egypt, though some scholars beg to disagree, suggesting there are at least two other candidates for Mount Sinai, which are not in Egypt at all but rather one in southern Saudi Arabia and the other in southern Jordan.

The Christian Bible, of course, is more, much more, than the Hebrew Bible. It also includes the New Testament, which was written AD—*anno Domini*. AD is not an acronym for "after death" as many people, myself included when I was younger, mistakenly believe. *Anno Domini* is the Latin for "in the year of the Lord." The "year of the Lord" also is not necessarily taken to mean that single calendar year in which Jesus was supposedly born, although that is precisely what it *might* mean. Some believe that we are still "in the year of the Lord"—meaning all the subsequent years from the birth of Christ until now and at least until the Second Coming— where the "year of the Lord" means "the *era* of the Lord."

Most scholars now believe Jesus was not born in 0 AD, as I was taught in my youth, but earlier, between 6 and 4 BC, and that Jesus died around 30 AD, leaving 0 AD representing not much of anything other than little Jesus beginning to learn carpentry from his father, Joseph, and learning patience from his mother, Mary.

The current calendar the world follows is the Gregorian calendar, so named after Pope Gregory XIII in 1582, who implemented its usage to better correlate what they, the Catholic Church, did not actually accept but nevertheless knew was the truth: that the annual transit of Earth around the sun is a 365.2425-day journey. The fudge factor added to the 365-day Gregorian calendar was of course a leap year every four years. Interestingly, a leap year is not an exact correction either. As a result, every once in a while whoever is in charge of these things—likely a man in a white coat wandering some dreary basement with drab gray walls and flickering fluorescent light bulbs—adds a few leap seconds in what is termed "Coordinated Universal Time."

The Gregorian calendar replaced the Julian calendar, adopted by Julius Caesar in 45 BC to reform the Roman calendar. Because of the lack of a leap year in the Julian calendar, it had hopelessly drifted—a digression especially noticeable when it came to the summer and winter solstices (the longest and shortest days, respectively) and the spring and fall equinoxes (where day and night are exactly equal). But the Julian calendar itself replaced the even *more errant* Roman calendar, perhaps dating back to 700 BC, which was so inadequate it required a Roman council to occasionally convene to correct its wayward and alarmingly useless timekeeping measures.

So, as I mentioned, it is a myth that Jesus was born in the year zero. First of all, there was no year zero, because the division between BC and AD had not been introduced. Back then, years were not counted like they are now; the months and days in the year were counted, but not the years themselves. At the time of the ministry of Christ, last year was just that, *last year*, and two years ago was just that, *two years ago*. They couldn't say things like I was born in 20 AD or my two-wheeled horse-drawn chariot is a 33 BC Model Charioette, not like we can say I was born in 1965 or my car is a 1963 Corvette.

It wasn't until 500 AD that the Roman monk Dionysius Exiguus, working backward from what he could glean from the New Testament, determined, incorrectly, that Jesus was born exactly 500 years earlier, taken to be 0 AD. Subsequently, he assigned everything before the birth of Christ as BC, *before Christ*, and everything after AD, *anno Domini*. To make matters a tad confusing, having the letter "B" stand for "before" was a rather odd but necessary choice at the time, as the Old English word *"beforan"* means "in front of," as in "in front of Christ," which sounds weird. The better choice than "before" would have been the Latin *ante*, as in *ante Christ* which would have meant "before Christ." But that had one glaringly obvious problem, the *ante Christ* phonetically sounds too similar to "Antichrist," the opposite of Christ—which would be the Devil—and that would not have gone over well with the surging Christian world. So, rather than "AC" for *ante Christ*, the good monk Exiguus assigned "BC" for *before Christ*.

Most if not all of the New Testament was written in the first century AD, after the death of Christ. The authorship of the New Testament is not as nearly in dispute as the Torah is. We know that the New Testament derived from several Jewish disciples of Jesus, including his followers Peter, Matthew, Mark, Luke, John, Jude, and Paul.

Which is a roundabout way of stating both the Old Testament and the New Testament make reference to diseases. In the book of Exodus, for instance, God reveals his great power through the ten plagues upon Egypt, the third plague of which was lice and the sixth that of boils: "… and have Moses toss it into the air in the presence of Pharaoh. It will become fine dust over the whole land of Egypt, and festering boils will break out on men" (Exodus 9:8–9). Of the ten plagues Moses wrought upon the Egyptians, four of them had to do with infection: lice, plague, pestilence, and boils.

Today we know there are any number of infections transmitted to humans by lice, especially typhus, which has likely killed more soldiers since antiquity than actual combat. The fourteenth-century Black Death plague wiped out 25% of the World population in eight years. And boils—well, they're bacterial infections. That phrase "toss it into the air" is quite interesting to the journey we're embarking on in this book, as it's an image that apparently made a lasting impression on one early physician from antiquity named Galen, who we will now visit with.

Galen of Pergamon was a physician of Greek ethnicity but Roman citizenship, born in 129 AD in Anatolia, modern-day Turkey. The reason why an ethnic Greek living in Anatolia could come to be born a Roman citizen in a country that was just previously under Persian control is the stuff wars are made of.

From about 10,000 BC to 2400 BC, Anatolia (modern-day Turkey) was pretty much Neolithic, a period in human development where humanity was switching from nomadic life to civilization. Basically, the time when humans were figuring out it was easier to stay put and not chase after wild game, a time marked by the domestication of plants and animals precisely so they didn't have to chase after wild game, a time where stone tools and weapons were refined and then discarded for copper tools and weapons, and then those discarded for iron tools and weapons. A time when humans became domesticated, especially women; men still struggle with the notion. Toward the end of the Neolithic age and the dawn of the first civilizations, early forms of written language were being developed, albeit in the simplest of forms. Actual *spoken* language, if you want to call it that, emerged in our ancestors perhaps 100,000 years ago. Learning to write the spoken word, *written* language, well, that took some time for our ancestors to master, perhaps first appearing in 3500 BC in Sumer, in southern Mesopotamia.

So how did the Greek Galen, a Roman citizen living in Turkey, come to exist?

For several thousand years, Turkey changed hands like my wife used to change her outfits before we went out on a Saturday night. Consider all these conquering events: In 2400 BC, under the command of Sargon I, the Akkadian Empire, which encompassed a Semitic-speaking people based in eastern Mesopotamia, invaded Anatolia, only to lose it around 2150 BC to a wandering band of Middle Eastern nomadic folks, the Gutians. Then like locusts devouring everything in their path, the Assyrians marched in from northern Mesopotamia, vanquishing the Gutians and laying claim to Turkey (so as to extract its silver). The Assyrian reign lasted several hundred years, until the Hittites moved in, themselves an ancient Anatolian people from the northeastern edge of the realm.

The Hittites controlled Turkey for 1,000 years—a remarkably long time for that critical region, critical because it was the land between the East and the West, or, more precisely, where the East and the Silk Road met the markets of the West. The Hittites finally relinquished control in 550 BC to the all-powerful, all-mighty Persian Empire of Cyrus the Great, also known as the Achaemenid Empire. The Persian Empire was the forerunner of modern-day Iran. Iranians are Persian, not Arab. While the two groups share a common religion, Iranians are Persian Muslims, not Arab Muslims—something too many Americans fail to appreciate. Religion and nationality are two different things.

Persian control over Turkey, as well as huge swaths of Greece, lasted a mere 200 years, until about 335 BC, when the imperial desires of Alexander the Great—who was from the ancient Greek kingdom of Macedonia—marched in and wrested control of Turkey from Persia. And soon after that Alexander the Great marched his army into the heart of the Persian Empire, cutting off the head of the snake. The rise of Alexander the Great and his empire kicked off what became known as the Hellenistic period of Greece (336 BC - 31 BC). The great Greek thinkers—Hippocrates, Socrates, Plato—predate the Hellenistic Period, and are known to us from the Classical Period (480 BC – 323 BC) with only Aristotle bridging Classical with Hellenistic Greece. When Alexander died unexpectedly in 323 BC—not in battle but from a bacterial infection, of all things—his vast empire began to slowly die right alongside him. Brick by brick it started to crumble in the far reaches of the empire, and with that collapse, the Seleucid Empire, which finds its origins in Babylonia between the Euphrates and Tigris rivers (modern-day Iraq), stepped in and grabbed control of Anatolia from the Greeks. Of course, that didn't last long either, the Roman Empire grabbed Anatolia and much of the Middle East by 190 BC and Egypt in 31 BC.

That is what was happening in the Middle East and on the eastern shores of the Mediterranean in the waning centuries leading up to the birth of Christ.

Meanwhile, across the vastness of the Mediterranean Sea and into the setting sun, another, even more formidable, empire was emerging: the Roman Empire. Soon enough it would engulf nearly the entire Western world in a conquest of several centuries, giving rise to the oft-repeated phrase "Rome wasn't built in a day"—but built it was.

Around 510 BC, Rome was little more than an enclave on the western shores of the Italian coastline, but by 220 BC it had expanded into all of what is now mainland Italy as well as absorbed the Mediterranean islands of Sicily, Sardinia, and Corsica, and even established outposts near Greece in what we would call today Albania. The emerging Roman Empire continued to push its way all along the northern Mediterranean coastline, including into Gaul, modern-day France. By 140 BC, Rome had destroyed the Greek city of Corinth, by 86 BC had seized Athens, and soon after and Greece a Roman province.

Also around this time, in 90 BC, Rome had expanded westward, conquering the Iberian Peninsula of current-day Spain and Portugal, as well as becoming more entrenched in Greece, and even settling outposts along the Maghreb of Northern Africa, notably laying siege and then destroying Carthage, modern-day Tunisia. Through the lust of the conquering Caesars (Julius, Augustus, Tiberius), a little over 100 years later—by 20 AD, corresponding to the ministry of Jesus—Rome had conquered nearly every land, every parcel, every bit of shoreline and coastline of the entire circumference of the Mediterranean Sea, including the Holy Land. If at that time in history you were to sail the whole coastline of the Mediterranean, like Ulysses must have on his ten-year odyssey to get back home to Ithaca, Greece after the ten-year Trojan War—as portrayed by Homer in his epic poem *The Odyssey*—every seaport, every seaside, every shipping lane, every seashell, every sandbar, that is to say: everything, was under Roman rule. The Roman Empire conquered everything, and as it conquered, it built roads for its army to travel on and maintain control, giving rise to another oft-spoken phrase: "All roads lead to Rome."

It didn't end there. By 70 AD, thanks to Caesars Claudius and Nero, Roman legions had crossed the English Channel into Britain, and Roman garrisons controlled the entire length of the Maghreb, from the Moroccan city-states of Marrakesh and Casablanca in the west, to Algiers, Tunis, and Tripoli in the middle, and Alexandria and Cairo in the east. Even the Caucasus, those lands encircling parts of the Black Sea—known today as Armenia, Georgia, and southwest Russia—were controlled by Rome.

Which is why Galen, a man of Greek ancestry born in 129 AD in Pergamon, Anatolia/Turkey, was a Roman citizen.

Taught medicine in the ancient Greek tradition of Hippocrates, Galen would go on to become the most recognized physician of his time. His teachings and beliefs—including ones that were wrong—dominated the medical landscape for well over 1,500 years after his death. In fact, to question Galen was practically considered heresy, almost like questioning the Pope. It took the European Renaissance (beginning in the fourteenth century), Scientific Revolution (sixteenth century), and the Enlightenment (eighteenth century) to bring much of Galen's well-intentioned but false medical dogmas crashing down.

Galen is most notably the father of the "miasma theory" of human infection. The miasma theory states that all infections, such as cholera, the Plague, and consumption (which you and I would call tuberculosis), are caused by "bad air." That's it. That's the entirety of the theory—bad air. The miasma theory was based on the partially understandable but significantly incomplete observation that many infections seemed to circulate in populated areas where rotting detritus, human and animal waste, and nasty vapors wafting through the air were in no short supply. Although infections do seem to go hand in hand with smelly environments, there is a lot more to it. Galen's miasma theory was, as we shall learn, a woefully simplistic conclusion. And speaking of medical education, we now turn to mine.

It was late in the summer of 1979 when I found myself a first-year medical school student at George Washington University in Washington, DC. The days were sunny and hot, I remember that. Of course, it was hot—it was DC in August. And it was humid. Of course, it was humid and muggy—it was DC in August.

First-year medical school was your basic grind—a palette of anatomy, microbiology, human physiology, biochemistry, and embryology. Our anatomy course, like all anatomy classes, consisted of a tedious drone in the lecture hall paired with the more rewarding, more enjoyable world of dissection in the anatomy lab. I liked human anatomy. In fact, anatomy was my favorite course—especially since in medical school we didn't have recess and gym class, which otherwise would have been my two favorites. Flourishing in anatomy lab likely explains why I became a surgeon. If I had floundered I'd probably would have become a sturgeon (school boy joke).

When my father, also a doctor, studied medicine in the 1930s, the anatomy course spanned two years. When I took anatomy fifty years later, it was one year. Not because there was more anatomy to learn in my father's era, but because there was less of the other stuff to learn. So, the 1930s medical student languished longer in anatomy while the modern-day medical student gets the privilege of stuffing the same amount of

"And, for those of you that want to relive your school days, our retirement community has a walking trail that is uphill both ways."

anatomy into their brain twice as fast. My pals and I had to cram human physiology, pharmacology, immunology, genetics, and more into our tortured brains—courses that did not exist or were in their infancy during my dad's day. Not to put too fine a point on it, but when my father started medical school, penicillin had yet to be discovered. When I started, penicillin was practically passé. In my father's era, he'd bemoan, he had to walk to school, ten miles, uphill, both ways. I would counter with a quip about how, when he got there, there wasn't all that much to learn.

Prior to Galen's era, before the rise of Christianity, early Greek, Persian, and Egyptian physicians studied anatomy through human cadaver dissection. If you want to be a physician who treats people, you quite likely need to know human anatomy. You know: "Toe bone connected to the foot bone / Foot bone connected to the heel bone / Heel bone connected to the ankle bone" and so on, as goes the spiritual song "Dem Bones," written by James Johnson in the 1920s, inspired by Ezekiel 37:1–14, which describes Ezekiel's visit to the Valley of Dry Bones. Although "Dem Bones" was first recorded in 1928 by the Famous Myers Jubilee Singers, the best-known versions are by Fred Waring and His Pennsylvanians, from 1947, and the Delta Rhythm Boys, released in 1950. But along with that song, your best bet to learn human anatomy is to dissect the corpse of a recently departed soul. Makes perfect sense, and for centuries before the rise of the Christian Church, that is exactly what ancient Greek, Persian and Egyptian physicians did.

Despite this core component of advancing medicine through dissection, would you be surprised to learn that, with the rise of Christianity in the first century AD, human dissection came to be considered an unholy act and subsequently forbidden in all of Christendom? It is odd, isn't it—or perhaps, in fact, not odd at all.

Human anatomic dissection was prohibited from about the time Galen was born in Pergamon to sometime around the Renaissance: a dissection drought of well over 1,200 years, which also paralleled, not so coincidentally, the fall of the Roman Empire, the rise of the Church, and the ushering in of the Dark Ages. What did the fall of the Roman Empire have to do with human dissection? Well, a great deal actually, as the fall of Rome and rise of the Church resulted in the squashing of many humanistic pursuits.

The Dark Ages are called precisely that due to the religious suppression of humanism and science, culminating in a dark period in history where not much advancement took place. With the rise of Christianity and concentration of power in the Church, a dark shadow was cast over the whole of the Western world. Coincidentally and incidentally, Earth at that time entered a mini ice age, partly due to excessive volcano activity around the globe. The ash from all those volcanos spewed into the atmosphere, blotting out the sun, making the skies darker, and subsequently dropping global temperatures a tad. For quite a few centuries, actually. But that global pall of ash darkness is not why historians refer to the Dark Ages as the Dark Ages. Just as volcanic ash blighted the sun, religious doctrine blighted humanism. The volcanic ash was merely coincidental—the Dark Ages were named after the blighting of humanism by religious suppression, not after a blighted sun. Emphasis was on the Church, not on the individual.

The fall of the Roman Empire—an empire more tolerant toward humanism—meant that there was no "enlightened empire" capable of keeping in check the overarching, overreaching will of the Christian Church. You might think it odd that totalitarian pre-Christ empires like the Greeks, like the Persians, like the Egyptians, like the Romans afforded some measure of protection for scientific pursuits, for humanism, but that is exactly what those civilizations did.

When the Roman Empire began to falter and in the fifth century went the way of the dodo bird—extinct—there was nothing formidable enough to keep in check the power and will and dogma of the Christian Church. Not much in the way of science, medicine, music, art, or philosophy advanced for well over 1,000 years—or, actually, longer. Not until the Renaissance began to rise from the ashes of the Dark Age, around the fourteenth century, did freethinkers begin to slowly crawl out from underneath the rock of religious suppression.

As for the miasma theory of infection (that is, bad vapors): Galen's shadow was allowed to reign supreme over Western medicine for nearly one and a half millennia. Many diseases and certainly anything remotely resembling an infection was blamed on smelly things. Not only was that a dogma in need of refinement to begin with, but the manner in which Galen and the other men of medicine who followed him could acquire such medical knowledge was restrained by the Church.

Galen's anatomical knowledge, his knowledge of pathology and how that knowledge applied to human disease, was gained entirely from ancient texts on human anatomy plus his own animal dissections, mainly swine and monkeys—and most animals do not share the same spectrum of infections that plague humans. That's right: Galen, the great physician of the second century, never actually dissected a human cadaver, nor did he even have a preserved human cadaver dissection from which to learn anatomy and correlate his theories on human disease. All of Galen's precepts, including his persistent and incorrect miasma dogma, endured for well over 1,500 years. Passed down from a man who never knew human anatomy firsthand.

Imagine today if you were going under the knife to have your gallbladder removed and your surgeon's anatomical knowledge was based on the dissection of a pig. That would not make for a comforting thought as you were being wheeled away on a squeaky gurney into the operating room. You just might jump off that rickety stretcher and flee for your life down the corridor with your hospital gown flapping wildly and your derriere bouncing in the breeze.

The relation between anatomy and disease is called pathology, that is, the ability to perform a postmortem autopsy and correlate a cause of death anatomically with the symptom presentation.

Galen was not always wrong. Contained within the passages of his medical doctrine are some very important underpinnings of doctoring. He established such things as the importance of monitoring the heartbeat and the color and smell of urine—fortunately not taste. He made the connection that stress, and anxiety cause some illnesses like intestinal irregularity, declared that arteries carry blood, and noted how eye color can indicate diseases, such as the yellow, jaundiced eyes of liver failure. Galen was good, but he would have been better if he had been allowed to perform anatomical dissections and autopsies.

And, certainly, Galen was wrong just as often as he was right. Besides his adamant belief in the miasma theory, he incorrectly declared that venous blood is produced in the liver and that blue venous blood is what carries nutrients throughout the body, which is the opposite of what we know today: arteries carry oxygen and nutrients, and veins carry spent blood back to the lungs (not liver) to get replenished with oxygen. (The only "vein" that carries blood to the liver is the portal vein, which leaves the intestines replete with recently absorbed nutrients, travels to the liver for processing those desirable goodies, and filtering out undesirable baddies). Galen elevated blue venous blood to a higher pedestal than red blood simply based on color. Blood from blue veins is darker and therefore must be richer, or so Galen incorrectly reasoned. But blue blood is blue because when a hemoglobin structure *loses* its attached oxygen molecule, the structure changes shape, turning from red to blue.

As for red arterial blood, Galen proclaimed it carried "spirits," both good and bad spirits—and I don't mean alcohol, although it can do that too. Galen was referring more to the realm of the supernatural. This concept of spirits, most particularly the bad ones, formed the foundation for the practice of bloodletting, a sanguinary way to rid one of evil spirits and bad humors and enable a return to good health.

Galen was a big promoter of Hippocrates's four humors. The concept—actually proposed by Alcmaeon of Croton 100 years before Hippocrates, and rather refined by Hippocrates into the notion of "four life essences"—held that the humors should be in balance. When they are out of whack, illness, infection, and disease most assuredly will follow. An overabundance of black bile (cold and dry) meant the person was depressed or melancholic, yellow bile (warm and dry) meant the person was quick-tempered or choleric, red blood (warm and moist) meant the person was passionate or sanguine, and phlegm (cold and moist), that is to say, the gross stuff we cough up, meant the person was apathetic and a bit on the nose: phlegmatic. And a disequilibrium in the balance of these four humors is what caused disease.

Religion, and not just Christianity, up until the Renaissance achieved, and is still attempting to achieve (can you say "evangelical"?), a dampening of scientific thought. Religious belief and science need not be mutually exclusive, although the two, all the way back to antiquity, have made for uneasy bedfellows. Religion tends to rely upon faith, revelation, and sacred traditions whereas science leans upon reason, fact, and empiricism. Socrates was forced to drink hemlock in 399 BC—known to history as the Trial of Socrates—when he refused to acknowledge the traditional Greek gods. Well, that and because Socrates also didn't care much for Greek democracy (he believed only the brightest philosophers should govern).

Even during the Renaissance, when some folks were trying to free themselves of religious doctrine, Galileo Galilei, like Socrates 2,000 years earlier, faced similar religious condemnation. The trial of Galileo in 1633, which resulted in his house arrest for the last decade of his life—at least he was not forced to drink Hemlock—was centered on Galileo's support of Nicolaus Copernicus's heliocentric theory of the solar system. That is to say, his support of Copernicus's bucking of religious doctrine when he placed the sun at the center of the known universe. This was decidedly not the view of the Catholic Church, which instead held the incorrect opinion that Earth is at the center of all things. Galileo was punished, and more to the point, *could be* punished precisely because he lived within the reach of the Church and the Church held sway over governance.

Copernicus, on the other hand, had the good fortune to be born in a part of Prussia—then called the Kingdom of Poland, now a part of modern-day Germany—that was beyond the reach of the Vatican. But it wasn't just scientists and philosophers who happened to be born beyond the Vatican's reach who were able to escape religious scrutiny. Even *religious leaders* were able to escape religious scrutiny in these places. It was no accident that the champions of the Protestant Reformation—Martin Luther in Germany, John Calvin in France, Thomas Cranmer in England—lived in places beyond the bounds of the Catholic Church's reign of religious persecution and control. They were able to ignite the Protestant Reformation precisely because they couldn't easily be punished by the Catholic Church.

I would be remiss if I didn't say that many modern-day Catholics are pretty cool and very tolerant religious folks, accepting of humanism and science. It needs to be said because, historically, this has not always been the case, and I wouldn't want to leave you, the reader, with a misconception. Laced throughout the lines of this manuscript is a general condemnation of religion in general and its habit of stifling human growth. Recorded history—that is, the Dark Ages—shows us how the Christian Church happened to be especially good at it, good at snuffing out humanism. Other religions have been and are just as guilty at stifling humanism. Or to quote John Dominic Crossan:

> My point, once again, is not that those ancient people told literal stories and we are now smart enough to take them symbolically, but that they told them symbolically and we are now dumb enough to take them literally.

Martin Luther and John Calvin were driven by religious doctrine, but Thomas Cranmer, the Archbishop of Canterbury at the time, had a different tormentor—he was driven by the lustful appetite of England's King Henry VIII, who desired to dispense with his first wife, Catherine of Aragon of Spain, so he could marry his next wife, Anne Boleyn. The Vatican did not easily grant marriage annulments and were certainly not about to offer one to a lustful king in a far-flung country. With Cranmer's help, the newly created Protestant-leaning Church of England granted the annulment. Henry's first wife, Catherine of Aragon, received a nice pension after the divorce; Anne Boleyn did not receive such a nice farewell when her time came, eventually losing her head over the marriage.

"His Highness is changing his relationship status."

In Facebook terminology, King Henry frequently and famously changed his relationship status. He had had six wives in total, with wife number two, Anne Boleyn, and wife number five, Catherine Howard, losing their heads. Wife one, Catherine of Aragon, was the daughter of the King Ferdinand II of Spain, so her head was safe by association. Even Henry was not about to give her the literal ax. Instead, he created the Church of England, a Protestant church, almost solely for the purpose of securing himself an annulment. Seems like an awful lot to go through for a divorce, right? Or maybe not. Wife three, Jane Seymour—his favorite—died while giving birth to what would be Henry VIII's only male heir, the frail King Edward VI, crowned at age nine and dead at fifteen. Henry's fourth wife, Anne of Cleves, bowed out of her own accord to avoid the chopping block, and wife six, Catherine Parr, outlasted King Henry himself.

We don't think of the Catholic Church, or at least not back during the Middle Ages, as ever granting a religious annulment, but they in fact did on the rarest of occasions. As long as you weren't British. The classic example is when Pope Eugene III, in 1152 AD, voided the marriage of King Louis VII of France and Eleanor of Aquitaine, based on the revelation that they were distant cousins—a fact discovered after they were married, and had had children. The divorce was granted, and each went on to remarry even closer cousins, making that annulment by the Pope rather farcical, divorce a distant cousin so you can marry a closer cousin.

One of the most fundamental doctrines of the Protestant Reformation was that the relationship between man and God was direct and unique—there was no intermediary, such as the Pope in Rome standing between God and man (and woman too, of course). In other words, being beyond the reach of Rome not only allowed major scientific advancements to unfold, not only ushered in humanism in science, art, music, and philosophy, but it also allowed the Protestant Reformation to be born.

It is not easy to reconcile religion and science. Borrowing from the ideas of famed American paleontologist Stephen Jay Gould, we can say that religion and science are in two separate areas of inquiry; by being placed in distinct domains, their individual teaching authority similarly should not overlap nor compete. Science and religion should not enslave each other.

Life is not that simple, though; it never is. Conflict arises when one domain uses its tools to quell or squash the other—fervid religious beliefs attempting to rein in scientific empiricism, scientific empiricism using facts to vanquish spiritual beliefs. As Bronze Age man emerged from the Neolithic period, religion and science arose independently of each other as two lines of inquiry that early humans used to explain their world. If you were to go back far enough, back before antiquity, and attempt to understand why our Neolithic ancestors created both the spiritual world and the empirical world, at the heart of it—besides a basic human thirst to understand—was an eagerness to make sense of misery and destruction. When a bad thing happened, like a drought destroying all your crops or a volcano leveling your seminomadic community, you sought either a spiritual answer or a scientific answer. You can't undo human history: our spiritual world and empirical world evolved simultaneously, but independently of each other. Yet conflict invariably arises when one world attempts to crush the other world.

I once went out on a few dates with a girl who believed in the biblical description of Noah's ark, and who also believed dinosaurs never existed. I didn't care to engage in any conflict (although the fact she was perhaps a shade gullible—open to suggestion—did pique my curiosity). It wasn't worth the fight, mainly because I was not about to change her spiritual beliefs and she certainly was not about to change my empirical facts, and thus there was no road to any reconcilable place that would see us dating.

Not all religious flocks are equally unsympathetic toward the sciences. Today wide segments of Catholicism, Christianity, Judaism, and Islam believe in evolution, even though vocal fundamentalists believe in the biblical story of creation. Many religious leaders simultaneously embrace both faith and empiricism. Consider this passage from the book of Genesis (1:31): "And God saw everything that he had made, and behold, it was very good. And there was evening and there was morning, the sixth day." Rather than read it literally as God creating the world in six days, a scientifically enlightened Christian might interpret that a day in the life of God is, say, one billion years, rather than our own twenty-four-hour day, which we puny humans measure by Earth spinning once on its axis.

Neither is it the case that the sciences never borrow tools from religion and faith in order to navigate the empirical world. Perhaps nowhere has that intersection been more apparent than during the creation of the first atomic bomb by the United States during World War II, where religious-based ethics, morals, and even faith figured heavily into the equation of creating and then dropping it.

The Manhattan Project, led by J. Robert Oppenheimer, harnessed the strong nuclear force of the atom—actually, he harnessed the strong nuclear force of a gazillion atoms all at once—to create the first atomic bomb. But that scientific discovery didn't come with a manual on ethics. Creating such a weapon of mass destruction, and then deciding to drop it on Hiroshima and Nagasaki in Japan, was a decision derived as much from America's religious underpinnings of decency and morality as from the goal of winning a war. There was a weighing of alternatives, such as the Japanese lives that would be lost by a nuclear explosion against the American and Japanese lives that would be saved by ending the war sooner.

For Oppenheimer, his moral transformation progressed from being a pure empirical scientist at the outset of constructing such a weapon—singularly focused on succeeding in building an atom bomb—to harboring a profound moral objection to what he had helped create, which presented an ethical conundrum for him and others. In the end, Oppenheimer regretted his role in creating the atomic bomb, borrowing his famous statement from the Bhagavad Gita: "I am become Death, the destroyer of worlds."

And speaking of worlds, small worlds, tiny worlds, atomic worlds, nuclear worlds, that is where we now venture.

The strong nuclear force, which allowed Oppenheimer and pals to unleash "I am become Death," a power our world had never seen before, except for few asteroid strikes here and there, is just one of the four fundamental forces in the universe. There four fundamental interactions in the universe that cannot be further reduced: electromagnetism, weak nuclear force, strong nuclear force, and gravity. The first three fundamental interactions are fairly well understood, but the "how" of gravity still remains largely a mystery—somewhat taken on faith. Scientists have yet to prove the seat of gravity, but we certainly believe in it. Just slip on some ice.

Electromagnetism describes such things as why electrons will travel endlessly down a wire looking for something positive in its life, or the interaction between two charged particles, the "why" behind such phenomena as a positively charged sodium ion (Na^+) binding to a negatively charged chlorine ion (Cl^-) to form sodium chloride ($NaCl$), otherwise known as table salt—a compound that has been a part of a rich, sometimes violent, world history.

In use even and coveted by Neolithic humans, salt has been an important ingredient in human history: as a necessary dietary supplement, as a spice, as a preservative for mummies in ancient Egypt, and as seasoning for our salads (which is a word derived from "salt"). Even the Roman expression "worth one's salt," meaning a person is deemed to be competent, is based on how Roman soldiers were often paid in salt, rather than coin, it being such a valuable commodity.

Electromagnetism working with the weak nuclear force is probably what keeps electrons in orbit around a nucleus. Earth stays in orbit around the sun mostly due to gravity, whereas an electron does not stay in orbit around its nucleus for the same reason. Electromagnetism working with its friend the weak nuclear force, sometimes referred to the electroweak force keeps those electrons in their shell, in their defined orbit. It is also why the rules of Newtonian Mechanics—body in motion stays in motion unless influence by another force or by gravity—falls apart in the atomic and subatomic world where particles follow Quantum Mechanics.

As a brief aside, at one time electricity and magnetism, like those magnets on your refrigerator with the grocery list, were thought to be two different forces at work. Electricity and magnetism are one and the same. Whether it is a steady flow of electrons down a wire which we call a current, or the negative end (north pole by convention) of a magnet attracted to certain metals which we call magnetism, or an electrical coil with electrons going around and around producing a magnetic field in the center which we call an electric motor, or two atoms combining which we call a molecule, all boils down to charged particles attracted or repelled by each other. Although several nineteenth-century scientist contributed to the runup to the concept of electromagnetism, names like Hans Christian Ørsted (1777–1851), André-Marie Ampère (1775–1836) and especially one of my personal favorites Michael Faraday (1791–1867), the unification of electromagnetism we owe to James Clerk Maxwell in 1865.

What did these electrified scientists do? Hans Ørsted the Danish scientist discovered that an electric current running down a wire produced a magnetic field. He was also greatly inspired by the eighteenth-century Enlightenment philosopher Immanuel Kant (1724–1804). The Frenchman André-Marie Ampère's contribution to electricity is in his name, the ampere or simply amp, which is a measure of electrical current over one second, the coulomb. But for our purposes Ampère also demonstrated that a circular current produced an even more profound magnetic field within the center. Michael Faraday, the dapper Brit contributed significantly to our understanding of electricity and chemistry, pioneering work in electrolysis (plating metals onto a surface using a current of electricity to force the chemical reaction: electrochemistry), but for our purposes Faraday elucidated the law of electromagnetic induction. What does that mean? Sending a current

around and around in a frenzied circle spool of wire will magnetically spin an internal metal rod which then turns things, that is, if that rod spinning inside the electromagnetic field is connected to something, the basis of the electric motor. You have one in your refrigerator, the alternator in your car is basically an electric motor, and if you own a Tesla, you have a slick electric motor and clever battery storage.

But as for Clerk Maxwell (1831–1879), he did for the fundamental force of the universe, electromagnetism, what Isaac Newton had done for the laws of motion. He demonstrated, and I will say this as delicately as possible because I too struggle with the concepts, that the waves of electricity, the waves of a magnetic field, and the waves of light are all fundamentally the same. That underneath it all is a basic fundamental property of the universe, that joins the weak and strong nuclear force and gravity as the four fundamental forces. When Einstein was asked if he stood on the shoulders of Isaac Newton to come up with his out-of-this-world theory of special relativity, he quipped: "No, I stand on Maxwell's shoulders."

The weak nuclear force describes the short distance force that keeps or does not keep certain subatomic particles from parting ways and it also plays a role in radioactive decay. A decaying atom is quite different from splitting an atom—don't confuse the two. The force that keeps protons and neutrons together in the nucleus is the strong nuclear force, not the weak nuclear force. Just split an atom and it goes Boom! If you drill down deeper into what protons, neutrons and electrons are made of, it is the weak nuclear force that helps those subatomic particles with weird names like quarks, leptons and bosons, combine so they make protons, neutrons and electrons.

Another example of the weak nuclear force—elucidated around 1900 by three great scientists: Wilhelm Röntgen, Henri Becquerel, and Marie Curie—is the naturally occurring decay of radioactive uranium and similar elements found on Earth. An atom's nucleus is composed of protons and neutrons, around which electrons orbit in defined shells. A perfectly balanced atom, say your everyday carbon atom has six protons, six neutrons, and six electrons. The number of protons define the "atomic number" of the atom, it's identity, so for carbon it is six, but the combined mass of the protons plus neutrons for an atom is its "atomic weight", so for carbon it is twelve. The number of electrons always equals the number of protons, always, but the same cannot be said of the neutrons, sometimes there are more.

Take uranium for instance, it can appear in several varieties, due to extra neutrons in the nucleus—not extra protons, not extra electrons, but extra neutrons—making for an unstable configuration of that nucleus that endeavors to become stable. The more neutrons the likely the atom is more unstable. This is important: If you add an extra proton to the nucleus, which can't be done, unless you also add an extra neutron and an extra electron, all you get is the next element along the periodic table.

Normal carbon has an atomic weight of 12, written carbon-12. If you add an extra neutron you get carbon-13 and add another extra neutron you get carbon-14, which is unstable and due to decay jettisoning the extra neutrons in an exact half-life, carbon-14 is the poster child for radiometric dating really old things, termed carbon dating. But if you add an extra proton to carbon-12, it will not happen unless it also adds an extra neutron and an extra electron, and what you get is the next element on the periodic table nitrogen N with seven protons, seven neutrons and seven electrons and an atomic weight of 14.

But extra neutrons can and do arise, the atom desiring balance is then achieved by jettisoning the extra neutrons, and each neutron ejection releases some form of radioactivity. That decay is most often in the form of an alpha particle (helium nucleus), beta particle called a positron (anti-electron) or gamma radiation (short-wavelength photon). That is the basis of how that chest X-ray you got last year works: the ejected radiation does or does not pass variably through your body tissues, depending on the different densities encountered—bone absorbs the radiation more than your lungs—and the end result is imaged on receptive film. The weak nuclear force plays a role in jettisoning those extra neutrons, that is, radioactive decay.

The strong nuclear force is the force that keeps protons bound to protons and neutrons bound to neutrons, as well as protons and neutrons bound to one another. If an atom is split by fission—if protons and neutrons are ripped from each other—it goes *kaboom*: an atomic bomb.

The theoretical basis for the strong nuclear force was hidden in Albert Einstein's famous equation $E = mc^2$, which in its simplest explanation means that energy and mass are two versions of the same thing and that there is a lot of energy stored inside *them there* atoms that make up all the visible mass of the universe. Although we'll get more into it later on, let us appreciate at this juncture how, prior to the Big Bang 13.8 billion years ago, there was nothing but pure energy—no mass, nothing. Due to some fluctuation, which some would call the finger of God, that pure energy formed the entire universe, that is, all the subatomic particles, which then formed all the protons, all the neutrons, all the electrons, all the photons of light, and also formed the four fundamental interactions.

The literature is a bit unclear as to what discovery came first: the existence of the electron or the existence of the proton. That there was something inside an atom that carried a positive charge—later named the "proton"—was demonstrated in 1886 by the German physicist Eugen Goldstein, but his experiment was limited to proving a positive charge, not the existence of the proton. The actual identification of the proton occurred in 1917, observed by one Ernest Rutherford, a New Zealand–born British physicist, when he demonstrated that the hydrogen nucleus has a single proton and no neutron—just that singular, lonely proton. The single proton hydrogen nucleus does have a single electron in orbit around it either keeping it company or driving it crazy, so there is that.

In 1897, another British physicist, J. J. Thomson (he preferred being called J. J. to Joseph John), demonstrated the existence of electrons in cathode rays, electrons passing from one end of his cathode ray tubes to the other. Electrons do that, they try to find that positive thing in life. It would take the work of James Chadwick, a dapper Englishman working out of the Cavendish Laboratory at the University of Cambridge, to arrive at the discovery, in 1932, of the neutron—that subatomic particle that, when paired with the proton, makes up the seat of the strong nuclear force.

To summarize the four irreducible, fundamental interactions of the universe, let us consider table salt, otherwise known as sodium chloride, or in science speak, NaCl. The weak nuclear force is what keeps the subatomic particles—quarks, leptons, bosons—bound to each other to make protons, neutrons and electrons so that atoms such as sodium and chloride can exist. The strong nuclear force is what keeps the protons and neutrons inside the nucleus of an atom of sodium Na bound to each other and the same for an atom of chloride Cl. Going back to the weak nuclear force it also keeps the complementary electrons in their orbital shell, and the charisma or temperament of those electrons—the electromagnetism—is what then allows the sodium atom Na and chloride atom Cl to pair bond giving us salt to sprinkle on nearly everything.

As for gravity, as I mentioned, it is still not fully understood—thou heartless biatch! Scientists have no idea what gravity is. Actually, more correctly, they have a *lot* of ideas of what it is, none of which has outgained the others at this point. Gravity was demonstrated by such Renaissance folks as Galileo around 1590, with his experiments that involved dropping cannonballs off the Leaning Tower of Pisa—a moment from science history that conceivably every elementary school student has been taught. Galileo determined that two cannonballs of different mass fell at the same speed from the top of the tower. One hundred years later, in 1687, gravity was given a mathematical equation—of which I will spare you the details—by yet another genius of geniuses, Isaac Newton.

Then along comes another genius of geniuses, Albert Einstein, with his general theory of relativity, which, among many things, correctly proposed that those two cannonballs of Galileo didn't quite fall at exactly the same speed. In the macroscopic Newtonian world, those cannonballs did fall at the same speed; in the cosmic world of Einstein, they did not. Which means, among other things, that gravity—although everywhere and all-knowing, all-seeing—is actually the weakest subatomic force of all. The most pervasive force, yet the weakest.

Newton postulated that gravity—famously portrayed by the unlikely story that he was sitting under an apple tree when an apple fell on his head—is an instant interaction among all objects in the universe. When Earth tugs down on you, you are tugging up on Earth. That is the gravity you and I know. To Einstein, however, gravity was not a tugging of two objects on each other so much as a curvature of space that affects them both. Sort of like placing a bowling ball on a trampoline: the spring-loaded mat curves slightly downward due to the weight of the bowling ball. Mass does that to space—it curves it—and if space is curved, matter and time are both affected.

The Earlier, Less Publicized Discovery of Gravity by Cornblatt

But what the heck is that supposed to mean: the curvature of space? And for that matter, what is the curvature of time? Newtonian mechanics works great for us in the macroworld on Earth, but it begins to fall apart when it's applied to what's happening out there in the universe, or down there in the inner recesses of an atom. Before we move on to dicing dogma in the next section, what exactly did Einstein mean when he said gravity is the curvature of space-time?

Consider this: if you go out into the most remote spot in the universe, somewhere where there is nothing around for billions of lightyears (it may take you a little while to get there, so pack some snacks, Oreos are the best), there will be no measurable gravity. Out there in that dark and lonely spot, space is empty, time doesn't really exist, gravity is nada—it's all just very, very flat. If you take an apple out of your pocket (don't eat it as your snack, because you need it for the experiment) and place it softly into that emptiness, it won't fall anywhere. It will just sit there motionless, because nothing is acting on it—no gravity, no curvature of space, no force you gave it.

To conduct this experiment, you'll also need to retreat to a safe distance, to ensure your mass does not influence the apple quietly sitting there, lonely and alone. If you then take another apple out of your other pocket and place it near the first apple (but you yourself must stay far, far away!), they'll ever so slightly tug on each other. That is Newtonian gravity. Somehow those two apples are communicating with each other over distance and time. Is there a secret signal between them (Newtonian gravity) or are those two apples slightly bending space (Einsteinian gravity) to give the *appearance* of an attraction to each other, or is it both? Since we're talking about Einstein's gravity just now, we'll go with his answer: space is bending.

Next in our experiment, you must take two identical Timex watches—or Rolexes, if you prefer—and place one watch on each apple. The tick of time will pass the same on either watch. Ticktock, ticktock. If you were to separate those two apples by a larger distance, but still within that empty bit of flat space, in Newton's world those two apples would tug on each other, albeit with less enthusiasm, and time would pass in the same way; in Einstein's world, those two apples would bend space but with less consequence on each other, and time would pass the same. All is well and good for both Newton's apples and Einstein's apples with their watches.

But what happens if other things are introduced into your bit of space out there, in the flatness of the emptiest part of the universe? Newtonian mechanics begins to fall apart, whereas Einstein's theory seems to hold fast. (I suspect Einstein's world is so different probably because he was an alien, not from this world. Although Newton, too, was likely an alien, but maybe just from a different neighborhood.) In Einstein's world, things change when you start adding new elements into the bit of space. If you were to take a nice big star, like our sun, out of your pocket and place it close-ish to the first Timex-equipped apple, but not the second Timex-equipped apple, and gave that first Timex apple a good nudge, several things would happen. The inertia of that first Timex apple coupled with the gravitational pull of that star, if the forces are balanced, would send the apple into a perpetual orbit around the star. This result is explained by Newton's first law of motion that a body in motion tends to stay in motion. If the inertia you give that first Timex apple is too great, it will saunter off into deep empty space forever, escaping the gravitational pull of the star, never to be heard from nor seen again. If the inertia you give that first Timex apple is too small, it will enter into a death spiral around and eventually into the star—also never to be heard from nor seen ever again.

Assuming that the first Timex apple enters into a successful orbit around that star, the other thing that would happen is that time on that now star-orbiting first Timex apple would slow down, while time on that second Timex apple, still way out there in empty space on its own, would not slow down. This is not Newton's universe; this is Einstein's universe. Space is being curved on that first Timex watch orbiting that sun, but not on the second Timex watch sitting all alone, meaning that the first watch takes longer to get from point A to point B in its curved space then the second watch, which is not moving in its flat space. The more gravity, the more the curvature of space, the longer the time—which is how us puny watch-wearing humans measure time: from point to point. This phenomenon is termed "time dilation."

The fun movie *Interstellar* (2014), which stars Matthew McConaughey—a rather talented actor—as a former NASA pilot turned deep-space traveler named Cooper, includes a scene that emphasizes how gravity is a product of the curvature of spacetime, in a way only a Hollywood movie can exaggerate and still drive home the point. Without too much in the way of an explanation, the backstory of *Interstellar* is that humans are killing Earth—big surprise—and Earthlings are exploring other planets in other galaxies on which to keep the human species alive, which they travel to using a wormhole through space. Cooper and crew have arrived at an unnamed planet somewhere on the other side of the universe. That planet happens to be near a supermassive black hole called Gargantua. The crew arrives there through that wormhole—an ultimate curvature in space. Putting aside all the scientific impossibilities of a Hollywood film, the movie reveals that, due to the gravitational curvature of time by the Gargantua black hole on the planet they're setting down on, one hour spent on that planet exploring for life is equivalent to seven years back on Earth. It is a Hollywood example of how mass curves space, causing time to slow down, otherwise known as the dilation of time.

I guess there must have been a great deal of gravity in my middle school third-period geography class, because time really, really, really slowed down, and the gravity of me not paying attention to my teacher, the gravity of me not doing my homework, came to bear upon me when I brought my report card home for mother's signature, as did the gravity of the paddle my dad walloped me with. Actually, my father never did that. But the gravity of my mother's disappointment was enough of a burden, expressed in the form of one week's grounding—house arrest in the black hole that was my bedroom.

A modern, everyday example of how gravity affects you and me, how it can affect time in our daily lives, are the many Global Positioning System (GPS) satellites orbiting 12,000 miles or so above Earth—a range termed "medium Earth orbit." GPS allows our iPhones and Androids, coupled with applications like Google Maps, to know exactly where we are on Earth, as well as enables Siri to tell us where the nearest coffee klatches are. But get this: the clocks on those GPS satellites above us tick ever so slightly faster than the clocks on Earth due to that decrease in gravity at 12,000 miles into space. Because of this discrepancy in the passage of time between an Earth clock and a GPS space clock—even though it is of a nearly imperceptible increment—NASA scientists have had to apply a fudge factor to those GPS satellite clocks, sort of like our previous discussion of the Gregorian calendar and its leap year. That way, when you ask Siri where the nearest Starbucks is, you don't end up next door at the mattress store where they always seem to be having a huge sale. Which might not be such a bad thing because, maybe, just maybe, you might need a new queen-size mattress at 75 percent off rather than a grande decaf light-foam vanilla latte for $3.75.

3 DOGMA IS NOT A MOTHER'S DOG

Having previously mentioned the concept of dogma permeating both religion and science alike, decrees made to look as if they came from God herself, or proclamations made to look scientific but often are either false or taken on faith, such as Galen's second-century miasma theory that smells—bad vapors—were at the seat of infection spread, we now turn to what exactly is dogma, using Galen of course.

The word "miasma" comes from the Greek word meaning "pollution," which is ironic considering how Galen's incorrect assumption polluted medical knowledge for so long. Galen's miasma theory began to unravel in the seventeenth century and was dealt a fatal blow in the nineteenth with the introduction of Louis Pasteur's germ theory. But a millennium and a half is a long time for a false theory to survive—maybe not in religion, but certainly in science.

To be fair to Galen, as previously communicated, the concept of "bad air" as the cause of human infection originated before him, coming to him in bits and pieces from earlier Greeks. Hippocrates and Aristotle also believed in the bad air theory, but it was Galen (who was also one of the personal physician to the Caesars of Rome) who codified it and put his stamp of approval on it, unleashing centuries upon centuries of dogma. The miasma theory wasn't entirely off: certainly colds and the flu with people coughing and sneezing droplets laced with viral particles all over each other transmitted infection through the air. But it wasn't the *smell* of vapors that caused illness, and bad air certainly wasn't the only mode of transmission of infections.

What exactly is "dogma"? We use the word so frequently but, really, what does it mean? Or, in the immortal words of Inigo Montoya to Vizzini in *The Princess Bride* (1987): "I don't think that word means what you think it means."

"Dogma" comes from the Latin, which came from the Greek, and means "opinion" or "tenet"—that is, that which one thinks is the truth or is told is the truth, when in fact it might not be true at all. Dogmas in religion are just that: unprovable "truths" handed down from a higher authority, year after year, century after century. Religious dogmas are not the sort of thing easily dismissed or disproven or discarded simply because they lack "independent proof." They're based on faith, not empiricism. Isn't that the whole thing about religious doctrine? It can't be empirically disproved because it's taken on faith, not on fact. And, as I stated previously, faith is a powerful form of inquiry into our world that arose independently from science and that should not so easily be cast aside like last night's guacamole or a doofus boyfriend.

In comparison, one would like to think that dogmas in science and medicine, especially if the "truths" are in fact untrue, are more easily disproven. And if you think that, you'd be correct—most of the time. But even in science, in medicine, sometimes dogmas are not as easily disproven as we'd like to think. Sometimes in medicine we do "things," believe "things," because we were taught those things, and when those things are disproven or come crashing down, it can still take time to dismiss that which we thought we knew to be true.

To fully understand the concept of dogma, let us leave the world of dictionary definitions and journey to the thought experiment of five monkeys, based on the research of psychology folks like Harry Harlow and G. R. Stephenson who plied their trade studying behavior modification in monkeys, what they gleaned from our distant ancestor, and how those lessons might be applied to humans…or not. One need only watch the hilarious 1978 classic movie *National Lampoon's Animal House* starring the irrepressible John Belushi who brought to life the equally irrepressible character John Blutarsky; you won't need to study monkeys to understand human behavior.

Five monkeys are placed in a cage where a banana sits atop a ladder. One monkey attempts to climb up the ladder to get the banana, and all five monkeys are drenched with freezing-cold water. After several more similar events of an ice shower, if one of the monkeys even attempts to go up the ladder to get that tasty banana, the other four chagrined monkeys, fearing an icicle bath, tackle the ladder-climbing monkey—drop it to the cage floor like fourth-period algebra. Eventually, after enough attempts have passed, after all five monkeys have been drenched with enough ice water, after enough days have gone by, after enough monkeys have gotten beaten up by the other monkeys out of fear of receiving an ice shower, no monkey ever again attempts to go up the ladder, despite a tasty banana perched at its top. Day after day after day, no monkey will touch that wretched ladder to get that inviting banana. Those five monkeys don't need to be drenched anymore, don't need to be beat up anymore, they just no longer even think twice about getting the banana.

Then one day, one monkey—and only one monkey—is removed from the cage and a new monkey takes its place. When that new monkey attempts to go up the ladder, the remaining original four monkeys repeatedly stop it, pulling it off the ladder like all hell is about to break loose, mercilessly beating up that new monkey. There is no longer any need for the ice shower—the bewildered new monkey soon abandons all thought of getting to that banana just because of the reaction of its cage mates.

This ruthless process goes on and on, with each of the five original monkeys being replaced with one new monkey until, one day, all five monkeys are new monkeys. Most important to understanding dogma is the fact that none of the new monkeys have ever been drenched with an ice shower. So, now you have a cage with a ladder with a banana atop it and five monkeys, none

"I can't remember. Are we for or against cloning?"

ever having been drenched with freezing-cold water, but all having been beaten up, and no monkey will dare climb up that ladder to get that banana. They know nothing of ice-water showers; all they know is they'll get beaten up if they go up the ladder.

Now, here is the dogma. If you were to ask any of those five new monkeys why they dared not climb the ladder to get the banana, all they could tell you is this: "We are not supposed to, and it's always been done this way." That is dogma.

Science in general, and for our purposes medicine in particular, should be adept at dispelling dogma, precisely because science relies on empiricism. Yet despite this fundamental underpinning of science— namely, proof—sometimes dogma still wins out. Which is why it becomes even more important that the empiricism inherent to the scientific method is designed to parse fact from fiction, truth from lies, the one from the many. This is where our story now turns: chapters a lot less about infection and medicine and a lot more about science and the nature of the scientific method, chapters in which we'll explore such things as the experimentation that led to the discovery of DNA and the process that gave rise to bacteria, the first earthly life forms from which all other organisms on Earth derived.

That phone rang.

A lot of things about the art of medicine, about the practice of medicine, can't really be taught, and dialing a patient over the phone is one of those innately unteachable lessons. And when I say, "dialing a patient over the phone," I don't mean literally dialing a phone. "Dialing a patient" in medicine means to turn the dials of health, like fine-tuning an old radio dial to perfect reception. What it describes is bringing all to bear upon patients to make them better. Dialing the ventilator settings, dialing the serum glucose levels, dialing the blood pressure meds and the heart contractility meds. "Dial" is from the Latin *dies*, meaning the course of a day and from whence we get "day," "daily," "sundial," and even "Dial soap." The soap brand used the word as part of its original marketing gimmick, with the claim that a single washing with Dial protected one's hands from germs around the "dial" of a clock; likely untrue, but it was a clever allusion marketing scheme, nevertheless.

It is difficult enough to dial a patient to good health when they're sitting right there in your medical office or lying on an ICU bed, but dialing a patient over the phone into better health is even more daunting, even more of an art. The process, devoid of observing the patient, relies frighteningly on the strength of words. When a physician has the opportunity to see the patient, to perform a history and a physical exam, a lot of things come together. Sometimes in a lickety-split-second a physician can simply look at a patient, lay eyes on a patient, not even ask a single question or perform a single exam or order a single laboratory test, and still have a strong hunch whether the patient is fine or not fine, whether the patient is on the road to recovery or has fallen off the curve. When talking over the phone to a patient, especially through an intermediary like a husband, as when Alan was conveying Deborah's illness through Barbie to me—well, a great deal can get lost in the translation. It is a classic "I heard from a guy, who heard from this other guy who knows a guy…" situation.

As I described in a previous section, "falling off the curve" in medicine refers to a patient who is no longer following the normal predicted sequence of events to recovery. The winter cold in particular follows an expected, predictable progression to resolution. When the normal resolution of cold symptoms does not occur, the wayward path the patient has gone down instead is called "falling off the curve," well illustrated in the next story.

Once, I moonlighted at an urgent care clinic, places where physicians are euphemistically dubbed "docs-in-the-box." I was "popping up" there because I needed extra bank to provide for my young family during those long, lean years of toil on a surgery resident's salary, with a wife and two young children. Anyways, one sunny day I walked into this urgent care clinic for my shift and was greeted by the receptionist, who looked uncharacteristically ragged. I knew something was brewing—my neck hairs stood on end, my stomach got a little queasy. The receptionist went on to explain as best she could, using nonmedical language, that the physician whose shift I was about to relieve had been buried neck deep in an exam room with a single patient for several hours. This was an urgent care clinic, not an emergency room, and, as such, the visits always went one of two ways: the diagnosis is made in short order and a treatment begun, or the urgency is in fact an emergency and the patient is quickly shipped off to the nearest ER. Either way, it shouldn't take gobs of time to decide which way the wind is blowing. An urgent care clinic is not a place to dilly-dally.

But on that day in that exam room, that physician was gobbling up precious time dilly-dallying, ordering all kinds of tests, many of which had to be performed off-site. That is, a blood test not performed on the premises, a complete urinalysis not performed on the premises, a microscopic exam not performed on the premises, and yet still all this time was being used up—or, to quote Mercutio in Shakespeare's *Romeo and Juliet*, that doctor was "burning daylight." Texans like to claim "burning daylight" is a Texas phrase, a cowboy phrase, it is not. It is pure Shakespeare.

Before I even made it into the exam room, that sunny day had started to turn appropriately cloudy. The nurse stopped me outside the door, looking even more uncharacteristically ragged than the receptionist had, and proceeded to explain and exclaim, with pleading eyes and in medical terms, that the doctor who I was about to replace was eyes deep with an ill patient, and that the physician had been unable to decide which way the wind was blowing. Now my neck hair was *electrified*, my heartbeat picked up tempo, my eyes narrowed, my brow furrowed, my pupils constricted, and my teeth clenched as it dawned upon me that I was about to walk into a room with a whole lot of hurt going on. The receptionist was on edge, the nurse was over the edge, and now I was approaching that edge and I hadn't even stepped foot in the exam room. What was once a sunny day was now a storm inside that doc-in-the-box.

I took a deep breath and opened the door. I gazed at the patient, gazed at the doctor, scanned the room, and then returned my focus back to the patient. I didn't ask a single question of the patient or the doctor, I didn't look at a single one of the lab tests that had come back. I simply stared at the patient for several moments, perhaps tilted my head slightly to the side, and then I turned to the nurse and told her to call 911. It was clear to me but apparently not to the other doctor that what we had before us was a patient who had fallen off the curve—way, way off the curve and was circling the drain. What the doctor thought was an urgency was an emergency, what he thought was something he could diagnose was for his skill set undiagnosable, what he thought was a patient he could dial in an urgent care center was undialable. He couldn't see the forest for the trees. He couldn't see the trees, either.

The ambulance arrived within ten minutes and lightning quick whisked the patient to the nearest hospital emergency room, sirens blaring. In the end, the patient was suffering from a particular type of heart attack—the symptoms of which can mimic the flu. I don't relay this story to brag about my keen sense of observation; to the contrary, it's more about the lack of a keen sense in that other doctor. Hold on to this vignette, because it figures later on in another story, and in subsequent volumes, a common recurring theme in medicine.

That phone rang.

So, there I was in Florida, trying to parse where the toad of truth sat, speaking to Alan over the phone about Deborah's supposed relapsed upper respiratory infection. Handicapped as I was because I could not lay eyes on her, let alone examine her, I listened carefully to glean whatever clues I could. Because even when you are able to lay eyes on a patient, as the previous story just conveyed, you don't always see what you need to see, and worse yet, you can get caught up running down a path you simply believe to be true. You start to see what you want to see, your eyes betraying what you *need* to see: a patient in a doc-in-the-box clinic circling the drain. Was Deborah circling the drain?

One day as a medical student rotating in the ER at Denver General Hospital, I was assigned to perform the initial evaluation of a woman on Gurney #3, a pleasant middle-aged woman complaining of flu-like symptoms. Her chief complaint was something on the order of "I don't feel well, I think I have the flu"—and, as you might have guessed, this is key to the story. The primrose path begins: the patient mentions "the flu" and we start to be taken down a pathway that indeed might be the wrong trail.

Gurney #3 was describing some nausea, sweating, shortness of breath, and an overall sense of not feeling well. In medical parlance it is termed "malaise." The triage nurse had assigned her symptoms, "possible flu," on the Big Board—that dry-erase board where the entire panorama of the ER unfolds. Today most ERs use a big-screen monitor to track the ins and outs and goings of patient care for all to see. Not in my day. Back then it was a huge dry-erase board, and even an old chalkboard at some hospitals. On that dry-erase board, complaints were listed, orders and labs scribbled, with emergency orders written in big red letters, progress revealed—basically the one place chronicling everything unfolding in the microcosm of the ER. A board that no doubt looked like chaos to the casual observer unfamiliar with the workings of an inner-city emergency room.

There is a saying in medicine: "Don't let the patient get in the way of the diagnosis." The corollary being, don't be led down the primrose path. A key aspect of the scientific method, even for a wet-behind-the-ears medical student like I was that day, is that you're supposed to rise above false information, put the pieces of the puzzle together, and not get distracted by false leads. You would like to think that a medical student with all that medical school knowledge neatly tucked away into the crevices of a bursting brain could easily zero in on a patient's ailment. And if you thought that you'd be wrong. Book knowledge without experience is only half the story.

Apprenticing medicine is just like apprenticing plumbing: this hands-on experience is where the medical student merges book knowledge with clinical knowledge and becomes the resident, becomes the healer, by taking hand grenades in those trenches. I was a lowly medical student drowning in the sea of a busy city ER. I was hardly the healer yet, but I was beginning the transition from working only with books to working with patients.

There's a joke about never wanting to show up ill at an ER the first week in July because you might be cared for by a first year intern fresh out of medical school. The internship and residency programs in medicine, it is true, are designed to further advance the clinical skills of the young physician. But the thing is this, whether it is a fourth year medical student, an intern or a young resident managing the point of service of an ER patient, they're not drowning in a sea of frailty. They have senior residents and attending physicians keeping their skiff sailing fairly smoothly, a few waves here and there.

"Actually, you're my *second* patient if you count that cadaver in med school."

That particular day with Gurney #3—due to the fact that once I was assigned to her, I was required to take a full history and physical exam as part of my learning exercise—felt like I was just burning up precious daylight. Everyone assumed we had all the time in the world with Gurney #3. After all, that patient just had the flu, or so the patient said, and as such the triage nurse directed her to the medical student as a learning experience. Real-time ER docs don't burn daylight gathering a bunch of minutiae like taking a complete history and physical exam; instead, they zero in on the problem, a pertinent history and targeted physical exam. Unless a small detail is germane to the ailment, ER docs are laser focused on the obvious problem at hand.

But us medical students assigned to an ER rotation wearing the short white coats had to, despite the emergency setting, ask about things like social history and family history, even though those inquiries had little if anything to do with the emergent ailment. So, there I was, being a dutiful medical student, asking an ill woman where she was born, what she did for a living, when she immigrated to the United States, what her husband did for a living, how many children and grandchildren she had. I might as well have asked what her favorite color or favorite flavor of ice cream was, because that has just as much to do with whatever medical problem was ailing her.

On that day, in addition to being a wet-behind-the-ears medical student, I was also what you might call a short-white-coat medical student. The length of the white coat a physician wears is like the insignia that a military officer displays on their uniform. Despite some TV shows like *Scrubs* (2001–10) showing medical students and residents only in scrubs, we also wore white coats, just like the seasoned docs. Not just because the coats were required, but because they were a practical vestment. With big pockets full of every imaginable thing that a medical student or resident might need to navigate through a day on the hospital wards, our coats were the difference between success and failure. In residency I weighed perhaps 160 pounds, but with my white coat on, I easily topped 180. Stethoscope, the day's to-do list with all the patients under my charge, cheat sheets, extra IVs, extra syringes, that cute nurse's home phone number scribbled on the lab report for the patient in room 737, a half-eaten candy bar, an unopened can of cola. Those short white coats' pockets were stuffed to the brim with everything and anything that could fit into them.

As for the "white coat protocol": medical students had to wear a white coat to the waist, no longer; residents wore mid-thigh length, no shorter, no longer; attending physicians—the bosses—wore those long ivory tower scholarly white coats that flow below the knees. White coat length was the caste system of the teaching hospital.

My white coat, like the coats of the all other medical students and residents, was a hazmat-condemnable item. Our coats were not provided by the hospital like our scrubs were, meaning not only did we have to buy them ourselves but also take them home and wash them ourselves. Like that was ever going to happen with any regularity, given our overwhelming workloads. White coats covered in blood, covered in vomit, covered in urine and poop, covered in food, covered in aerosolized viruses from hacking patients, covered in every microbe that could walk into or out of a hospital. At day's end, when we were finally able to go home for a quick meal and some much needed sleep, our white coats ended up back in our hospital lockers, where all the defilement could percolate until the next shift began. And maybe even some mold would join the fray. If you were organized and assuming you did laundry with any consistency, you could bring in replacement armor. But, most likely, you were donning that same white coat for the next day of assaults, and for many, many days after that.

The chief complaints of Gurney #3, my middle-aged female patient, combined respiratory flu-like symptoms with what seemed to be some intestinal flu-like symptoms sprinkled in. This should have been a red flag. It was not. For most people, the flu is a respiratory condition and, technically speaking, the flu is just that: the flu. But there is also such a thing as the "stomach flu," although its correct nomenclature is "viral gastroenteritis" (the medical world frowns on the moniker "stomach flu"). Usually a person doesn't have both a respiratory flu and a stomach flu, since they are caused by totally different pathogens. Children with respiratory flu often have "stomach symptoms" because, well, they're children, but adults usually—not always, but usually—don't have stomach complaints with a respiratory flu.

I took a history that was not significant other than the flu-like complaints with a stomach component. But since I was a medical student required to take a more complete history, I came to know a great deal about Gurney #3, information that would otherwise not be part of a normal ER visit. Information that would make me look good to the attending physician to whom I reported, in short, who was grading me. Information that was, in retrospect, a waste of time, focusing on the trivial when perhaps there was a monster lurking. Burning daylight.

That phone rang.

Time ebbs away. I was burning daylight with Gurney #3 and didn't even know it. That physician whose shift I relieved in the doc-in-the-box anecdote was certainly burning daylight. Was chatting over the phone with Alan about Deborah's illness burning daylight, too?

The physical exam of Gurney #3 was not much more revealing than the history I took. Her vital signs were acceptable. I listened to her heart and it was beating, which is about all a medical student can tell from a stethoscope exam. Her lungs were quiet, no noisy sounds like the ones that accompany pneumonia. I didn't get much from the abdominal exam either.

I dutifully reported my findings to the attending physician, doing my part to advance the well-being of Gurney #3—or so I thought. The attending asked me questions, some of which were pointless, just to see if I had done my duty. He would ask me what her temperature was and then something like where in Mexico she was born, just to test me, to see if I had performed a social history on her.

Which is where our story now turns: one essential aspect of science is the asking of questions.

Science perhaps more than any other area of human inquiry is an endeavor that builds upon knowledge derived by those who came before (like Einstein quipping he stood on the shoulders of Maxwell, not Newton). Understanding how science builds and grows, adds to the appreciation of the scientific method. The scientific method is how we can tell nonbelievers the difference between fact and opinion—like the difference between normal global warming and abnormal human-driven global warming—fact versus opinion, a concept lost on too many to even begin to count.

Chemistry is perhaps the poster child of how science builds upon science, and so we'll take a moment here to explore the scientific method using chemistry. Don't fret—the following passages will be sheared of medical jargon. There will be no test at the end.

Classical Greece had Aristotle and his four elements of the universe: earth, air, fire, and water (not to be confused with the 1970s R&B band Earth, Wind & Fire, purveyors of such hits as "September" and "Shining Star"). Today there are 118 known elements listed on the periodic table, the first ninety-two of which exist naturally on Earth, starting with hydrogen (H) at number 1 and running up to uranium (Ur) at number 92. But for Aristotle there were not ninety-two elements, there were only those four.

In the course of a little over 2,000 years, we went from Aristotle's four elements to the current ninety-two natural elements. Actually, it was a much lesser-known Greek, Empedocles, who proposed the four visible elements—earth, air, fire, water—with Aristotle giving it his stamp of approval, as well as Aristotle adding a fifth element that could *not* be seen: ether. Ether—not the anesthetic gas—was thought to be an invisible element permeating the entire universe that allowed for such mysterious things as how light from the Sun and stars propagated to Earth if not for a medium in which to propagate. Consider sound, there is no sound in outer space because there is no medium for the sound waves to move within.

Perhaps the ancients didn't know much, but they knew a wave needed a medium in which to, you know, be a wave, and to them light was no exception. From Aristotle's day up until Einstein, scientists knew light behaved like a wave, and like all waves it needed something to propagate through. As a consequence, this fictitious "ether" was invented as such a medium. Einstein would later propose that light had both wave properties and mass properties even though they're sort of massless, made up of those bosons mentioned earlier, and, as such, doesn't need a medium in which to propagate; it is its own medium. But that is another story we'll get to in a bit.

The remaining twenty-six elements after uranium at ninety-two have not been discovered naturally on Earth—including plutonium (Pu) at position number 94. These twenty-six unnatural elements are termed transuranium elements, meaning they come after uranium. All transuranium elements are byproducts of nuclear reaction in the laboratory setting, are highly unstable, sometimes only exist for a few fleeting moments, and decay into other more stable natural elements. It is controversial if plutonium (Pu) at spot 94 on the periodic table, is produced on Earth or not; its presence on our world might be solely as a byproduct from nuclear reactors. It is not as though you can go do dig up a little bag of plutonium ore like you can uranium ore. After uranium, the last twenty-six elements on the periodic table—the element of surprise not one of them—do *not* occur naturally, at least not on our world. These transuranium elements have been synthesized in the proverbial physics laboratory using nuclear reactions and particle accelerators. If any of those last twenty-six elements do exist naturally in the universe without the help of a Shake 'n Bake laboratory, apparently it is not here on Earth. Maybe on some other planet in the universe, but not this one.

Why do larger and larger atoms like oganesson (Og) at 118 not exist naturally, definitely not on Earth and perhaps nowhere in this universe? Simple: stuff more protons and neutrons into a nucleus, and that nucleus becomes increasingly unstable. The strong nuclear force keeping those atomic particles packed together, let alone the complementary electrons, is just not doable. Not doable on Earth, anyways, and maybe not doable anywhere.

Of the naturally occurring elements, element 85—astatine (At)—is so rare that at any given moment there is only a spoonful of it throughout the entire Earth's crust. That's it, perhaps twenty grams. It could be even less, just a handful of atoms of astatine coming in and out of existence at any given moment. Francium, at number 87, is considered the second rarest natural element on Earth, with perhaps thirty grams present at any given time. As precious as we think diamonds are, if a guy really wants to show his love for his fiancé, give her an astatine engagement ring. Sure, there might be just a few unseeable atoms in the ring setting, if that, but what a way to show—or rather, not show—your undying devotion.

If any of the twenty-six synthetic elements, numbers 93 through 118, ever naturally occurred on Earth—plutonium at 94 might occur naturally but we just don't know—it was in the planet's early days, and those elements having long since disappeared.

Perhaps the first true chemist in human history, or at least one of the first, was the eighth century Persian polymath Jābir ibn Hayyān. His father was Syrian, and his mother was Persian, and in typical Roman fashion his name, Jābir, was Latinized to Geber. He studied and apparently knew everything—a Renaissance man nearly 1,000 years before the Renaissance. Geber was familiar with astronomy, geography, math, and chemistry. As a chemist, he introduced systematic observation and experimentation to the laboratory, an important addition to the Aristotelian method of observation and deduction (without experimentation). Experimentation is one of Geber's key contribution to the advancement of science.

Geber, living in Persia as he did, was beyond the reach of the Christian Church, beyond the conditions imposed in Europe during the Dark Ages. As such, chemistry was able to advance in parts of the Middle East in a way it couldn't in the whole of the Western world. Scientific empiricism did not find its footing in the West until the early seventeenth century, when the British scientist Francis Bacon formally introduced the scientific method, which basically asks the question "How do we know something is true?" Bacon's question is one crucial example of when science cleanly broke from dogma and faith-based lines of inquiry.

Francis Bacon was born in 1561 along the Strand, London. The Strand is a three-quarter-mile thoroughfare in central London, running east from Trafalgar Square to Temple Church, where it becomes Fleet Street. One of my personal favorite bits of London is the Savoy Hotel along the Strand. As for Fleet Street, it continues eastward to London Wall, a fortification wall built between the second and third centuries AD during the Roman occupation of Britain.

Bacon was a child prodigy who entered Trinity College Cambridge at age twelve. When I was twelve not wearing the "dunce cap" in 6th grade was a good day. So precocious was the young lad that royalty, who pretended not to know that anyone below their social stratus even existed, knew of the child genius Francis Bacon, including the Queen of England. Upon meeting Queen Elizabeth I as a child, and wowing her with his brilliance, the enthralled Elizabeth dubbed him "the young lord keeper."

In England the "lord keeper" is an officer of the Crown who keeps the Great Seal, which essentially means they have physical custody of the royal seal, those wax seals, that were affixed to public documents from which the power and authority of the Crown flowed. Francis Bacon at twelve was the "little lord keeper;" when I was twelve, the only sort of "keeper" title I held was when I played the children's game "keep away," where the two keepers throw a ball back and forth and the players in the middle attempt to intercept the ball, with the honor of joining the ranks of the keepers when they do. The game ends when there is one lonely sad-sack middle player left, the DFL, and acronym for "dead ****** last." But keep away is not nearly as brutal as dodgeball, where multiple bruises, minor concussions, hurt feelings and total annihilation abound. There were a few boys who had flunked down a grade or two, and even sported facial hair—they were men among sixth-grade boys—and they could really hurl that dodgeball, leaving welts and bruised feelings.

When Francis Bacon introduced the philosophy of scientific empiricism to the world, he was laying down the foundations of the scientific revolution that was soon to unfold. That was 450 years ago. Yet today we still have "science deniers" who do not understand the difference between fact and opinion. These are indeed two different concepts: fact can be proven, opinion is a judgment or a view; fact can be verified with evidence, opinion cannot be verified with evidence; fact is objective, opinion is personal; fact does not generally change over time, opinion differs according to different people and over time; and, finally, others can prove fact, others cannot prove opinion.

Science deniers interchange fact and opinion, or conflate the two, as if these two concepts have one and the same meaning. Consider this, when a scientist states the world is 4.543 billion years old, that is fact; it has been proven with such scientific accuracy as radiometric dating (analyzing radioactive decay—that unstable extra neutron thingamajig previously discussed—of elements which tick along like a clock) or proving the age of earth with the redshifting of star light from distant galaxies (the further the moving away the more light source shifts into the red spectrum, the older the universe must be). When a believer in Genesis says the world is a little over 6,000 years old, that is opinion, not fact. A person is welcome to their own set of opinions, but they are not welcome to their own set of facts. Not everything is subject to the dominion of "opinion," and that is what some science denier folks don't understand. In the simplest example I can offer, saying the world is flat—there are some today who state that—is not even an opinion one is welcome to. It is so far off what we know to be true that there is no room for opinion on the subject. When someone says the world is flat they are not welcome to that opinion. It is even an insult to the definition of the word "opinion."

Bacon's scientific method is the foundation upon which we have been able to prove fact. It lets us know the difference between fact and opinion. The scientific method involves at least six steps that tend to unfold naturally, so listing them here seems a tad silly, yet is it silly?

- **Step 1**—Ask a question: Why does my mother sometimes boil plain water and at other times add salt to the water?
- **Step 2**—Gather information: My mother adds salt when boiling water for pasta, but not when making hard-boiled eggs.
- **Step 3**—Make a bold hypothesis: My mother adds salt to the water for pasta because the water will come to a boil faster.
- **Step 4**—Design an experiment to test your hypothesis: I will measure the time it takes for salted and unsalted water to boil (or forgo the scientific method and just ask mother why she adds salt, and then go outside and play keep away.)
- **Step 5**—Analyze the results: Salted water takes more time to reach boil than plain water, not less.
- **Step 6**—Present a conclusion: My bold hypothesis was wrong.

Further experiments would have to be designed to reach the fact of the matter: your mother doesn't add salt to the water to boil pasta faster or slower, she adds it because salted pasta tastes better, as it satisfies our human craving for that particular mineral. As for why salted water takes longer to boil, not less, that will be explained momentarily.

Whether a hypothesis is proven right or wrong is both hardly the point and precisely the point of the scientific method. Proven hypothesis or disproven hypothesis—the scientific method allows science to advance onward. The scientist needs to look at the results of their experiment objectively and arrive at a conclusion, even if it is not the conclusion desired or expected. Proving a hypothesis wrong is just as important as proving one right. Both are learning experiences, and both allow for progress. To paraphrase Thomas Edison when asked about all his light bulb failures (and putting aside whether Edison truly invented the light bulb—because he did not, but he did invent the best light bulb filament at the time), Edison quipped: "I have not failed 10,000 times. I have not failed once. I have succeeded in proving that those 10,000 ways will not work. When I have eliminated the ways that will not work, I will find the way that will work."

The scientific method also involves the scientist taking what others have done and moving the knowledge base a bit further down the road. This process can be appreciated from looking at how we arrived at the universal properties of water, an understanding derived from experiments by many scientists, each adding their own bit to the puzzle. It is known that life is possible without oxygen; it is believed, however, that no matter where you go in the universe, life is impossible without water—which is where we our story now turns. Why *are* the properties of water so essential to life? And to understand that we now return to our experiment with boiling salted water.

The reason is takes longer for salted water to boil is that the added salt raises the boiling point of water and how it accomplishes this is quite simple. Water molecules love each other and are disinclined to separate. Adding energy, like heat on a stove, to water eventually forces water's molecular love affair with itself to be broken, causing it to enter the vapor state. Adding salt to water adheres those water molecules to each other even more, amping-up the love affair, and so more energy is needed to overcome the molecule-to-molecule attraction water has for itself—and therefore takes longer to reach a boil. Why? The temperature of boiling salt water is higher than boiling plain water: it's not your typical 212°F or 100°C. The more salt added the higher the temperature needs to be to reach a state of boil, in turn, the longer it takes to reach that point. Oddly, if you throw your pasta in that superheated boiling salt water it just might cook faster since the temperature is higher. But that is not why your mother adds salt—it's the flavor, baby.

Having said that salted water takes longer to reach its boiling point, oddly the ability of salt water to absorb heat is faster than plain water. You'd think that if salted water gets hotter faster it should boil sooner, but it doesn't. The ability of salt water to retain heat, its heat capacitance is lowered so it gets hot faster, its heat capacity is not a reflection of the heat needed to force water molecules to boil; heat capacitance is an independent variable to boiling point. An extreme example would be a hunk of metal, like copper. It has a low heat capacitance, so it gets hot quickly, which is why metals are used for saucepans, they heat quickly and transfer that heat to the food you're trying to cook. But throw a hunk of copper into a saucepan and good luck trying to get it to melt, let alone boil.

It goes the opposite way, too. Salt destabilizes water molecules in the frozen condition, preventing those same water molecules from maintaining a solid frozen state, lowering the melting point. The sodium Na and chloride Cl ions tug on the hydrogen H of frozen water H_2O, breaking the water molecule-to-molecule bonds at lower temperatures and the ice melts, which is why people in snowy climates sprinkle salt on their icy driveways and why municipalities salt their roads, often using a magnesium chloride salt, since it is less corrosive than table salt (sodium chloride). Same concept, different salts.

So why isn't salt added to a car's cooling system if it both raises boiling point and lowers freezing point? First off, it doesn't raise boiling or lower freezing points enough to be of much use as a coolant for an internal combustion engine. But the other reason is corrosion. Salt would rapidly interact with the metal of the radiator and rust it away before you could drive your new car home from the dealership.

So, instead, antifreeze is the additive used for an engine's cooling system. The most common is ethylene glycol, which, like salt, achieves two seemingly opposite things: it greatly raises the boiling point of water (wouldn't want your engine to boil over, would you) and it also greatly lowers the freezing point of water (also wouldn't want your engine to seize into a frozen state). Antifreeze does this by stabilizing water molecules in both directions while not causing corrosion. Why do the internal combustion engines of most cars need a radiator cooling system with antifreeze-doped water? Besides antifreeze raising the boiling point and lowering freezing point by altering how water molecules interact with each other, water itself, *independent of the coolant*, is good at absorbing heat—that heat capacitance mentioned a moment ago—water absorbs the heat from the combustion of fuel and blows that heat off as the water percolates through the baffles of the radiator. To summarize, you need the water in and of itself—its heat capacitance—to draw off heat from the internal combustion, and you need the ethylene glycol coolant to keep that water from boiling over (and from freezing solid) so it can do its thing.

Water has one of the higher heat capacities, which is a measure of how much heat a liquid element or molecule needs to absorb to raise the temperature one degree. Take something like the element mercury (Hg), which is liquid at normal earthly temperatures. Mercury has a very low heat capacity, meaning it does not take much heat to raise the temperature one degree, which in turn expands the mercury volume, making it exquisitely sensitive as a thermometer. With the mercury warmer it expands, having nowhere to expand to but up the thermometer glass tube in precise increments calibrated for each degree rise (or fall) in temperature. At the other end of the spectrum is ammonia (NH_3), which has a huge heat capacity. Water is right below ammonia.

Things like oils are closer to low middle, closer to mercury, and things like alcohol are closer to high middle, closer to water and ammonia. Metals (like mercury) have a very low heat capacitance—metal atoms are close together—meaning they heat pretty quickly, which is good, because metals want to rid themselves of that heat, transferring the energy from the stove through the slick metal Calphalon sauce pan into the salt water for mom to boil the pasta. Gases, like the ones of Earth's atmosphere—78% nitrogen N_2, 21% oxygen O_2, 0.93% argon Ag, 0.04% carbon dioxide CO_2—have a very high heat capacity since relatively speaking the gas molecules are far apart—meaning they can handle a great deal of energy input with little rise in temperature. But it's not endless, not without consequences—which is the issue at the heart of global warming. And too much greenhouse gases—carbon dioxide, methane, nitrous oxide—and you have your global warming ready to happen. And that is about all the capacity I have to discuss heat capacities.

Let's move on and push into more fundamental principles of water and what makes it essential to life. The love of water molecules for each other, their cohesiveness, is evident in nature besides the boiling point and freezing point scenarios. Take, for instance, a tree. How do you suppose water makes it all the way to the top of a giant sequoia, a tree that can reach 350 feet, as high as a thirty-story building? There are no pumps in tree roots, nor does a tree have a heart somewhere in its trunk to propel water all the way up to 350 feet. We humans have a beating heart so as to push blood into every nook and cranny of our tissues, assisted by arteries and veins that are lined with muscles that squeeze, ushering the blood along. Trees and plants have no such muscle-bound system, no heart-pumping mechanism. So how does it work? How does even a single water molecule make its way from the roots deep in the ground all the way up 350 feet to the sequoia canopy, let alone hundreds of gallons of them?

Consider this: because water molecules are so attracted to each other, because they love each other and hate to part, when a single water molecule evaporates from a leaf way, way up at the treetop, as it becomes vapor it tugs ever so slightly on the water molecule next to it, which then tugs the next water molecule below it, which then tugs the next water molecule below, and so on all the way down to the tree roots. Theoretically, a single water molecule evaporating at the treetop can exert a pull that eventually extends 350 feet all the way down to the roots. That is how the watery sap, also termed xylem, makes it to the top of a sequoia. There comes a point, however, where a water molecule at the top of a tree can't tug on a water molecule in a root, where a tree can't grow any higher, where the burden of gravity overcomes the cohesiveness of water. Otherwise, Jack's magic beans might very well have been able to grow a beanstalk to "a land high in the sky" in the real world, not just in make-believe.

"Jack and the Beanstalk" is a fairy tale that originated in England, believed to come from 5,000-year-old folklore, handed down generation to generation. The first known published version of "Jack and the Beanstalk" is from the 1730s, which was followed by numerous variations, the most common of which was published in 1890 by Joseph Jacobs. Several moral lessons can be

"And that is why GMO's are bad."

extracted from the tale, such as do what your mother tells you to do, don't buy something different when your mother sends you to the market, don't buy something you can't afford, what is yours is yours and what is theirs is theirs, and, finally, if you know something is wrong, don't do it.

To paint the full picture, I should add that it's not just water molecule evaporation in the leaves that pulls xylem up from the roots. It's more complex than that. Warmer days, as the spring gives way to summer, with warmer air and warmer ground, heat the roots, trunk, and branches, producing relative swelling, which squeezes xylem to flow upward. As the fall gives way to winter and cooler days, the process is reversed—squeezing from thermal swelling stops—allowing the nutrients to flow back down into the trunk, into the roots to be preserved till spring. If you have access to some maple trees, the best time to tap is late winter or early spring, when the xylem begins to flow out of the roots where it has been stored and concentrated all winter. It is safe to tap a maple tree of about 7 percent of its mapley xylem syrup—as if anyone actually measures that percentage or can measure that—and it will be hardly missed at all. This is coincidentally about the same amount of blood a person can donate at one time: one unit of blood is 470 ml, which is about 8 percent of total blood volume. There must be some mysterious physiologic connection there, like the Bermuda triangle, the intersection of lines, but we need to move on.

Another example of water cohesiveness, or for that matter liquid cohesiveness in general, is the phenomenon of siphoning. I'm still amazed that in my youth my friends and I could use a flexible rubber hose to siphon gas out of my mother's car—never my father's, he'd know it was missing—by sucking out that first mouthful of gasoline to get the gas flowing and, presto, you have enough gas to cruise the streets in your pal's Econoline van on a Friday night. Siphoning works because the initial suck—after spitting out that mouthful of octane—creates a vacuum. There's nothing nature abhors more than a vacuum, and so it endeavors to close the vacuum down by having one molecule of siphoned fluid chase the next in a wild frenzy that's helped along by a gravity assist. The siphoned fluid continues to flow downhill—the gas can must be lower than the car's gas tank—until the supply is either exhausted or the seal broken.

What was even more amazing than the principle of siphoning gas from my mom's car, more amazing than cohesive water molecules reaching the top of a sequoia, more amazing than a vacuum chasing its tail, is that my friends and I were so poor as teenagers we had to siphon gas in the first place, at a time when gas wasn't much more costly than Bazooka bubblegum. It is with grim disappointment that I report gasoline was that cheap in my youth and yet I still couldn't bankroll even a few gallons of high octane. Octane, by the way, is the chemical name of gasoline, with a chain of eight carbons—*oct-* from the Greek for "eight"—to which eighteen hydrogens are attached: C_8H_{18}. When ignited using oxygen as the fuel to accept electrons, the hydrogen atoms peel off the carbon chain and in so doing release energy stored in the chemical bond.

As for the phenomenon of boiling, it is merely enough energy being added to a liquid to break its intermolecular bonds, sending those molecules spinning into the vapor state. Freezing is lowering the energy state of that liquid until a point is reached where the molecules no longer have the energy, the willpower, to slide about each other in their liquid state and instead become locked into a lattice, otherwise known as the solid phase.

Now here we turn our attention to water's indispensable role in sustaining life, especially with regard to what happens with water at the moment of freezing. All liquids as they become colder become more and more dense, their molecules arranging into a tighter compact phase, a tighter lattice, until, voilà, the molecules are frozen, which in molecular terms implies they are not moving about each other. As far as we know, nearly all liquids in the universe do this—become denser, more compact in the frozen state than the liquid phase—all except one. And that would be water. This unique property of water is, indeed, one property that makes it essential to life wherever you might venture in the universe. Life doesn't necessarily need oxygen, or carbon-based organics, or Godiva chocolate, but life does need water. If life exists anywhere in the vastness of the universe, water is there.

Having said water is the only molecule that is less dense when frozen, that is not entirely correct. Acetic acid which is white vinegar you sprinkle on your salad is less dense frozen. But acetic acid and other like-minded less dense molecules lack other properties of water that would make them SOL as fundamental to life.

While it is intuitive to think that the solid form of water, which would be ice, would be denser, more compact, than the liquid form, that is not the case. One merely needs to pour a scotch and soda on the rocks to appreciate that ice floats. How is this so, and more importantly for life, *why* is this so?

As water initially approaches freezing, it, like all other liquids, becomes denser, and as it becomes denser, the nearly-frozen-but-not-quite-frozen water molecules sink toward the bottom. More densely packed means it is heavier. Take a pond, for example: as the surface water *almost* freezes, it settles toward the bottom of the pond. Nearly-frozen-but-not-quite-frozen water is more dense than liquid water.

But something unique—something astoundingly strange—happens right at the moment water fully freezes. In the words of Rod Serling, host of *The Twilight Zone* (1959–64), the freezing of water is an event "beyond that which is known to man," "a journey into a wondrous land whose boundaries are that of imagination. Your next stop, the Twilight Zone!"

That Twilight Zone moment for water is this: at that exact moment of freezing, its molecules suddenly reconfigure into an orientation that is both solid yet less dense. The formerly denser not-quite-frozen ice crystals becomes less dense totally-frozen ice, and because the icy water is less dense, it floats to the top of the denser near-freezing water. It is a beautiful arrangement in chemistry of the universe that apparently water possesses, and few other molecules or elements.

Why is it important for life that ice, or solid water, is less dense than liquid water? If at the point of freezing the icy water stayed dense, did not rearrange its molecules and become lighter and float to the surface, ponds and lakes and even shallow parts of the ocean would freeze from the bottom up, killing all marine life as we know it. Everything in the fossil record never would have come to be if ice didn't float, as early life on Earth began in the soup kitchen of the primordial oceans. All life on Earth never would have arrived, or if it had arrived somehow, it would have perished soon after doing so, snuffed-out once the ice water sank on its head. Especially during one of Earth's five major ice ages. But that's not what happened, because that's not what happens.

Frozen water sits on the top of ponds, lakes, and oceans, protecting the marine life down below. Add to the equation the heat from Earth's core, and marine habitats remain toasty warm for the long haul. The "long haul" might not be just one miserable Chicago winter—the long haul might be one miserable extended ice age of a few million years. The other great thing about ponds and rivers freezing at the top is that it gave Joni Mitchell a frozen river that she could "skate away on" for her 1971 classic melancholic hit, "River" off the *Blue* album.

All these facts about salt and water and boiling points and freezing points and floating ice came to be understood through the scientific method. They came to us not from a book written several thousand years ago by well-meaning sheepherders wandering in the Sinai desert but rather through scientific empiricism, through one scientist adding their bit of the story to the last scientist's bit of the story and making that available to the next scientist to add their bit. Science building on science.

In addition to the six previously mentioned scientific steps laid out by Francis Bacon—ask a question, gather information, make a bold hypothesis, design an experiment to test your hypothesis and perform that experiment, analyze the results, present a conclusion—there are four more important actions to take:

- **Step 7**—Repeat the experiment: So as to be sure.
- **Step 8**—Share the results: So that others can learn, critique, criticize, and grow the body of knowledge.
- **Step 9**—Encourage others to repeat the experiment: So as to prove or disprove the hypothesis, making fact reproducible (whereas opinion is not).
- **Step 10**—Create a new hypothesis based on the results: So as to continue the chain of science—science building upon science.

And perhaps in the field of biology, the scientific method and scientists standing on the shoulders of those who came before, and adding their bit to the saga, is nowhere more illustrative than the elucidation of DNA, which is where we venture next.

5 RNA OR DNA, A CHICKEN-EGG WHICH CAME FIRST

A classic example of the scientific method—and one worth retelling—is the discovery of the double helix structure for deoxyribonucleic acid, better known as DNA, the genetic code, a discovery that could not have happened without the work of the other scientists who came before. You can go really far back in chemistry to find the start of the story of how DNA was discovered. That is, you could discuss such things as how chemists developed the assay techniques needed to determine the structure of other compounds, from simple to complex, from proteins and fats to carbohydrates, techniques then adapted for nucleic acid assays for DNA. Oh my, where to begin a story about DNA?

In Lewis Carroll's *Alice in Wonderland* (1865) the White Rabbit is the herald or trumpeter for the trial of "Who Stole the Tarts." You know, the trial where the Queen of Hearts repeatedly yells "off with his head" or "off with her head". No one's head actually ever gets chopped off in Carroll's books; otherwise the book would not have made very good bedtime story for children. When the White Rabbit is eventually called to the witness stand, he puts on his spectacles and asks of the King of Hearts "Where shall I begin, please your Majesty?" The answer from the King was straightforward: "Begin at the beginning…and go on till you come to the end: then stop". And if the White Rabbit is not familiar to some of you through Carrol's masterpiece, perhaps you are familiar with Grace Slick, the Acid Queen and her "White Rabbit", a vignette to be discussed at the end of this chapter before moving on, I promise.

A story about DNA cannot possibly begin at the beginning. So, for our purposes, we won't journey that far into the past, and unlike the White Rabbit when called to the witness stand beginning at the beginning, the station along the timeline for us is 1869, when the Swiss physician and biologist Friedrich Miescher isolated what would eventually become known as DNA from some discarded bandages soaked with pus. Yuck! Up to that point, scientists pretty much knew life was composed of three distinct organic compounds—proteins, fats, carbohydrates—as well as one other critical component: water. The key chemical structure of all proteins is the amino acid NH_2, the key structure of all fats is the carboxyl acid $COOH$, and the key feature of carbohydrates is right there in the name: hydrated carbon, or CHO.

The material that Miescher extracted from the pus were none of those things, yet he knew this new organic material was apparently somehow essential to life. He surmised this because the new stuff was extracted from the nucleus of the bacteria cells he was studying. Which is why Miescher termed this new compound "nuclein"—it came from inside the bacteria cell's nucleus region (bacteria cells don't possess a nucleus, but they can boast a *nucleus region*). In a classic unfolding of the scientific method at its best, others would stand on Miescher's shoulders and take his discovery further down the road. Way further down the road. More on that in a moment.

We talk about the designations "organic" and "inorganic," as in he hated inorganic chemistry in college but hated organic chemistry even more, but what is the difference? The basic definition of "organic," as in organic life, organic chemistry, is a compound where carbon (C) atoms are the key element involved and essential in the compound's structure. When scientists mumble "organic life," they mean carbon-based life, carbon being much like the ringleader of a big-top circus. All life, at least on Earth has carbon as the central player. "Inorganic" is everything else on earth that is not life nor derived from life, like your table salt, the water you just sipped, the oxygen in the atmosphere you breathe, and the silicon dioxide rock that makes up most of Earth's crust.

Just to be clear, whereas water is *essential* to organic life, water itself is not organic. And carbon steps in as *ringleader* of the organic life made possible by water. Besides carbon, the three other essential elements, the main players in organic life are hydrogen, oxygen and nitrogen. Carbohydrates and fats might only have three—carbon, hydrogen, oxygen—but proteins have all four. So do nucleic acids, the building blocks of our genetic material.

About ten years after Miescher's discovery of nuclein, in 1878, the German biochemist Albrecht Kossel determined that bacteria nuclein was not protein, was not carbohydrate, was not fat, but rather was a new organic compound (meaning, of course, that carbon was still the ringleader). Kossel speculated that nuclein, later named nucleic acid, was a central player in all biological life. Its exact role was unknown, but it was clear it was key, nonetheless. Kossel also showed that there are five distinct nucleic acid bases in organic life on earth: adenine, guanine, thymine, cytosine, and uracil. There are other nucleic acid bases, but they are not part of the circle of life. At least not life on *our* planet.

In 1919, the American biochemist Phoebus Levene identified three subcomponents of the five nucleic acid rings discovered by Kossel, which were a sugar group and a phosphate group attached to the nucleic acid ring. Levene further suggested that this new compound, later to be called DNA, formed a long string of nucleic acid bases connected by the phosphate group. It was like a string of pearls, where the pearls were the nucleic acid rings and the connecting thread between each pearl was the phosphate group. As for determining the three-dimensional structure of this string of pearls, not much else happened for nearly two decades.

Then, in 1937, British physicist and molecular biologist William Astbury used X-ray diffraction studies—that is, he sent X-ray beams through those isolated nucleic acid compounds—and was able to image, or begin to image, how this nucleic acid compound refracted the beam. (How X-rays were discovered in the first place and then how the X-ray machines that image our lungs also allowed Astbury to X-ray nucleic acids is another huge bit of science to be discussed further down the road; but, once again, it is a case of scientists using the work of other scientists, often not even in the same discipline, to advance the cause.) Astbury had already provided X-ray diffraction analysis of the protein keratin a few years earlier, which is what allowed the American chemist Linus Pauling to determine the alpha helix of nearly all proteins. Astbury then demonstrated that these newly discovered nucleic acid molecules, like protein molecules, were likewise helical. That turned out to be an important bit of information.

Pauling's alpha helix is the most common secondary structure of all proteins. This helical spiral is always right-turned, never left-turned. Why? Only God and Linus Pauling knew why right-turning spiral proteins survived in the world of organic life while the left-turning spiral protein was relegated to obscurity. While the secondary structure of all proteins is the right-turning spiral, the *primary* structure of proteins is the various recurring carbon-based amino acid chains of molecules. What is all that I just said? Well, in other words, the secondary structure, the helix arises from the fact that the primary structure, the sequence (string of pearls on a necklace) forces a right-turning alpha helix (the pearl necklace twisted into a single helical spin).

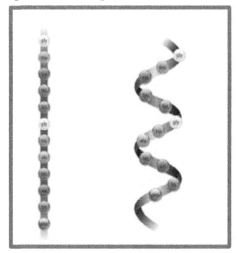

As an aside, the primary amino acids of protein—not the nucleic acids of DNA—go by such names as lysine, glycine, glutamine, proline, arginine, alanine, cysteine, and asparagine; as you can see, a lot of amino acids end in "-ine," but there are a few non "-ines" for good measure, such as tryptophan and aspartic acid, both amino acid bases. From *The Big Bang Theory* sitcom Sheldon Cooper's favorite amino acid is glutamine, episode "The Friendship Algorithm" 2009.

DNA, however, as compared to protein, is an entirely different structure altogether—built differently, yet still helical. And as you can guess, the sequence of nucleic acids—the primary structure—is what forces their helical nature—the secondary structure.

Actually, up until the 1940s, before the discovery of DNA, it was believed that proteins were the seat of genetic material. This scientific dogma had been around for 100 years or so, and then was disproved—another classic example of how science critiques itself, corrects itself, and moves on. It was becoming apparent to biologists and organic chemists that the scales were tipping toward these newly discovered nucleic acids being the true seat of genetic material. How did that happen?

In 1944, Oswald Avery, Colin MacLeod, and Maclyn McCarty, in their famous Avery-MacLeod-McCarty experiment, determined that these nucleic acid structures were indeed the basis of genetic material, and not proteins, when they demonstrated that nucleic acids were responsible for bacterial transformation. Bacterial transformation is the ability of one bacterium to take up genetic material from a second bacterium to form a third novel bacterium—how scary is that, viruses do it all the time—and this transformation occurs at the level of nucleic acids. To prove this phenomenon, the Avery-MacLeod-McCarty experiment involved exposing a live, nonvirulent *Streptococcus* bacterium to a dead, virulent *Streptococcus* bacterium's genetic material—its isolated DNA—which produced a novel virulent *Streptococcus* bacterium. Nothing else changed in the analysis; no protein changed, just the nucleic acid. Fortunately, that novel virulent strep bacteria species that the three scientists produced didn't crawl out of the lab on the bottom of one of their shoes. At least we don't think it did. But the trio did prove DNA was the seat of genetic instruction, not proteins.

You might think that with names like Oswald Avery, Colin MacLeod, and Maclyn McCarty that they were key characters in the 1995 film *Braveheart* alongside Mel Gibson's William Wallace, or perhaps were three scientist lads working somewhere in Scotland or Britain, out of the Cavendish Laboratory at Cambridge or in Edinburgh, perhaps. And if you thought that, who could blame you? But you would be wrong. Oswald Avery and Colin MacLeod where Canadian Americans and Maclyn McCarty was American, and all three performed their pioneering nucleic acid structure research out of the Rockefeller University Hospital on New York's Upper East Side.

After combining the Avery-MacLeod-McCarty experiment in 1944 with knowledge gleaned from the 1937 Astbury X-ray diffraction study showing that nucleic acid structure is a regular pattern with a spiral shape, all of a sudden everyone was focused on the hypothesis that nucleic acids, the seat of genetic material, were helical. Protein amino acids were helical, so why not nucleic acids? The race was on to see who would be the first to crack the three-dimensional structure of what would become known as DNA, of our genetic material. A Nobel Prize was most certainly waiting at the end of that journey.

I once met Linus Pauling when I was in college at Antioch College which is nestled in the quaint hamlet of Yellow Springs, Ohio. I attended a lecture Pauling gave at nearby Wilberforce University. Pauling was perhaps the leading organic chemist of his day, but, for reasons that certainly escape me, he was stuck on a triple-helix model for genetic nucleic acid. You see, up until that time, no one had produced clear enough X-ray diffraction studies of spiraling nucleic acids to determine what *type* of helix existed. All that was known for certain was that nucleic acids *were* helical. Was it a single helix, a double helix, a triple helix, a quadruple helix?

The prediction was that if nucleic acids were truly helical, X-ray diffraction studies would reveal some type of recurring X-shaped pattern on radiographic film. Take a rope for instance: if you were to X-ray a rope, the overlapping, intertwined pattern that would be imaged would reveal some type of crisscrossing X pattern, and that image would reflect how many strands—two, three, four—were wrapped around each other to make the rope, assuming the X-ray images were clear enough. In other words, a characteristic X pattern would emerge depending upon how many strands were intertwined. But X-raying a rope is far easier than X-raying submicroscopic DNA.

As an aside, the "X" in X-ray was not because the first X-rays made an X pattern or some such thing; "X" assigned by the discoverer of X-rays in 1895, the German scientist Wilhelm Conrad Röntgen, who, when he happened upon them, in true mathematical physicist form, assigned the unknown element he encountered the letter "X." You know, all those tortuous algebra equations from high school where "X" equals something other than that girl's phone number, which sadly was also unknown. On the other hand, the X patterns of proteins and nucleic acids and rope are something entirely different—they are literally the crisscrossing XXXXX pattern of an X-rayed spiral.

The reason Pauling, and separately the molecular biologists Francis Crick and James Watson, predicted an X-shaped pattern was not a guess, nor was it a leap from the alpha-helical shape of proteins. The three of them, but especially Pauling, were gifted chemists. They already knew about the inherent 3D shape chemical bonds must make in space, and after adding to that the repeating pattern of nucleic acids—the primary sequence forces the secondary structure—those three chemists were able to mathematically arrive at some type of helical configuration without necessarily needing an X-ray image to prove it. But, oddly, that is exactly what they needed—an X-ray image of DNA—to put the final pieces of the puzzle together.

Take something as simple as the water molecule: we write it as H_2O, but it is really H-O-H, where oxygen has two hydrogens attached to it. Water is not a straight molecule like it appears on this page, H-O-H; rather the hydrogen atoms form an exact 104.5-degree angle off the oxygen atom, which is partially based on how their outer electron orbits interact, that James Clerk Maxwell electromagnetism we talked about. If you were to give a gifted chemist like Pauling a complex compound, with a bunch of carbons and hydrogens and oxygens, he could figure out the angles between them based on how atoms and groups of atoms and long chains and outer electron orbits respond to each other: sequence forces structure. Now imagine someone like Pauling figuring out the exact angles and structure of something as complex as intertwined, spiraling nucleic acid strands. Sounds impossible, but that is exactly what those three scientists did—or almost did— over late nights and the pots of coffee and vitamin C tablets that they, or at least Pauling, certainly consumed.

Consider this: there is likely no silicon-based life in the universe—silicon sitting right below carbon on the periodic table—perfectly positioned to theoretically be the central ring-player in organic life. Silicon like carbon could hypothetically form organic-like compounds with hydrogen and oxygen. Pauling and his ilk were such gifted chemists that if you gave them a sequence of silicon-based nucleic acids, that is, replaced all the carbon atoms (C) with silicon atoms (Si), leaving everything else the same, Pauling could build you your silicon sequence and helix structure for those aliens living out there on their lonely, unfillable longing in space in the Andromeda galaxy, a life form that is silicon-based life, not carbon-based. Why is there probably no silicon-based organic life out there in the vastness of the universe, or if there was or is, why wouldn't it advance very far? Silicon its atomic weight 28 compared to carbon at 12, would be make for relatively unstable organic compounds that just couldn't hold it together when the going got tough.

Yet it remains a fact that DNA is so chemically complex that, despite their collective knowledge of chemistry and molecular relationships and angles, Pauling and Watson and Crick still needed an X-ray image to guide their work.

The 1999 film *Notting Hill* stars the irresistible Julia Roberts as the actress Anna Scott and Hugh Grant as the regular, but very handsome, British bookstore owner William Thacker—a delightful rom-com if there ever was one. But Notting Hill is important in our story for another reason: an often overlooked key player who helped unravel the structure of genetic nucleic acid, Rosalind Franklin, was born into a prominent British Jewish family in London's Notting Hill neighborhood in 1920.

Franklin received a top-shelf education from childhood, starting with the private day school Norland Place in West London. She continued her education at two boarding academies, first the Lindores School for Young Ladies in Sussex and then St. Paul's Girl's School, also in West London. From there she did her college and graduate work at the University of Cambridge and her postgraduate work in Paris, where she learned X-ray crystallography. Franklin then returned to London and joined the faculty of King's College. In the run-up to elucidating the structure of DNA, Rosalind ended up having a falling out with two other important X-ray crystallographers, John Randall and Maurice Wilkins—a falling out over a very, very special photo: Photo 51, which turned out to be the final clue to the structure of DNA.

Photo 51 was really called that. It was likely the fifty-first photo taken in a series of X-ray crystallographs in Franklin's lab and should not be confused with another famous "51," namely Area 51, a highly classified facility of the U.S. Air Force in Nevada. The U.S. Air Force acquired what has become known as Area 51 in 1955 primarily to test the Lockheed U-2 aircraft, a high-flying reconnaissance plane used by the military and by the CIA. But due to the high security of the "secret" facility, it spawned conspiracy theories about unidentified flying objects—your basic UFOs piloted by aliens. The highway leading to Area 51 is affectionately known as the Extraterrestrial Highway. Feeding into the conspiracy was the 1947 crash of a military weather balloon near Roswell, New Mexico, which many believe to have been a crashed UFO, that was then taken to Area 51.

Photo 51, taken in Franklin's lab, was the clearest-yet X-ray diffraction of genetic nucleic acid. Doctoral graduate student Raymond Gosling produced the actual image in 1952, but it was Franklin, Gosling's doctoral supervisor, who appreciated the significance of Photo 51 among a slew of other X-ray diffraction photos. She showed the image to senior researchers at King's College, notably the above-mentioned John Randall and Maurice Wilkins, who, without Franklin's permission, showed Photo 51 to fellow Brits Francis Crick and James Watson.

Immediately, or perhaps close to immediately, Crick and Watson analyzed Photo 51 along with all their other data, following the scientific method, and realized that the genetic nucleic acid was a double helix—not a single helix, and not the triple helix that Pauling was fixated on, but a double helix. It is said that if Pauling had been shown Photo 51 at the exact same moment as Crick and Watson—like a TV game show where you ring a bell—Pauling would have figured out the double helix instantaneously (*Ding ding ding!* We have a winner), faster than Crick and Watson, based on Pauling's singular and superior knowledge of chemistry. Not that Crick and Watson weren't stellar—they were. But Pauling was practically godlike in the world of organic chemistry. He was *more* stellar.

But Pauling didn't see Photo 51. He was an American whose lab was at the California Institute of Technology, in Pasadena, better known as Caltech—also home to the heroes of Chuck Lorre's smashing science sitcom *The Big Bang Theory* (2007–19). Meanwhile, as Pauling was toiling away at Caltech, that vital X-ray diffraction work was being performed in London. Watson and Crick *did* see Photo 51, and from all the information they had at hand, in a paper published in 1953, they determined our genetic material is deoxyribonucleic acid, DNA, two strands of nucleic acid sequences intertwined in a double helix. In 1962, Watson and Crick, as well as Franklin's lab mate Wilkins, were awarded the Nobel Prize in Physiology and Medicine for their pioneering work on DNA. Sadly, Franklin did not receive a Nobel because she had died four years earlier in 1958, age thirty-seven, from metastatic ovarian cancer. The Nobel is not given posthumously, but Franklin surely performed Nobel-worthy work.

Don't feel too sorry for Linus Pauling, though. He won the 1954 Nobel Prize in Chemistry eight years earlier for his brilliant work on the nature of the chemical bond, a body of work that allows a trained chemist to figure out structure and chemical reactions, why they are the way they are, and why they occur the way they occur. Pauling also won the 1962 Nobel Peace Prize for his outspoken stance against the then bourgeoning nuclear arms race between the US and USSR. Linus Pauling died in 1994 at age ninety-three,

"We ran a full DNA test, STR and Mitochondrial analysis... and Bob here 'Googled' it just to make sure."

Francis Crick died in 2004 at age eighty-eight, and, as I type these words, James Watson is still alive at age ninety-one.

James Watson has recently been stripped of nearly all his academia honors other than his Nobel (because you can't be stripped off a rightfully earned Nobel) for his outspoken racist comments over the previous decade or more. The gist of his racism centers on his belief that Black people are genetically less intelligent than white people. Watson may have discovered the 3D shape of DNA, but we all discovered he's a bona fide racist.

This story of Watson's racism reminds me of another brilliant scientific mind who was also shockingly racist: William Shockley. Shockley, born in London in 1910, became one of the Silicon Valley originals before Silicon Valley even existed, when he perfected semiconductors and transistors, the former of which are still at the heart of computers and the latter still at the core of many electronics. He did most of his pioneering work after he immigrated to America, working for the famous Bell Labs in New Jersey, after which he became a professor of electrical engineering at Stanford University in California.

But our attention is directed at Shockley's other work, specifically his interest in eugenics: the study of the genetic quality of the human population, which, as a byproduct, parses humans into groups based on race—groups that are then judged to be superior or inferior to one another. Shockley proposed that African Americans—like Watson's opinion of Black people—were intellectually inferior to European Americans.

Not only did Shockley shockingly propose white people were superior, he then put himself on top of that delusional heap. Thinking he was some sort of gift to humanity, he promoted the creation of a sperm bank where deposits came from men like him—"geniuses"—so that women desiring intelligent children could make a withdrawal and end up with guaranteed little Einsteins, or in this case, little Shockleys. The sperm bank's founder Robert Klark Graham called it the Repository for Germinal Choice, located Escondido, California, opened in 1980 and went defunct 1999. The 2007 pilot episode of Chuck Lorre's *The Big Bang Theory*, the one where Leonard (and I) fall in love with Penny, begins with the genius scientist Leonard Hofstadter and even more genius Sheldon Cooper—combined IQ 360—making an unsuccessful visit to one such genius sperm bank. I can only assume Lorre was being sarcastic. Because the point of it is this: what if a woman makes a withdrawal from the Repository for Germinal Choice sperm bank hoping she'll give birth to a genius only to discover, years later, that she's in fact been saddled with a racist.

The discovery and elucidation of DNA, beginning with the Miescher isolating "nuclein" in 1869; Kossel in 1878 determining nuclein was in fact nucleic acids; Avery, MacLeod, and McCarty demonstrating in 1944 that nucleic acids are the seat of genetic material; in 1952, a guy we haven't talked about yet, Erwin Chargaff at Columbia University in New York, showing that of the four nucleic acids, adenine always paired with thymine and guanine always paired with cytosine; followed by Pauling, Watson and Crick homing in on some type of helical structure; and finally Franklin identifying the DNA twirl in Photo 51 in 1952, all allowed Watson and Crick in 1953 to elucidate that genetic nucleic acid was DNA. That, folks, is the scientific method at its best— scientists standing on the shoulders of the scientists who came before them, peering a tad further into the gray horizon.

Watson and Crick and company showed that nucleic acid is a double helix and, just like proteins, is always a right-turning helix. Chemists are not sure why life chose right-turning protein helices and right-turning DNA helices over their left-turning counterparts, even though I previously quipped only God knows. But actually, there are some unproven hypotheses. At some level it appears to have something to do with the way electrons spin within an atom—not the electron orbit around the nucleus, but the electron spin about itself (like a spinning top)—those electrons in atoms seem to favor right spins over left spins which then might force right protein helices and right DNA helices. And why electrons might favor right spinning over left no one knows. Yet. And that is about all I have to say about that.

That above-mentioned base pairing sequence of Erwin Chargaff, a Hungarian Jew who trained in chemistry in Vienna and immigrated to the US in 1925 for graduate work, is exact and incontrovertible: adenine always pairs with thymine, and guanine always pairs with cytosine—sometimes abbreviated as A/T and G/C—spiraling into a beautiful double helix carrying the genetic code for all cellular life on Earth.

We've discussed the elucidation of DNA and haven't said peep about RNA—ribonucleic acid—which predates DNA on Earth's chemical evolutionary timeline, an especially important discussion since it was RNA that gave us DNA that then gave us life. And which is where we now venture.

RNA is built slightly differently than DNA. For starters, RNA is a single-stranded nucleic acid sequence and not necessarily helical, not double helical like Big Brother DNA. RNA is composed of ribose sugar ring nucleic acid (the "R" of RNA) rather than deoxyribose sugar ring nucleic acid (the "D" of DNA). The difference is in the name—"deoxy"—which means for DNA each ribose loses an oxygen atom making it *deoxy* DNA that translates into being able to twist snuggly into a helical structure. The ribose of RNA, on the other hand, does not permit tight helices making RNA's nucleic acids more vulnerable, which for life on Earth is a good thing. This instability makes RNA unable to form double-stranded helical rings but still able to carry its complement of the genetic blueprint in single helical strands. RNA's other difference is that one of the nucleic acids is different from those found in DNA: a fifth nucleic acid, uracil, takes the spot of the thymine. This difference is the story of genetic life on Earth—precisely how life evolved into DNA.

Because RNA is single stranded, it is quite vulnerable, not robustly helical, as its nucleic acids do *not* pair up to form a much more stable double helix, as do adenine/thymine and guanine/cytosine do in DNA. For RNA, its four nucleic acid groups are essentially unprotected flapping in the breeze, whereas the four nucleic acid base pairs of DNA are ensconced in a protective double-helical arrangement. This is important because the four bases of DNA are our genetic code and being inside that safe double helix means that the code cannot be messed with, or at least not messed with easily. With RNA, the code is external and exposed, and as such is more easily damaged. Sort of like the difference between wearing a cape versus a buttoned-up coat—one is going to do a better job of keeping you warmer than the other. Or, more convincingly, the difference between racing along a dangerous winding highway in a 1967 Ford Mustang convertible with its top down against a 1967 Ford Mustang hardtop.

Before we delve further into RNA and DNA, you might be wondering what exactly RNA and DNA code for, or, more to the point, what is so gosh darn important about them. Well, simply put, DNA codes for RNA, and RNA codes for protein enzymes. Why is there all this ballyhoo over DNA and RNA if all they do is to eventually code for protein enzymes? Well, to give you a little spoiler for a subject we'll encounter later on, the chemical evolution that led to life on Earth—accounting for everything from the color of your hair to the height of a giraffe to the size of a peapod—all boils down to chemical structures, which in turn are nothing more than chemical reactions. Those chemical reactions that built you and built me and built that alligator on the fifteenth fairway in Tampa could not, cannot, proceed without those necessary protein enzymes that make structural and functional proteins and also build fats and carbohydrates that—when combined with water and a dash of other ingredients, when assembled in just the right order, in just the right concentration—make you and make me. I'll avoid putting too fine a point on it, but DNA unravels and codes for RNA, which then codes for protein enzymes, which then go about making the ingredients and substrates that make you.

When the road to life first began on Earth—that is, the road to life that would use nucleic acids to carry genetic code—RNA was the original template, not DNA. DNA did not exist. Since RNA is single stranded and cannot base pair to form a protective double helix, by its very single-standedness nature, RNA exposed its underbelly genetic code to damage—to flapping in the breeze in the Mustang convertible. Not only that, but near the beginning of that road, RNA chose uracil over thymine as its preferred nucleic acid, because it is easier for nature to make uracil than thymine. That is, not as much energy is expended in building uracil as in building its cousin thymine. Despite its lack of stability and lack of thymine, RNA could and can still code for the protein enzymes necessary to life as we know it.

At this point in chemical evolution, before DNA arrived on the scene, we have RNA floating about in the primordial seas, but we still don't have life on Earth yet. We have a bunch of RNA on the *road to life*, but not life itself. We also have a bunch of protein enzymes, mostly coded for by the RNA, also floating about in the primordial stew. But as for life, not so much. At some point, however, the road switched directions from RNA to DNA by evolving two key changes: removing the oxygen atom (hydroxyl group) from ribose to make the sugar deoxyribose, and by substituting uracil with thymine, which pairs nicely with adenine whereas uracil does not. These two chemical evolutionary changes allowed the nucleic acid sequence to form two tight interwoven strands, or a double helix, or DNA, thereby protecting the genetic code from harm. Voilà, we now have the Mustang hardtop.

Now, we still don't have life on Earth yet. The story of how bacteria (life!) formed from the four basic ingredients of A, T, G, and C will be told later in this story. But, for now, what we do have are a bunch of ingredients—proteins, fats, carbohydrates, nucleic acids—all waiting around for Godot. But unlike *Waiting for Godot*, Samuel Beckett's 1949 play where the titular character never shows up, life on Earth did eventually arrive.

How does DNA unzip its fly to expose the protected genetic code to then code for RNA? Enzymes, of course—enzymes that the DNA coded for in the first place. We understand that changing out uracil for thymine allowed base pairing, allowed the formation of the double helix. What's the big hullaballoo about switching from ribose sugar to deoxyribose sugar? The deoxyribose of DNA as just mentioned is missing one oxygen atom on every base pair, and in missing that one, single, solitary oxygen atom, it not only facilitates helical transformation, that deoxy site within the sequence is also the lock that allows certain protein enzymes to act as keys, whereby they attach to that exact site, unzipping the DNA into two single strands. Those two single exposed strands of DNA then code for complementary RNA messengers, which in turn do the actual coding of the protein enzymes. Enzymes unzip DNA, which codes for messenger RNA, which codes for protein enzymes, which construct structural and functional proteins, fats, and carbohydrates—the stuff you and I are made of.

In a genetic world where the desire for stability and durability is worth the expending of energy—to be a hardtop and not a convertible—switching over from uracil to thymine and dumping the oxygen atom to change from ribose to deoxyribose made perfect sense for DNA. These changes provided protection for the code and the unlocking of the code for transcription by securing it on the inside of a double helix, thus protecting the base pairs that carry the building blocks of life for the reasonable price of a small increase in expended energy.

So, life kept both RNA and DNA for good reason. In the same world where you need to protect DNA, you also need a bucketload of RNA transcribed from the unzipped DNA to make all the protein enzymes that make life possible. Whereas DNA is better at carrying and protecting the genetic code, RNA is better at mass-producing protein enzymes.

Where exactly does all this DNA unzipping and RNA production of proteins happen? DNA lives inside the nucleus of every biological cell (except in the case of bacteria, which have no nucleus but rather a "nucleus region"). Inside the nucleus or the nucleus region, the DNA unzips—that unzipping is only partial, not the entire strand—and codes for various RNA: along with messenger (mRNA), there's transfer (tRNA), ribosomal (rRNA), and small nuclear (snRNA). The assembly of proteins by mRNA happens in the cytoplasm outside the nucleus or nucleus region. Once DNA has coded for the various RNAs, for the most part all those various RNAs leave the nucleus and enter the cell's cytoplasm in order to do what they do best: assemble protein enzymes.

To describe that in more detail, what happens is that the messenger RNA, transfer RNA, and ribosomal RNA all leave the nucleus and enter the cytoplasm. Then the transfer RNA finds floating in the sea of cytoplasm the exact amino acid building block it's looking for and delivers it to the messenger RNA for the assembly of proteins, a template, and the ribosomal RNA helps the transfer RNA align those amino acid building blocks for the messenger RNA to accept. Think of it this way: you're at a Catholic high school dance in the gym. The messenger RNA are the girls lined up on one side of the basketball court. The transfer RNA is Sister Mary who goes to the other side of the gym and finds a boy amino acid to dance with a girl nucleic acid. Sister Mary brings the amino acid boy back to the line of nucleic acid girls with Sister Madeline's (ribosomal RNA) help. Of course, in this pairing of an amino acid boy with a specific nucleic acid girl, enough space is left between the two for the Holy Spirit. None of that mushing his chest with her chest. How sad is that?

As for small nuclear RNA, it actually stays in the nucleus and helps the unzipped DNA assemble all the RNAs in the first place. That is, small nuclear RNA finds the correct nucleic acid building block floating in the nucleoplasm—not *amino* acid in *cytoplasm*, but *nucleic* acid in *nucleoplasm*—to assist in the DNA's translation into RNA.

And I think that's enough detail on that.

Protein enzymes, still within the cytoplasm, then code for structural and functional proteins, as well as carbohydrates and fats. Once assembled, the proteins, fats, and carbs either stay inside the cell, doing whatever bit of cellular task is required of them, or are transported out of the cell and travel to wherever else they might be needed. Cytoplasm is the fluid inside a cell; nucleoplasm is the fluid inside the nucleus.

In the world of chemical genetic nucleic acid evolution, uracil is best for RNA's purposes, because what you're after is quantity, and it is much cheaper to make uracil for RNA's purposes. Where it's dependability that you need, fidelity—DNA's main concern—thymine is best, because it allows double-helix base pairing, the protecting of the code.

Another advantage of the double helix of DNA, besides durability, fidelity, and protection of the genetic code, is that more information can be stored per unit length. That is to say, the double helix has stunningly efficient data storage. RNA, which is not nearly as coiled or not coiled at all, and not very long chains, does not enjoy the same amount of genetic storage capacity per the three-dimensional space it consumes. A single human DNA strand has roughly six billion base pairs of adenine-thymine and guanine-cytosine steadfastly coiled in a total length of only six microns, which is one millionth of a meter. That is a lot of data in a very small package. It is more unit data than the current smallest computer processor Intel makes for commercial use. Think of your Mac or Windows computer, or your iPhone or Android phone, and all the storage they accommodate. DNA still easily outperforms whatever chip is currently at the top of the heap for data storage.

If we were able to take that tight, highly efficient human DNA chromosome coil with its six billion base pairs and unwind it and pull it into a straight line without breaking any chemical bonds, it would be roughly two meters in length. Maybe two meters long doesn't sound all that impressive—but that's 500,000 times its original length, which is surely an impressive figure. Another way of looking at it is this: our DNA fits nicely into the nuclei of our cells at six microns, but uncoiled and stretched, it would be six and a half feet in length, the average height of an NBA basketball player.

Now, that six and a half feet of DNA is for just one human cell chromosome and each cell has forty-six chromosomes. When the math is done, the human body has about thirty-seven trillion cells, and except for a few cells, like red blood cells, that jettison their nucleus in order to be who they are, each cell has a nucleus containing the exact copy of that organism's DNA. If you were to uncoil from one human body all those 46 chromosomes in each of thirty-seven trillion cells and place them end on end—6.5 feet x 46 chromosomes x 37 trillion cells ÷ 5280 feet per mile—well, that would stretch to about two trillion miles of DNA. In comparison, Pluto on a good day is just over three billion miles away. The nearest star, Alpha Centauri is twenty-five trillion miles across the cosmos. All the DNA in a single person uncoiled and placed end on would make it beyond the Oort Cloud into interstellar space, that part beyond the gravitational influence of our sun.

Oddly—and then I will say no more about this—most of the DNA base pairs are garbage: they don't actually code for anything like protein enzymes. They are almost like fillers or placeholders or only God knows what. It is estimated that only 1 to 2 percent of the entire strand of DNA base pairs that make up the human genome chromosome actually codes for anything worthwhile. Some of the junk non-coding DNA might assist in how the coding DNA conducts itself, but still, there appears to be stretches of silent DNA along a chromosome. Go figure.

The other advantage of uracil being in RNA but not in DNA has to do with recognizing errors in the DNA sequence, errors that can occur at any moment. If the unzipped DNA strand has an error, it usually results in the cytosine degrading to uracil. A type of RNA whose sole purpose is to recognize DNA errors registers that degradation of cytosine to uracil, because it isn't able to match up correctly. The nucleus is then alerted and goes about DNA repair—the mechanisms of which I'll spare you the details of.

The take-home lesson is this: on Earth, RNA came before DNA; RNA then evolved into DNA to protect the genetic code it carries and to more easily store a gazillion bytes of information in the tiny space of a cell's nucleus. DNA is the genetic code for cellular life on Earth. It is a double-stranded nucleic acid helical polymer; that is to say, DNA consists of two very, very, very long chains of nucleic acids intertwined in a precise double-helical structure, down to every single atomic detail. The DNA sequences—which we call "chromosomes"—contain the genetic instructions for whatever organism it lives inside the cells of. All cellular life on Earth has DNA as its genetic code; viruses, on the other hand, can be either DNA or RNA, because quite simply, viruses are not cellular. Depending on what definition you use to define "life," viruses might be life, but they are not "cellular life."

Genetic adaptation with favorable heritable traits is the stuff we are made of. Earth is 4.543 billion years old. When bacteria first appeared on Earth about 3.8 billion years ago, many variations of nucleic acid RNA existed, and possibly several variations of DNA as well. At some point, one of those bacterium progenitors had the favorable DNA sequence of adenine (A) always paired with thymine (T) and guanine (G) always paired with cytosine (C), while the other competing bacteria did not. Those bacteria that possessed this ideal base-pair DNA code moved on to the next stage of biological evolution, and the other competing bacteria candidates died out.

The last universal ancestor (LUA) was a bacterium that lived about 3.5 billion years ago, and it advanced onward to form all cellular life on Earth—all plants, all fungi, all protists, and all animals, as well as all viruses. How the simpler bit of genetic spit viruses evolved from the much more complex bacteria will be told further along this journey. And it was from LUA that, through a series of evolutionary steps, sometimes violent ones, *Homo sapiens* arrived on the world scene around 300,000 years ago. The extant subspecies *Homo sapiens sapiens*—which is modern-day humans, that is, you and me—arrived on stage maybe 70,000 years ago. To go from the last universal ancestor bacteria to humankind as we know it, it took a mere 3.5 billion years.

Now as for that Grace Slick story I promised before moving on. Grace Slick wrote a song "White Rabbit" while with the San Francisco band The Great Society, a composition she brought with her when she was then asked, in the fall of 1965, to leave The Great Society band and join the Jefferson Airplane band, another up and coming San Francisco group. Jerry Slick, her husband, who started The Great Society was not invited to join the Jefferson Airplane, just Grace. Grace Slick also brought another composition "Somebody to Love" 1967 along with "White Rabbit" also 1967. Both songs helped place the Jefferson

Airplane on the map of Rock 'n Roll fame, and likewise earned Grace Slick her other name, The Acid Queen.

The Jefferson Airplane version of Grace Slick's "White Rabbit" is a crescendo song, similar in style to Ravel's *Boléro* 1928 orchestral piece. Maurice Ravel was a late nineteenth-century, early twentieth-century French composer. Ravel's *Boléro*, a movement that Bo Derek's character Jenny Miles liked to have sex to in the 1979 romantic comedy movie *10*, featuring Dudley Moore, Julie Andrews and Bo Derek, is a composition that begins slow, quiet and in a straight forward, deliberate manner, then crescendos in tempo, decibels and thickness; just as quietly as it begins, it loudly and suddenly ends. In similar fashion to Ravel's *Boléro*, Grace Slick's ability to crescendo "White Rabbit" with her voice was and is simply unmatched.

Grace Slick's song "White Rabbit" uses imagery found in the fantasy books of Lewis Carroll's *Alice's Adventures in Wonderland* 1865 (sometimes titled simply *Alice in Wonderland*) and *Through the Looking Glass* 1871, to describe the world of psychedelic drugs, with images changing size, shape and form after one swallows a psychedelic acid pill, an acid trip. That is the song. The real White Rabbit is a main character in *Alice's Adventures in Wonderland* by Charles Lutwidge Dodgson, but you and I know him as Lewis Carroll. It tells the story of a girl, Alice falling down a rabbit's hole into a fantasy world populated with animals anthropomorphized with human qualities. It is not just a children's book having been enjoyed these many, many years later by adults, too.

On the banks of a gentle river, Alice discovers a White Rabbit dressed in clothes, and together, they go down and subsequently fall down the rabbit's hole. Alice sees a bottle that says: "Drink Me" and drinks from it; she sees a cake that says: "Eat Me" and eats of it, allusions to the fantasy that then unfolds before Alice's eyes. And no doubt these passages of "eating" and "drinking" something in order to enter a fantasy world gave inspiration to Grace Slick's acid-fueled song "White Rabbit".

Grace Slick was also likely being ironic when she composed "White Rabbit" as she was decidedly not promoting acid trips. To the contrary, parents for generations since Lewis Carroll's books were published have read and encouraged their children to read *Alice's Adventures in Wonderland* and *Through the Looking Glass*, all the while glossing over the fact Alice was entering a world of fantasy. Lewis Carroll was surely not promoting drugs, that's not what he meant. But taken on the nose (or in the nose) one can't help to make the association between swallowing something and getting high. The irony for Grace Slick no doubt was these same parents, on the one hand promoting a book extolling the virtues of fantasy by swallowing something and on the other hand, becoming alarmed and shocked when their own children did precisely that, started experimenting with drugs, especially during seminal years of the 1960s. Whether Grace Slick regularly, or if at all, used acid is hardly the point; the point is "White Rabbit" earned her the title the Acid Queen, as well as being known as a slick singer. And as for Lewis Carroll, a mathematician at heart, he did not do drugs, and seldom drank. The drug-fueled countercultural movement of the 1960s most probably misappropriated *Alice in Wonderland* as extolling drugs.

The thought of such an acid or drug trip never appealed to me but certainly the use of mind altering drugs or getting "high" is not peculiar to recent human history. Getting high or obtaining an altered state of consciousness dates back to the earliest of human recordings. Our troglodyte ancestors were quite familiar with the fermentation of grapes and other carbs into alcohol, and the secrets of the opium plant off the upper Asian belt of Afghanistan, as well as the euphoria of chewing the coca plant from South America were well known in Neolithic times.

That phone rang.

Deborah was your classic Southern California girl, like her sister, my wife, Stacey and her other sister, Shelby. You know, the Beach Boys kind of California girl: long blond hair, hangs out at the beach.

> I wish they all could be California girls
> I wish they all could be California
> I wish they all could be California girls

To be fair, Brian Wilson and Mike Love, who wrote "California Girls" in 1965, were not just talking about California girls being the best. They thought girls all over the world were the best. They just wished they were all *in* California so a guy wouldn't have to travel too far to find them. Although truth be told, Wilson and Love were especially fond of born-and-raised California girls. And who could blame them? Brian Wilson's two talented and beautiful California daughters Carnie and Wendy Wilson are two-thirds of the Grammy Award winning Wilson Phillip with Chynna Phillips, another California girl, the daughter of John and Michelle Phillips of the highly successful The Mama and the Papas.

Deborah was a California girl, but for reasons that escape me, she found her way to Denver, Colorado, for college—to the Colorado Women's College, now defunct —which is where she met her husband, Alan. Alan was a student there too, and one of few male students at a supposedly women's-only college. Talk about having the deck stacked in his favor. Alan had his pick of girls to date and he had the good fortune to land Deborah, who you can imagine was Colorado Women's College resident California girl and no-doubt the prettiest girl on campus.

Think about what it'd be like to be the only guy at an all-girl's college. Worse yet, think about being the only guy at an all-girl's college and not be able to land a date. That would be a blow to one's ego. That would be like a guy walking into a women's prison clutching a handful of pardons hoping to get lucky and not getting any takers.

Recall that story about me walking into that doc-in-the-box shop, where the patient had fallen off the curve and the physician from the earlier shift was mired in ordering all kinds of useless tests, unable to see the forest for the trees. Even to this day I cannot fully understand—although I have a better handle on it—how a doctor with the same basic training as me possessed little in the way of common sense. There was nothing terribly special about my ability to walk into the exam room that day and see the situation for what it was; rather, there was something terribly unspecial about that physician's inability to see it for what it was. But it happens.

Whether it's born of training or an innate common sense or both, *most* doctors can tell fairly quickly when the numbers are not adding up, when something is wrong with the picture. Perhaps this skill also partially arises out of a fear of being wrong—being on the wrong side of a medical decision. In the end, and however it is a physician acquires medical common sense, that doctor that day did not have the right stuff. He was most assuredly unlike the Mercury Seven, those seven fighter pilots who became the right astronauts with the right stuff for the first American manned spaceflights, the Mercury missions—first manned flight May 1961—as chronicled in Tom Wolfe's 1979 book *The Right Stuff*. Nor for that matter was that doctor like the '80s boy band song from the New Kids on the Block ("You got the right stuff, baby / You're the reason why I sing this song").

Looping back to Rod Serling, master of ceremonies of *The Twilight Zone*, "imagine if you will" trying to dial a patient over the phone when all you have to go on is what the patient is telling you from some other room, far, far away. Worse yet, try dialing a patient through an intermediary, or even two intermediaries—Serling's "middle ground between light and shadow"—where neither the patient nor their doctor is on the phone, and instead two nonmedical intermediaries are conveying what they *think* they heard. With each degree of separation and uncertainty added—dialing a patient over a phone, dialing a patient over the phone with one intermediary, dialing a patient over the phone with two nonspecialized intermediaries—the disasters waiting to happen also exponentially increase. I heard from a guy who heard from a guy whose first cousin once removed heard from a guy…!

To add to the misery, imagine the burden of trying to dial a patient over the phone through two intermediaries who are, in fact, your patient's family members, who are also your own family members—all of you pulled abruptly into a saga no one requested nor desired. That's what was unfolding before me when I was vacationing in Florida and Deborah was laid out ill in her Los Angeles home, me the reluctant physician trying desperately to not get involved in a situation I had no control over. It's sort of like Michael Corleone's lament in *The Godfather Part III* 1990: "Just when I thought I was out, they pull me back in!" Yes it is true, men quote The Godfather movies with much abandon: Clemenza: "That Sonny's runnin' wild. He's thinking of going to the mattresses already," in which "going to the mattresses" means going to war with rivals using ruthless, brutal tactics. Although I suspect the brutal part is not exactly what Joe Fox played by the ever-skilled Tom Hanks meant when he said to Kathleen Kelly brought to life by delightfully beautiful Meg Ryan in the wonderful 1998 romcom *You've Got Mail*: "Go to the mattresses. You're at war. It's not personal, it's business."

But that phone rang.

As we were getting ready to leave for that New Year's Eve dinner, Alan called again and asked to speak to me. He had not taken Deborah to the ER as I had strongly suggested a few hours earlier. Perhaps he thought she was feeling better. But I knew from the previous call that Deborah was in need of emergency care, feeling better or not. Alan, not being trained in the art of medicine, apparently did not see it that way.

Now, the following is key—or at least it's been key to my sanity all these years later. I declined to talk to Alan over the phone on that second call. No intermediaries, no discussion. I flat-out refused to insert myself into a situation I suspected was spiraling out of control 2,500 miles away, a situation that offered me no means to figure out where the toad of truth sat, no means with which to proceed. It was my fervent belief in that microsecond when I emphatically refused to talk with Alan over the phone, refused to talk through an intermediary, decisively refused to get further involved—other than to urge through the proxy of my mother-in-law that he take Deborah to the ER—that my inaction would impress upon him the firmness of my resolve that his wife needed to go to the ER. Let me restate that again: by refusing to get involved, I thought it would make such an astonishingly galactic impression it would force Alan's hand to actually go to the ER.

I felt a bit just like the unsurpassable Al Pacino as Michael Corleone: "Just when I thought I was out, they pull me back in" except I was not about to be "pulled back in." Not into a situation with Deborah's failing health that I could not control, especially when I knew her life might be in the balance.

My refusing to take the call put in place a firewall. Not only did I refuse to talk directly to Alan, I also refused to talk to him through an intermediary, our common mother-in-law, Barbie. We might as well have been speaking two different languages. It might as well have been the biblical story of the Tower of Babel.

The Tower of Babel is a myth that follows on from the great flood myth (Genesis 6:9–9:17). According to the Bible, the great flood was due to God's disappointment in mankind, upon which realization he decided to flood the earth, returning it to a near pre-creation world. Incidentally, the biblical narrative of Noah's ark resembles the Epic of Gilgamesh, a non-Hebrew Sumerian flood that predates the Bible's flood. I smell copyright infringement.

"Well, the weather was terrible, but the buffet was AMAZING!! They had EVERYTHING!!"

"No, the Tower of Babel wasn't built for better phone reception."

But as for the Tower of Babel: after the great flood, all that remained of mankind spoke one language and they decided, against God's wishes, to build a tower to the heavens so as to join God. God, not wanting any visitors—a bit of a recluse, like me—decided to confound their common speech into many languages. No longer able to understand each other, the workers on the Tower of Babel stopped building in concert with another, and the tower crumbled back down to the earth, eventually leading to the scattering of people to all corners of the world. The biblical lesson from the Tower of Babel narrative is to do what God commands. But the between-the-lines story is that humans, in speaking many different metaphorical languages, can never truly work together in concert. That's a sobering thought.

So, there I was on December 31, 1995, with my own Tower of Babel, Alan relaying incoherent information about Deborah to Barbie, Barbie relaying incoherent information to me. As I attempted to not get any further involved, the one salient point that rose to the surface was that Deborah, like our two intermediaries, like the workers on the Tower of Babel, had become incoherent. Literally incoherent.

"Coherent" is from the Latin *cohaerere* and means "connected, consistent, together," as in a person's thoughts, words, mannerism, things that are flowing normally or flowing together. "Incoherent" means the opposite, especially in terms of language, where thoughts expressed are incomprehensible, confusing, unclear, make no sense.

Incoherent? What about "incoherent" in a patient would not be a red flag that Deborah was not just falling off, but had *already* fallen off, the curve? A red flag is a universal warning of danger. At the Beach Boy's Redondo Beach, a red flag would mean don't go in the water. On a first date when a thirtysomething man reveals he still lives with his parents, that should be a red flag for the girl to use the "emergency phone call" trick. In medicine, it means there is something wrong with the picture.

Through the intermediary at my end, I learned that Alan still wanted me to call in antibiotics. Of course, through Barbie, I shook my head no. My people were talking to Deborah's people. I refused to talk to Alan, I refused to call in antibiotics, I refused to insert myself into that incoherent world. I drew a second line in the sand, having drawn the first line hours earlier, by telling Alan to dial 911. This second line in the sand was a real obvious line, a line so deep, so wide, you would fall into it strolling along Santa Monica Beach. With no ambiguity, with no hesitation, with firmness of resolve, I told Barbie to tell Alan to call 911. That placing Deborah in their car for a drive to the ER was a ship that had sailed. This was a 911-dire situation—no time left to fire up their Volkswagen Karmann Ghia.

I stood firm and I stood tall and I stood immovable, listening to make sure Barbie conveyed precisely that to Alan: dial 911. Deborah's infection had evolved into a beast of a different order. To be perfectly fair to Alan, it is unclear if he truly understood the urgency in my voice through an intermediary, or if he perhaps thought part of my reticence was just not wanting to care for a family member. It is true that physicians can lack perspective and judgment when caring for family members, their professional ability to render care hampered by a difficulty in being objective. And speaking of compromised objectivity …

Gurney #3 and the primrose path, a reference to Shakespeare's 1602 smash hit *Hamlet*, when Ophelia speaks to her brother, Laertes: "Himself the primrose path of dalliance treads / And recks not his own rede." She is warning her brother not to take the attractive, easy path of sin to hell. Loosely translated into modern-day English, it means not to be fooled, because the inviting easy path will bring disastrous consequences. In short, "don't be led astray," "don't be led down the primrose path," as perhaps we had been by Gurney #3's initial complaint: "I think I have the flu."

I was a medical student at the University of Colorado School of Medicine, which is a really fine institution with a collective of five rotation hospitals that participated in the medical school and residency training programs. Together those five hospitals covered every type of hospital a teaching program could possibly need, making the education of its medical students and residents rather complete. We didn't call those five hospitals by their proper names. Instead, we had our own language, or own babel, just like how Jane Goodall observed a unique language among the Tanzanian chimpanzees. That is not to say we were chimps—perhaps chumps, but not chimps. With the long and demanding days of medical school and even longer and more demanding days of residency training, a titch of humor, laced with moments of levity, went a long way in staving off insanity, or depression. As a resident and medical student, even as a doctor, and nurses too, we all need whimsy and amusement

"There are some things they don't teach you in medical school. I think you've got one of those things."

to survive in the trenches, taking hand grenades. One needs only watch the wonderful 1970 dark comedy war movie *M*A*S*H*—the intermingling of seriousness and comedy—to appreciate the human proclivity to humor in order to block the horrors of death and dying. We're not taught humor in medical school as a coping mechanism, so it must be a cultural thing. M*A*S*H is the acronym for <u>m</u>obile <u>a</u>rmy <u>s</u>urgery <u>h</u>ospital and although the setting was the Korean War (1950-1953), the film was subtextually about the Vietnam War.

Denver General Hospital, the community hospital of the City and County of Denver (now called Denver Health), was nicknamed "the Dog House." This wasn't out of any disrespect, nor because rotating at Denver General was so horrible it was like being sent to the proverbial "dog house." Rather, Denver General was called the Dog House simply because its acronym, DGH, sort of looks like DOG. In fact, the Dog House was and is a stellar city hospital, well run with devoted nurses and physicians and, for us medical students and residents, a great place to learn medicine.

When I say, "a great place to learn medicine," what that means is, like other inner-city hospitals, the Dog House saw and sees its fair share of illness and injury and disease. It has the very patient population that, for better or for worse, exposes a student of medicine to a wide breadth of knowledge and experiences. It is a fact of life that the sickest of the sick, the most injured of the injured, also tend to be the poorest of the poor. In contradistinction to Galen's miasma theory, whereby infection was seen as the providence of the poor because they smelled more than their rich counterparts, disease and injury is in fact more often the providence of the poor for no other reason than that they are poor. It is precisely the ill poor, that demographic group, from which a great deal of medicine is learned, medical students and residents grateful and humbled for having had the opportunity. To those we learn medicine from we owe you a debt of gratitude that can never be repaid.

The Denver General Hospital emergency room also had a special nickname: "the Knife and Gun Club." It earned that moniker not only because the Dog House ER saw more than its fair share of knife and gun violence but because a 1989 nonfiction book by Eugene Richards chronicling the Denver General ER was titled just that: *The Knife and Gun Club*. To be sure, the Denver Dog House ER was not and is not the only candidate in America that might hold a Knife and Gun Club designation. Every single major hospital in every single major US city, from LA County Hospital to Grady Memorial in Atlanta, and Detroit Receiving to Bellevue in New York, has its own credentials for a Knife and Gun Club ER title. But the Denver Dog House ER was just the first so designated because of that book.

When I was interviewing for a residency program at Emory School of Medicine in Atlanta, Georgia, which has Grady Memorial as one of its teaching hospitals, the resident taking me on the grand tour of the program hospitals—your typical dog and pony show—told me a story that could have been true or was maybe merely urban legend. What I was told was this: One hot, humid Atlanta night, the type of night where tempers easily flare, a gangbanger who had been shot up was brought into the Grady Memorial ER. Despite multiple bullet wounds, he was expected to live.

Word got out on the street that he was not dead, and would not die, so his assailant hearing this soon ran into the ER to finish him off right there sprawled out on the gurney, much to the horror of the doctors and nurses attending to him, who quickly backed off, lest they become a statistic, too. And then, just as quickly as that assailant had run in, his magazine now emptied into the victim, now most assuredly dead, fled into the darkness of night. What had been a bullet-riddled alive patient was now a more-bullet-riddled dead patient.

The other hospitals within the University of Colorado teaching program had their nicknames too. The Denver Veterans Administration Hospital was called "the VA Spa" simply because it rhymed. It was hardly a spa-like rotation. Across the street from the VA Spa was the private Jewish hospital Rose Medical Center, known affectionately as the Rose Bud, not because of Bud Light beer, but because, like a rose, it was an easy, sweet rotation, and it had the best cafeteria food, too. The Children's Hospital Colorado was simply known as that: Children's. I'm not entirely sure why Children's Hospital didn't have its own whimsical moniker, but my guess is out of respect to the young patients to which it tends. However, during my plastic surgery training in Cleveland, the children's hospital at Case Western University did have a special name: Rainbow Babies and Children's Hospital. We just called it the Rainbow.

Finally, there was the University Hospital, known during my tour of medical school and residency alternatively as "the Mecca" and "the Mother Ship," simply because the University Hospital was *the* center of all that was our teaching program, all that was us.

Although "Mecca" with a capital "M" names the holiest of holy cities for Muslims, which is located in Saudi Arabia, the lowercased word "mecca" simply means "a place that attracts people of a specific group"—like how Fifth Avenue in New York is a "mecca for material girls." The Mecca, when we used it in our training program, meant the Mother Ship hospital. The meaning of "the Mother Ship" needs little further explanation, popularly used in science-fiction Star Trek lore. Quite simply, the four other hospitals in our program—the Dog House, the VA Spa, the Rose Bud, and Children's—were all smaller, satellite spacecraft of the Mother Ship.

The first two years of medical school is pretty much classroom drudgery, where you shove as much information into your brain as possible and then vomit it out during exams. As the years pass on, you forget a good deal of what you learned. Or do you? The brain is a strange thing. It remembers bits and pieces of information that you cannot possibly recite upon command but are oddly retrievable when taxed or tickled or prompted.

In medicine, common things happen commonly, and specialists specialize, so as the years go on, that body of information a general physician and especially a specialist knows and uses on a daily basis is a fraction of what was learned in medical school. The most knowledgeable bookworm in all of medicine is the poor sad sack who just finished medical school, a beleaguered, brain-popped shell of a once vibrant human who knows a lot about a lot of things. But as the days turn into weeks and the weeks into months and the months into years, that medical student can only recall a fraction of what was once an encyclopedic arrangement of information spanning an array of subjects.

Even internists and family physicians and ER docs, who see the widest gamut of illnesses, sheds so much of what they once knew. For those of us who specialize, and worse yet, subspecialize, we quietly work away in our little niche of the world of medicine, and that niche of knowledge is just a titch. But this is the kicker: every once in a while, a patient comes along with an oddity that you haven't dealt with perhaps since medical school, or you overhear some doctors in the lounge talking about a disease you've long since forgotten, and yet somehow your brain finds some stored bits and bobs that come floating to the surface. In the words of Forrest Gump, "just like that" the information, long buried, percolates into your consciousness. After all that stuffing of information into your brain during the years of medical school and residency, whatever synaptic neurons have been storing it, ignite the long lost information that go ahead and fire.

A subtle shift of weight. A tug of fabric. A sudden draft. The moment he first felt comfortable calling himself a plumber.

Pretty much from third-year medical school on through residency training, book learning gives way to hands-on learning, or what some educators term experiential learning. It's really not much different than plumbers and electricians who learn through an apprenticeship. The evolution of a medical student through residency into professional practice is a necessarily slow, tedious apprenticeship, the aim of which is to ensure that the final product who enters the "practice of medicine" is up to the task. And hopefully, unlike the plumber, no visible butt cracks are on display.

My microbiology class in medical school was taken right out of the textbook, practically verbatim. There was really no need to even attend lectures. One merely needed to memorize the book (which is what I did), skip many lectures (which is also what I did), and study for exams (I also did this, but sometimes before exams I also went running to clear my overstuffed head).

What makes this book you're holding—a book whose backbone is human infection—especially strange is that during medical school I, its author, didn't have a keen interest in the world of human pathogens. Microbiology, as well as a few other medical subjects, held very little interest for me. Perhaps that was because devastating human infection, the kind of infection that saps lives, was for me in medical school a remote subject, far removed from anything I had experienced in my life at that point. Sure, as a kid I had chickenpox and a few colds here and there, but raging infection was foreign to me. As you advance in medical training, as you start taking care of patients in third-year medical school, and especially during residency, all of a sudden those boring, distant infection subjects come to life in a crash-zoom close-up shot. They're no longer obscure factoids in a book or lecture hall but frighteningly, tangibly real events that hit home with all the crashing violence of an unsurfable ocean wave.

7 THE BIG BANG

In a previous chapter, we arrived at the realization (with all due respect to the book of Genesis) that the first life on Earth was a bacterium that had the right kind of DNA to advance, making us humans, along with every other living thing, descendants of that one last universal ancestor affectionately called LUA, that unicellular microbe that wiggled 3.5 billion years ago. All these eons later, we humans—derived straight from that first bacterium, from LUA—maintain a relationship with modern bacteria, also descendants of LUA—ancestral cousins—a relationship that can best be described as a love-hate affair.

One type of bacteria we love is what scientists term "beneficial bacteria": probiotics. These little microorganisms live on our skin, in our gut, and within our bodily orifices and help us digest things or keep a balance between other warring microbes. It's a symbiotic relationship best described as "mutualism." We provide the probiotics a home and food; they help keep that home running smoothly. It is a slick mutual arrangement.

Certain gut species of the bacteria *Bacteroides* and *Lactobacillus*, which are also found in cultured yogurt and probiotic pills, help us digest foods we normally would not be able to, such as plant fiber and dairy. There are the *Bifidobacterium* and the *viridans streptococci* bacteria species, also in our intestines, who, like soldiers standing guard on a wall, kill other unwanted intruding bacteria we don't like or need. The bacteria *E. coli* that lives along the length of the intestine helps synthesize the vitamin K we need for blood clotting, while yet other types of *E. coli* can run amok in our gut, causing gastroenteritis. A different type of *Lactobacillus* than the yogurt variety is remarkably beneficial for vaginal health—and no, depositing plain cultured yogurt into the vaginal vault does not do any good to maintain a healthy flora at all, since it is the wrong species of *Lactobacillus*.

As long as the balance of the beneficial bacteria remains copacetic in and on our bodies, we hum along just fine.

These are just a few examples of the love affair humans have with bacteria. We are also all too familiar with their ancestral cousins, our ancestral cousins, of the more sinister variety that cause human infection. The bacteria we dislike, the infection-causing kind, likewise share a common ancestor with us. Can ancestors get any ruder than that? It is both interesting and a bit disconcerting to ponder that the LUA bacteria gave us both "us" and also gave us strep throat, staph abscesses, and bacterial pneumonia. Family…what can you do?

If you're wondering why some bacteria are beneficial and others cause infection, and, for that matter, why viruses cause infection, that's a great question. There are billions of bacteria living on our skin, munching away at dead skin cells, helping to keep our skin healthy. There are gazillions of bacteria in our gut helping to digest foods into component parts for absorption, foods we cannot break down ourselves. Because so many of them live on our skin and in our gut, those "beneficial" bacteria species decided long ago that even though we kill some of them with soap and evacuate others from the sanctity of our bowels when we go boom-boom, it's still an okay arrangement, as there are always plenty more beneficial bacteria to take their place. Subsequently, these beneficial bacteria don't wage war against us—unless they get out of balance—don't generally lead to infection, even though we wipe out some of them with Purell. They take their losses and remain mutualistic with us, their host.

But other bacteria—as well as pretty much all viruses that plague humans, mainly cold and flu viruses and viral gastroenteritis—determined long ago that they needed a surefire way to extend their lineage, to make sure their species did not die out at the end of a lonely road in some host they parasitized. And what better way to make sure your progeny gets spread to all corners of the earth then through inducing the symptoms of infection, inciting coughing & sneezing and vomiting & diarrhea as a way to keep moving on to the next host? Most human infections—colds, the flu, strep throat, cholera, tuberculosis, VD, you name it—always include a mode of transmission. Symptoms caused by infection are usually the mode of transmission, the avenues that ensure the survival of pathogens that decided long ago to take up a war stance against humans.

Before we delve into some of the more familiar human bacterial infections, there is a nonmedical bit of science worth exploring first. We humans are derived from the last universal ancestor, but where on earth—or not on Earth—did LUA come from? It's a form of the same age-old question humans have pondered for millennia—looking up into the cosmos, with our dimly lit, uncomprehending brains, asking, "Where do we come from?" It's the same sort of question some women ask themselves about their husbands: "Where on earth did he come from?"

Consider these statistics. According to recent polls, in the United States 35 percent of people hold the view that God in some way created humans, a percentage that has been dropping steadily for years. In Britain, the number of believers in creationism is more like 15 percent. Creationism is straight out of the biblical book of Genesis. If you parse these numbers out even further, or if you ask the question differently, some interesting facts emerge.

Older people tend to believe in creationism, and younger folks believe in evolution. If you restrict the question of creationism to the literal interpretation that God created the world about 10,000 years ago over six twenty-four-hour days, the numbers really drop. On the other hand, if you open up the question to encompass God having created the world, but maybe not 10,000 years ago and maybe not in six twenty-four-hour days, the numbers rise. If you add the idea of "intelligent design" to the equation, as Ridley Scott's 2012 sci-fi film *Prometheus* does, the numbers rise further. If the question asked implies that the universe began on its own, but God had a hand in it, then the numbers go up even more.

But one thing appears certain: no matter how open, how liberal, you make the question, the majority of people in America, Canada, Britain, and most other Western countries don't believe in creationism. We're talking anywhere from 60 to 80 percent of people in the Western, largely Christian, world do not believe in any variation of creationism.

Most of those who believe in God but not in the literal interpretation of Genesis are at a loss to offer an alternative to creationism that includes God but not Genesis. And many of those who don't believe in creationism or God are equally flummoxed by the question of "Where do we come from?"

The narrative that follows traces the scientific explanation for how LUA, the last universal ancestor bacteria, came to be. Apart from knowing the phrase "the Big Bang," many people are unable to provide even a vague concept about what the Big Bang is other than to state, "In the beginning there was a big explosion and the universe was created." Actually, it wasn't an explosion, no proverbial Boom!, despite being known as the Big Bang. We'll investigate that momentarily.

We are descendants of LUA, but where on earth did LUA come from? And where, for that matter, did Earth come from, where did our solar system come from, where did our galaxy come from—indeed, where did the universe come from? How, from the Big Bang, did I arrive at this computer screen, typing away—all those atoms traveling billions of years across billions of light years to make me "me" so I can sit here trying to explain the Big Bang to you, without glossing over the facts nor boring you with overwrought detail? Whether or not the hard facts and stats put us in danger of a yawning fit, it is a story worth telling—and it is actually not that complicated at all. Hopefully I'm able to dispense with the usual audience-terrorizing science jargon—and thus the yawns—that often accompany attempts to explain the Big Bang narrative, also referred to as an "initial singularity."

Our solar system began to form about 4.6 billion years ago, with Earth taking shape soon after, at 4.543 billion years ago. Both arose out of the gravitational collapse of and subsequent rearrangement of the molecular cloud that had formed in our little nook of the Milky Way galaxy, an atom-filled celestial cloud that was the result of the Big Bang, some 13.8 billion years ago. Simple atoms like hydrogen and helium formed from that initial singularity 13.8 billion years ago, a singularity that created our entire universe. This story will be fleshed out in the pages that follow, but, for now, in a nutshell: those simple atoms from the Big Bang underwent two fundamental processes that created more complex atoms and even created a few simple molecules (that is, atoms shacking up with one another), such as little package deals of water and breathable oxygen.

The next part of this CliffsNotes version involves one of the very first fundamental process ever to occur in the known universe after the Big Bang: star formation. Hydrogen and helium collapsed together to make stars, beginning about 200 million years after the Big Bang singularity, thus becoming the nascent universe; those early stars later on exploded, which led to the forming of more complex atoms and a few molecules. The second process, termed "nucleosynthesis," was similar, but bypassed the first step of star formation, as out there in the hot, hot, hot soup of space smaller atoms were able to form more complex atoms and also some molecules without the stars' help. Going from simple atoms to more complex atoms is your basic nuclear fusion, which combines atoms (as opposed to fission, which splits), a process that created all the naturally occurring elements on the periodic table (except, again, the element of surprise).

Simply stated, all the hydrogen, nearly all the helium, and some of the lithium in the universe formed from the Big Bang. All the other elements formed from nuclear fusion, mostly through star explosion and some from nucleosynthesis in the hot cosmic soup. How hydrogen, helium and some lithium came about from the Big Bang will cover momentarily.

Our sun formed about 4.6 billion years ago from the gravitational collapse of mostly hydrogen, and a bit of helium. What the sun didn't grab became all the other planets, moons, asteroids, and comets in our blossoming solar system. Earth is thought to be 4.543 billion years old, forming 9.3 billion years after the Big Bang, making it a little bit older and a little worse for wear than Keith Richards. Galaxies coming in and out of existence, solar systems coming in and out of existence, planets and other solar stuff coming in and out of existence, have all been going on since soon after the Big Bang.

That initial singularity of the Big Bang was a point nowhere in time and nowhere in space where there was nothing around but pure energy. There was no mass, no space, no time. Then something odd happened—what physicists term a "slight fluctuation" in that pure energy. That fluctuation set into motion the conversion of pure energy into all the mass of the universe, as well as created space and the flow of time as the universe expanded. And created those four fundamental forces of the universe.

Those 13.8 billion years ago, pure energy possessed the potential for all the mass and space-time of the universe, and that initial singularity unleashed it all. Before the Big Bang, everything was crunched into a speck of energy. That speck might have been the size of a football field or a small town—no one really knows for sure. It likely had some sort of dimension, although it did not have mass. I know that is a total Twilight Zone statement, but it was what it was.

Then this strange thing happened: that quantum fluctuation—also termed a "vacuum fluctuation," which are big words that simply describe how a single ripple triggered a chain reaction—prompted the Big Bang to unfold, which then allowed for the creation of all elementary subatomic particles—quarks, leptons, bosons—which then formed all atomic particles—protons, neutrons, electrons—which then formed the simple atoms hydrogen, helium and lithium, as well as photons of light, plus the creation of space and the creation of time.

If you believe in God and you want to believe in the Big Bang and want to believe God had a hand in the creation of our universe, this would be the moment for you. Perhaps God caused that quantum fluctuation, or God *was* that quantum fluctuation. But to be sure, it was 13.8 billion years ago, not a mere 10,000 years ago, as written down by Moses and the wandering sheepherders in the Sinai desert. How do we know how old the universe is? We'll turn to that in a moment.

"O.K., YOU'VE CREATED THE UNIVERSE, WHAT'S YOUR EXIT STRATEGY?"

This Big Bang resulted in the rapid inflation of the universe. "Inflation" in this regard does not mean the universe cost more like the inflationary costs of buying a new home, inflation in this setting means going from the complete nonexistence of space and time to, after that singular moment, the universe all of a sudden being upon us, and expanding rapidly, inflating rapidly—very, very rapidly—creating space in the sense that us three-dimensional-handicapped humans are able to understand it. That initial period of time in the universe is called the Planck epoch and roughly corresponds to an infinitesimally short time span of only 10^{-43} seconds. That's right—the entire quantum fluctuation Planck epoch lasted only 10^{-43} seconds, there and done. These basic elementary particles—euphemistically called "the God particle," but more correctly known as the Higgs boson—then formed those quarks, leptons and bosons needed to form protons, neutrons, electrons and photons and so on, were all created from pure energy during the Planck epoch. Simply put: all the matter in the universe formed during the Big Bang.

You're probably wondering words like quark came from. The American physicist Murray Gell-Martin coined the term "quark" in 1963 from James Joyce's book *Finnegan's Wake* 1939: "Three quarks for Muster Mark! / Sure he hasn't got much of a bark / And sure any he has it's all beside the mark." Although a "quark" is derived from the German and means "dairy product," it was the "three" in the first line that inspired Gell-Martin, as it reflects the way subatomic quarks fit in nature, three pairs giving six total quarks: up and down, charm and strange, and top and bottom. It all has to do with their charge, mass and spin giving the six flavors as physicists call it.

Lepton was coined by Belgian physicist Léon Rosenfeld in 1948 from the Greek *leptós* meaning "fine, small, thin." He chose lepton because at the time the only known lepton was an electron and the mass of an electron was small consider to the mass of protons and neutrons. Since then other leptons have been discovered, six total, some larger than a proton, but their purposes need not concern us here other than their flavored names: electron, muon, tau, electron neutrino, muon neutrino, and tau neutrino.

As for boson, its origin is much simpler. It was so named in 1947 by British physicist Paul Dirac to honor the contributions of Satyendra Nath Bose, an Indian physicist at the University of Calcutta. Bose collaborated with Einstein in the 1920s developing the Bose-Einstein equations and proposing the Bose-Einstein condensate, which in its simplest definition I can offer means the lowest possible quantum state a boson can occupy as temperatures approach absolute zero. What I think that means is once you approach absolute zero the boson is reduced to not much more than its dynamic properties which can then be observed, in layman's terms, the background noise is gone. There are four bosons other than the God particle Higgs boson: photon, gluon, Z boson, and W boson.

Now fast forward from 13.8 billion years ago to 4.6 billion years ago, when it is believed our solar system formed. The universe was huge by this time. In our little neck of the woods, all the elements, from hydrogen (H) at least through uranium (U) and possibly plutonium (Pu), needed to build a solar system—or at least our kind of solar system—existed. If they didn't come from the Big Bang itself, then they arose out of star explosions and nucleosynthesis. Now, call to mind that previously reviewed Timex-wearing apple thought experiment and how gravity works. With that in mind, in our little neck of the universe, our little niche of the Milky Way galaxy, a cloud of atomic dust collapsed to form our sun, mostly hydrogen and some helium. The atomic gases and more complex atoms and whatever other molecules that escaped being captured by the sun then collapsed into our planets and moons and asteroids and comets and other general space debris.

The sun's core, by the way, is mostly hydrogen gas. Due to the immense temperature and crushing pressure exerted by our sun, a fairly standard star, the hydrogen fuses with itself, forming helium, and in so doing, releases a bucketload of energy—one form of which we call sunlight—which life on Earth so desperately needs and without which would not exist. Get this, at the sun's core when two hydrogen atoms combine to form helium and release energy photons, it takes somewhere between 10,000 and 200,000 years for that photon to reach the surface of the sun. Why so long? Too many collisions with other stuff within the interior of the sun. It is mind-boggling. But once that photon reaches the sun's surface, escaping its gravity and heads towards earth, at 186,000 miles per second needing to travel the 95 million mile distance, that photon reaches earth in eight minutes. That photon once formed took 200,00 years to get from the core of the sun to the sun's surface, but a mere eight minutes to reach earth.

Leaving for a moment the sun out there in our nascent solar system and directing our attention to those collapsed gases the sun did not grab, gases containing parts of or the entire spread of the periodic table, plus several already formed water and oxygen molecules, they then condense into sufficient masses of balanced force and began orbiting the sun. We call these our eight planets—nine, if you include poor Pluto. Those other collapsed objects of smaller dimension but with sufficient mass and balanced force began orbiting the planets that captured them. We call these moons. Still, those other even smaller collapsed debris caught between Mars and Jupiter also started orbiting the sun and became the asteroid belt. And further still, then there are the TNOs, the trans-Neptunian objects, some of which make up the Kuiper Belt, a region in the solar system beyond Neptune where tiny planets, asteroids, and comets zip about, mostly orbiting the sun. Finally, there is the unobservable Oort Cloud, a region beyond Pluto that apparently contains a bunch of ice objects and other cosmic schmutz.

The Oort Cloud has never actually been observed from Earth but is believed to be home to long-period comets, comets that take more than 200 years to make one orbit around the sun. The Kuiper Belt, closer to us, near Pluto, contains short-period comets that take a mere 200 years or less to make a solar run. The most famous of these, Halley's Comet, is visible to the naked eye from Earth every seventy-six years or so. It has been known to astronomers (ancient Babylonian ones) since 240 BC but is named after the seventeenth-century scientist Edmond Halley, since he was the first astronomer to correctly determine its seventy-six-year periodicity. That is, Halley determined that a comet seen in 1531, 1607, and 1682 was the same comet. Halley's Comet was last visible in 1986 and it will next be visible in 2061, which puts me, but not Keith Richards, SOL.

Planets, and the moons they drag along with them, have enough Newtonian force, enough vitality, to remain in orbit around the sun (and, for the moons, around their planets). But the orbits of many asteroids and comets have less vigor, less vitality—which is to say that their unbalanced forces sometimes allow them to leave their assigned region in the solar system and either fly off into deep space, never to be heard from again, or come crashing into other objects, pelting vulnerable things like our own moon—which has a pockmarked, albeit friendly, face that no teenager would want.

The reason asteroids and comets sometimes hit things is because they orbit with less predictability, taking oblong and variable paths around the sun. Take, for example, that big asteroid that apparently hit Earth sixty-five million years ago off the coast of the Yucatán in Mexico, which condemned the dinosaurs to extinction. Errant asteroids also make for good Hollywood scripts, like *Armageddon* (1998), starring a tougher-than-nails Bruce Willis, a few space cowboys and a huge rock in a totally scientifically inaccurate portrayal of humankind attempting to destroy an asteroid heading for Earth. Between the average distance the asteroid belt is from Earth (180 million miles), the speed with which an asteroid travels (upwards of 100,000 mph), and the fastest a man-occupied spaceship currently can possibly travel (25,000 mph), well, the Bruce Willis team of deep-core oil drillers would be waving at the asteroid as it went zooming by. But I like the movie anyway, especially Steve Buscemi (any movie he's in he's good).

NASA and other space agencies attempt to track asteroids in the asteroid belt between Mars and Jupiter and are even more taxed to track those that are further out in the Kuiper Belt beyond Neptune—to quote from *Armageddon*, "it's a big ass sky." So, only a small percentage of asteroids in the asteroid belt and even fewer in the Kuiper Belt are actually trackable. Fortunately, and that is a soft "fortunately," despite the speed with which asteroids travel, if an asteroid out there in either belt is discovered and calculated to be slowly orbiting the Sun inward toward Earth, and is large enough to be a global killer, we would have us a few decades to figure out what exactly to do. That is because the asteroid is orbiting the sun and its path is slowly degrading inwards. It's not like an asteroid one day just decides to jump ship and head straight towards earth. The current best option is to send spacecrafts with multiple nuclear warheads to explode simultaneously next to or on the asteroid, altering its trajectory. Landing Bruce Willis on the asteroid to dig a hole in which to place said nuclear bomb ain't going to happen.

Recently, in July 2019, the Japanese spacecraft Hayabusa2 landed on the surface of the Ryugu asteroid 5.5 million miles from Earth, out there among what are termed the Apollo asteroids, or near-Earth asteroids, the first spacecraft to land on an asteroid. Hayabusa2 then blasted the surface of the asteroid—not to see if Bruce Willis's techniques are feasible—but to aerosolize the surface for scientific analysis, like, you know, searching for water or a girl's best friend: diamonds. But, for our purposes, the Hayabusa2 demonstrated that, if given enough time, we might just be able to blow up a global killer asteroid that is hurtling toward Earth, or at least blow it onto a different trajectory. Either option works for me. Just saying.

Most of the asteroids in our solar system are within that asteroid belt between Mars and Jupiter, orbiting the Sun through somewhat balanced Newtonian mechanics. The Kuiper Belt beyond Pluto, on the other hand, is mostly made up of comets, although they too are bound by gravity. What's the difference between the two? Well, asteroids are rock and metal, while comets are frozen ice balls of water, methane, and ammonia.

Now, if there is some huge asteroid out there beyond our solar system that has Earth's name on it, an asteroid that originated from another solar system, perhaps from the neighborhood of our next nearest star, Proxima Centauri, an asteroid not bothered or bound by the gravitational pull of our Sun into an orbit—well, that would not be a pretty picture. The fastest recorded object in the universe is something called the Runaway Star, which is traveling through our Milky Way galaxy at a slick twenty-six million miles per hour. An asteroid or another runaway star of sufficient size and similar speed as the Runaway Star originating from elsewhere in the Milky Way, not bound by orbiting the Sun, and heading straight toward Earth would first become noticeable as it neared our solar system's asteroid belt. That's because most astronomers looking for asteroids and such have their telescopes trained on a small bit of sky, and, even at that, it's a lot of sky to keep track of.

If we had a huge runaway asteroid or star traveling twenty-five million miles per hour heading straight toward Earth, and it was first observed emerging from the asteroid belt near Mars 300,000 million miles away travelling at 25 million miles per hour, then we would have twelve hours before impact—that is, if some astronomer happened to see it coming in the first place. Otherwise, we'd simply see a flash in the sky and then we would be gone—revelation and then annihilation becoming near simultaneous events.

The same would be true of an exploding star, a supernova. If a supernova exploded it would emit gamma radiation and other deleterious debris, like charged particles, that, if near enough to Earth, would clean the flesh right off our bones. If you think that such a star explosion ten light years away would give us ten light years' worth of time—that is, 137,000 years—to say our goodbyes, you'd be wrong. By the time we became aware of a nearby supernova explosion, it would have already happened 10 light years earlier, meaning its fallout would have reached us at the same time we became aware of it—we'd be aware and simultaneously ripped to smithereens.

Is earth in danger of being hit by supernova explosion radiation? It is estimated that a "safe distance" from a supernova, outside the splash zone of its wrath, is about fifty to 100 light years (bearing in mind that light years is a measure of distance, not time). The nearest known star of supernova potential, termed a "progenitor supernova," is IK Pegasi, at a relaxing 150 light years away. Betelgeuse, 1,000 times larger than our sun, is a supergiant star in the constellation Orion that is getting ready to go supernova—or maybe it already has. But, at 640 light years away, I think we're safe.

Having said that we are safe, and while a near-Earth supernova might not entirely wipe us out, any supernova within 1,000 light years—including Betelgeuse—can have a measurable effect on Earth's atmosphere and biosphere. These effects might include harming the ozone, altering the atmospheric gases, injuring the ocean plankton, and worse of all, interfering with our mobile phone reception, and other things like that.

Max Karl Ernst Ludwig Planck was born in 1858 in Kiel, in the Duchy of Holstein, now part of Germany. Planck's family, on both sides, was well educated. His paternal grandparents were theology professors in Göttingen, his father an attorney. His mother Emma (née Patzig) Planck was well-known for her lively participation in the local academic world. As a child, Planck's talent appeared to be in music, but he instead followed a path into physics at the Ludwig Maximilian University of Munich. Oddly, or perhaps not really all that oddly, the association between music and mathematics is not so trivial. The push and pull between melody, harmony, and rhythm certainly has mathematical underpinnings, which is why many great musicians are often gifted at math and many great mathematicians find learning music easy. I'm not sure where that leaves me. I can do some of the math, but I have absolutely zero musical talent. As Judy Garland said: "I got rhythm … / Who could ask for anything more?" (while Garland famously sang it, George and Ira Gershwin actually composed it for the earlier stage musical *Girl Crazy* (1930), with Garland picking up its refrains in the 1943 film remake by the same name).

What was rather strange, and singularly impressive, about Planck pursuing a career in physics is that his university physics professor, Philipp von Jolly, advised Planck to *not* go into physics. According to Jolly—and this was in the 1870s—everything in the field of physics, except for a small number of question marks, had already been discovered. Professor Jolly was of the opinion that physics was a dead-end career. He, as it turns out, was wrong, way wrong, and fortunately Max Planck didn't heed his professor's advice. After completing his physics studies, Planck eventually landed a professorship at Friedrich Wilhelm University in Berlin, now known as Humboldt University of Berlin.

Planck became fascinated with the concept of entropy, which to you and me in simple terms means a system running down, using up all its available energy, to the point that no apparent energy is left in the system. The key word in that line is "apparent." As the system loses energy, it heads into complete randomness. A perfect example of this type of entropy, of such disorder, is a teenager's bedroom.

For a theoretical physicist like Planck, however, "entropy" is something more complex. It is actually the measure of energy in a closed system that has no apparent energy but actually still has some energy—unavailable energy—as well as its degree of disorder and the uncertainty of that system. Another way of looking at it is this: a system that appears to have spent all its energy and is in a state of total randomness still in fact has energy. That is entropy in physics.

Thomas Edison has been credited with inventing the light bulb, but as has already been mentioned, he in fact did not invent the light bulb but rather came up with the best light bulb filament at that time, a carbon filament. Today, tungsten, not carbon, is the most common filament in a traditional light bulb. Whether the filament should be made of tungsten or carbon or something else was of great interest back in the late 1800s, as improving lighting efficiency was big business. Especially since those gosh darn filaments didn't last very long, heating up and then breaking. One German manufacturer, wanting to create a more efficient product, enlisted Planck's help in designing a light bulb that would emit the maximum light with the minimum amount of energy input into the system, ergo, less heat to burn up the filament but still have a worthwhile glow. At the time, Planck was known in the physics world for having done theoretical work on what is called the "blackbody," a theoretical body that absorbs all light and reflects nothing, sort of like a black hole out there in the universe.

Planck correctly theorized that the energy emitted from a blackbody was emitted in direct correlation to the temperature applied to the system and had nothing to do with the composition of the system. This theory became known as Planck's law. On a theoretical level, such a blackbody would be a highly efficient light bulb—but that is not where this story is taking us, nor is it where it took Max Planck.

While trying to invent a more efficient light bulb, the proverbial light bulb went on in Planck's head. In 1900, he went on to determine that the thermal energy emitted by a blackbody is emitted in packages of energy, and not as a smooth emission of energy as one might have thought, but in a step-pattern emission, like a staircase, which he called "quanta." This principle became known as Planck's postulate, and it ushered in the field of quantum mechanics, that is, the study of what goes on at the atomic and subatomic levels including, among other things, why energy is always delivered in discrete packages. And old Professor Jolly thought physics was already dead back in 1874! Professor Jolly didn't live to see his student Max prove him wrong, since Jolly passed away in 1884, long before Planck ushered in the field of quantum mechanics.

Who cared about this quantum of energy thingamajig? Apparently, a fellow German, Albert Einstein, pondering the universe while sitting at his desk job as a patent officer in Bern, Switzerland, cared about it. Picking up from where Planck, and others, had left off, Einstein in 1905 published three scientific papers of such magnitude that that year has been designated the *annus mirabilis*, or the "miracle year."

Earlier in these pages I waxed poetic about science building upon science, about one scientist standing on the shoulders of previous scientists to advance a theory, peering a little bit further into the unknown. But every once in a while, someone comes along and proposes or advances a theory that is such a radical departure, that is so utterly unlike anything that came before, that there were no applicable shoulders for them to have stood on. Einstein was such person. With moments like the one Einstein ushered in, the *annus mirabilis*, you are forced to take a moment of pause, rub your eyes, and wonder if such a person was either an alien or touched by a god. No one had a vertical leap like Albert Einstein did. Other than perhaps Isaac Newton. And Wolfgang Amadeus Mozart. And if Einstein did stand on some shoulders—Isaac Newton, Clerk Maxwell, Max Planck—he barely needed to.

The first Einstein paper of 1905 was on the photoelectric effect, a phenomenon that scientists already knew about observationally but to which Einstein added the theoretical underpinning. The photoelectric effect is also the basis of solar power panels, so each of us has some relation to it in our daily lives. The photoelectric effect states that when light strikes certain materials, especially metals, electrons are elevated to a higher orbit. If you have solar panels on your roof at home, this is how they work.

When it comes to solar energy, when a light photon strikes a solar panel—most panels are made of the element silicon (Si)—the photon's energy is converted into electricity by the emission of an electron into a higher energy state. The photon excites an electron, a one-to-one event: one photon, one electron. Those higher-energy-state electrons then leave the panel and are either stored in a battery for later application or are used directly to power an electric current, such as your home energy grid and your newly installed smart-home system. Even Alexa finds solar panels *exciting*.

Before Einstein weighed in on the photoelectric effect, the observation ran into theoretical problems. Most pressingly, it was observed through repeated experiments (the scientific method!) that with very, very, very dim light, absolutely nothing happened—no electrons are excited, no electrons hit a higher energy state, no emission of electrons. It's sort of like a kiss on the cheek from a girl at the end of a first date—it might be nice but it's not going to excite much compared to a kiss on the lips. And to be crude for a moment, there'll certainly be no emission of any sort.

The "aha moment" with the photoelectric effect came with Einstein's first 1905 paper, where he correctly theorized that light comes in packages of quanta. That is, Einstein posited that a photon behaves not merely as a wave but also as a particle, which would explain how one photon could excite one electron. Prior to Einstein, light was thought of as only being a wave, like a sound wave, and as having no mass. Einstein was proposing something different, something entirely off the charts. First, he was proposing that light had mass, and not just mass but quanta of mass. Like Planck's blackbody emitting quanta of energy or discrete units, a single light photon traveled like a wave, had mass, and possessed a quantum of energy. Without much explanation, a person can more intuitively imagine the wave nature of light—after all, we walk right through or into light all the time and it never knocks us over—but simultaneously appreciating that those photons of light striking us have mass as we sunbathe along Matira Beach in Tahiti is more difficult to accept. Which is where we now turn: the experiments to prove both the wave properties and the mass quanta properties of photons of light.

In 1801, the British polymath Thomas Young performed his famous double-split experiment where he showed light to be a wave. To understand Young's results, consider this: If a rock is thrown into a pond, it creates waves. If two rocks are thrown into a pond near each other, they both create waves, but some of those waves will amplify each other as they cross paths, and some of those waves will cancel each other out, and some of those waves will mess around with each other in a farkakte pattern, partially amplifying, partially canceling. Those waves are merely energy ripples traveling through water, a wave of energy.

Young's double-split light experiment is similar to throwing two rocks into a pond, but instead what you do is pass a stream of light through two parallel slits onto a screen, which allows you to observe the characteristic fringe patterns of amplification, cancelation, and superposition of overlapping light waves. Light is a wave, end of story. Or so it was thought.

If light was just a wave and had no mass, it couldn't possibly kick an electron into a higher orbit, because electrons don't absorb energy and get all excited—they need an actual bump. But, if light photons also had mass, if light was a particle, too, it could kick an electron in its booty. Einstein knew this, in his head, without the need for a laboratory experiment. He was the ultimate theoretical physicist—all of his experiments were performed in his head, were thought experiments. He never lowered himself to conducting an actual experiment. All he needed was his pipe (his favorite was Revelation pipe tobacco, how weird is that?), his blackboard, and a bit of chalk.

If Einstein performed only thought experiments to prove his theory that light is both wave and particle, what was the *actual* experiment that showed light behaves as a particle? Well, we already mentioned it, it was the already observed photoelectric effect, except no one realized that that was what was going on—no one until Einstein, who just sat at his patent desk in Bern, smoking a pipe, stamping patents, all the while making this stuff up in his head. True story. (I'm leaning toward alien.)

In that one paper, the photoelectric effect for which Einstein won the 1921 Nobel, he demonstrated light had mass, which was a more important revelation than explaining the photoelectric effect itself (maybe not as important as his Revelation pipe tobacco), which was the second thing the paper established.

Photons have mass and behave like a wave. What about electrons? We know electrons have mass and they have quanta. But do they also behave as a wave? After all, what is good for the goose is good for the gander, right? In 1927, Clinton Davisson and Lester Germer, working out of Western Electric, later known as Bell Labs, in New Jersey, also performed a double-split experiment of sorts, but in reverse. They shot electrons at a crystal that diffracted the electrons into a wave pattern, demonstrating that electrons, which are particles, also behaved as waves because they diffracted into a spectrum of sorts. Like a rainbow, but not a visible light spectrum, not like a crystal that bends light into colors, nor the water droplets in the sky after a rainstorm diffracting light into a rainbow.

Photons and electrons are particles and have quanta. Photons and electrons behave as waves despite having quanta. Strange things happen at the atomic and subatomic level. Or, to borrow a line from Jennifer Lawrence's Katniss Everdeen impromptu song in the third *Hunger Games* movie, *Mocking Jay—Part 1* (2014): "Strange things did happen here / No stranger would it be."

The second 1905 Einstein paper was on Brownian motion. Robert Brown was a Scottish botanist—those Scots sure are smart—born in 1773. In 1827, Brown noticed that pollen suspended in water and held in a glass beaker ejected tiny particles—now known to have been starch and lipid organelles—that then seemed to maintain a persistent jittery motion despite no energy being applied into the system. Subsequently termed "Brownian motion," this observation was subsequently made of all manner of solutions, although without any realistic explanation. Basically, Brownian motion is energy movement in a system with no apparent energy in or added to the system: it's hidden energy, or entropy. (Recall that entropy is the energy left in a system that has no apparent energy.)

Einstein postulated that the relationship of the molecules to each other, or for that matter, of atoms to each other—their attraction or repulsion—was the hidden energy driving Brownian motion. Einstein's 1905 Brownian motion paper not only offered an explanation for entropy—how in a system with no apparent energy there in fact exists intermolecular and interatomic interaction energy that drives motion—but also proved that molecules and atoms actually exist. I know that sounds odd, but prior to Einstein's hypothesis, chemists and really any type of scientist knew atoms and molecules *existed*, but they had never been "seen" in action. Einstein explaining Brownian motion provided the "seeing."

For at least a few hundred years, chemists and physicists spoke of atoms and molecules as if they knew them intimately—but of course it was just assumed they existed. Before Einstein's

Brownian motion paper, a scientist could not "see" a molecule, and certainly could not "see" an atom, so had to find comfort in merely knowing they were there. Einstein's explanation for Brownian motion was proof that the atomic and molecular levels existed since it showed that, in a closed-state, no-apparent-energy system, a dance is still going on, a dance between unseeable things. "Strange things did happen here / No stranger would it be." This dance could not, would not occur unless the unseeable strangers really existed.

It would take the electron microscope, which finds its origins in the 1930s and uses electrons rather than light photons to image the most miniscule of things, to directly prove the existence of molecules and atoms. The reason electrons, the mass of which is about the same as photons, can image down to a single nanometer and a photon cannot image below 200 nanometers is all about wavelength—the electron wavelength being much shorter than the photon wavelength. The shorter the wavelength, the better it can image a smaller target.

Robert Brown is perhaps best known for describing Brownian Motion which, to be blunt, was not as great of a discovery as one might think. It took Einstein to explain it. But what Brown should be remembered for is discovering and describing the nucleus of a cell in 1831. Although the Austrian Franz Bauer, a botanist, gave some passing mention of the nucleus of a plant cell in 1804, it was more an artistic drawing of what he observed under the microscope, without ascribing any significance to his observation. It was Brown who really described the nucleus as an important feature of the cell structure, that is, in terms of "jeez, this thing is probably important." "Nucleus" is from the Latin and means "kernel" as in a kernel of corn or a nut, and from which was derived nucleic acid, the building block of DNA, which is housed in the nucleus.

The third paper of the *annus mirabilis* is perhaps the most singularly defining paper ever written—ever: Einstein's theory of special relativity, formally titled "Zur Elektrodynamik Bewegter Körper," or in English, please, "On the Electrodynamics of Moving Bodies." That paper alone I consider perfect proof that Einstein was an alien, as there were few, if any, Earthling shoulders he could have stood on to see the horizon he saw. But truth be told, there were *some* shoulders he stood on. The tale of special relativity first requires a backstory, and that backstory begins with another possible candidate to prove that aliens walk among us: the seventeenth-century genius Isaac Newton, with whom you'll recall we've already had an encounter of some gravity.

Newton's laws of motion, first published in his *Philosophiæ Naturalis Principia Mathematica* from 1687, offer three fundamental principles for the observable macroscopic world, otherwise known as the world you and I see and live in:

1. A body in motion tends to stay in motion, if not subjected to gravity or some other force. For example, an object in outer space, once put into motion, will stay in motion.
2. Force equals mass times acceleration (F = ma).
3. For every action there is an opposite and equal reaction, like when a cue ball hits a billiard ball.

To those three laws of motion Newton added the concept of gravity. However, when it comes to the unobservable microscopic world—the atomic level—and even celestial motions, cracks start to appear in Newtonian mechanics, a phenomenon we saw earlier with the timekeeping apples floating in space and with how a GPS satellite clock ticks slightly slower than a clock on Earth. Subatomic particles do not always play by Newton's rules. Nor do celestial events always play by Newton's rules. They have a rule book all their own.

To further understand this, there is no better example to explore than the speed of light. A glance back through history shows us that various scientists since antiquity have messed around with trying to determine the speed of light, including the great Galileo, who, in 1638, suggested light travels ten times faster than sound. Galileo was way, way off—but what the heck, he's Galileo, we'll let him off the hook this time. The first scientist who came at all close to measuring the speed of light was the Danish astronomer Ole Christensen Rømer. In 1675, while observing a series of eclipses of Io, a moon of Jupiter, Rømer noticed a delay in the eclipse light reaching Earth based on the changing proximity of Earth and Jupiter to each other as they orbited the sun. That is, sometimes the eclipsed light reached Earth faster and then slower, then faster, then slower, all due to the changing distance between the two planets. Recording the change in time it took light to reach Earth, with the only altering variable the distance between it and Jupiter, allowed Rømer to calculate his value for the speed of light at 124,000 miles per second. Very impressive for the seventeenth century.

English physicist James Bradley in 1728 offered up perhaps the earliest more or less accurate observation of the speed of light when he used star aberrations, as they related to Earth's orbit around the sun, to arrive at a slick 187,000 miles per second. Think of Bradley's experiment this way: if you are standing in rain that's falling vertically (it's not a windy day) while holding an umbrella, you hold the umbrella straight up, and the rain comes straight down. If you run in that rain with that umbrella, then you'll now tilt your umbrella in the direction you are running, because it *appears* the rain is coming at an angle—which it isn't, it's still vertical. The faster you run, the less vertical and more horizontal *appearing* the rain coming down will be. If you know the speed you're running and the speed the rain is falling, you can calculate *the angle* your umbrella must be at in relation to any given speed you run so as to not get wet. Conversely, if you know the various speeds you are running and the difference in the angle you're holding your umbrella at, you can calculate the speed of the falling rain. This is that trigonometry we all found drenching in high school.

With star aberrations, the star appears further in front of where it actually is because Earth is moving around the sun. Bradley knew the speed of Earth's orbit, knew the change in angles of his telescope as he observed the starlight, which—like the rain you're standing in—was coming straight at him, leaving him only to trigonometrically calculate for the speed of the light based on the change in angles. I'm guessing when Bradley was in trig class at Westwood Grammar School in Gloucestershire, England in the early 1700s, he was paying attention and not passing notes to girls.

The Frenchman Leon Foucault came the closest to calculating the actual speed of light when, in 1862, rather than using stellar light he used an Earth-bound apparatus, a rotating mirror and a fixed mirror, to measure the change in the angle of light between them. From this setup he derived a value of 186,284 miles per second. Let me try to explain Foucault's experiment. You have a light source, a rotating mirror and a stationary mirror 90 degrees to the rotating mirror. You beam a pulse of light at the rotating mirror and when that mirror is, during its rotation, perfectly aligned at a forty-five-degree angle to both the light source and the stationary mirror, the beam of light—because it is traveling so very, very fast—will hit the rotating mirror, then the stationary mirror, and then back to the rotating mirror, then back toward the light source, all before the rotating mirror has moved, or has moved very far. But that is just it: the rotating mirror will have moved just enough that the light returning to the source will be ever so slightly off by an angle. That change in angle, plus knowing the distances the light traveled, is related to the speed of light.

The current accepted speed of light is 186,282 miles per second.

While scientists were getting closer and closer to the value for the speed of light, there were several problems with their calculations. Prior to Einstein, light was believed to be only a wave and not a particle, and therefore it needed a "medium" in which to travel. As we discussed earlier, a rock thrown into a pond creates energy waves, and those waves need the water to propagate along. Sound travels on Earth because there is air to carry the sound wave. There is no sound in outer space because there is no medium to carry sound waves. If you're an astronaut on the International Space Station doing a spacewalk to repair a busted antenna and you drop that $1 million wrench and it hits the tin can you call home, it won't make a sound. If it careens off into space hopelessly lost—$1 million floating away—your swearing and cussing will be heard through the microphone of your slick NASA space suit. However, if you take off your space helmet to scream even louder, there will no longer be any sound—not just because sound doesn't travel in space but also because you'd be dead.

Out of (morbid) curiosity, let's consider how death would unfurl if you found yourself in outer space without your nifty NASA spacesuit on.

The relative vacuum in outer space will pull the air from your lungs, the oxygen gas in your tissues will expand such that you'll balloon up to twice your size like a Macy's Thanksgiving Day Parade float, and the water in your tissues will begin to boil away while those very same tissues simultaneously freeze solid. You'll pass out within ten miserable seconds if not sooner and die of asphyxia, although you'll be unaware of it. But you won't explode nor implode, despite popular misconceptions. You'll just balloon.

Returning to the speed of light problem, because it was believed that light, as a wave, needed a medium in which to travel, scientists beginning with Aristotle ended up adding that fifth element to Empedocles's earth, air, fire, and water, a medium they called "ether"—not the anesthetic gas but rather that mysterious, unidentifiable substance that permeates the entire universe. Yet this made-up ether was unmeasurable, unidentifiable, unknowable; like religion, it was taken on faith. Think about that for a moment. Scientists took on faith this ether as an explanation for how light crossed the expanse of the universe. Faith and fact. Two strong phenomena. Ether is from the Greek *aithēr* and means "upper air."

We know that ripples in a pond need the pond water in order to ripple along, and sound waves need air, or water, or two tin cans connected by a wire, or whatever, to propagate along. But light—what the heck was it propagating in to reach us from the sun or from another distant star? What was out there if not this made-up "ether" thingamajig?

Here is a far-fetched example to show how it is that light can travel through the cosmos whereas sound can't. In the 1999 comedy *Austin Powers: The Spy Who Shagged Me*, Dr. Evil, played by Mike Myers, has installed a "death ray" on the moon and is going to blow up Earth unless he is paid a "kajillion bajillion dollars." As a way to stop Dr. Evil's plan, the president of the United States, played by Tim Robbins, wants to send a nuke into space to blow up the moon:

> **The President:** Jiminy Jumpin' Jesus, I can't believe we're gonna pay that madman. I got nukes out the ying-yang. Just let me launch one, for God's sake.
> **Commander Gilmour:** Sir. Are you suggesting that we blow up the moon?
> **The President:** Would you miss it? *[looks around the table]* Would you miss it?

Well, if the president did blow up the moon with nukes, we'd see it on Earth, we just wouldn't hear it. And by the way, the Chinese expression is "yin-yang," with a "yin" not a "ying." Yin-yang are opposites where, according to Chinese philosophy, *yin* means negative, dark, and feminine while *yang* means positive, bright, and masculine; the interaction of these opposites influences one's destiny. As for "ying-yang" with the "ying" it appears

to be American slang for having a lot of something, as in "she's got money up the …"—need I spell it out?

And we would indeed miss the moon—and not just because our blood boils over with romantic lust whenever its full, as extolled in that classic Dean Martin song "That's Amore": "When the moon hits your eye like a big pizza pie, that's amore / When the world seems to shine like you've had too much wine, that's amore."

The moon serves purposes besides being something to look at during clear nights or to help facilitate falling in love. The moon stabilizes Earth's spin and rotation. Without it, God only knows where our planet would end up spinning to. Without the moon, ocean tides would disappear, days would be shorter, Earth would be hotter, the poles would melt, some or all of life on Earth would disappear. Some scientists postulate that had the moon not existed from the get-go 4.543 billion years ago, life on Earth might have taken a different trajectory.

Without being able to prove the existence of this fictitious ether, and beginning to doubt it even existed, scientists were at a loss to explain how light traveled across the universe. That is, at loss until Albert Einstein correctly proposed that light is both a wave and has mass, that light has quanta. And if light has quanta, that means it only needs itself to reach us from the far reaches of the universe: light is its own medium within which to propagate, and Einstein explained this to us. But the weirdness of light doesn't end there—the story of light gets even more bizarre, also thanks to Einstein. "Strange things did happen here / No stranger would it be."

Before Einstein's 1905 paper on special relativity, nearly two decades earlier, in the summer of 1887, the physicist Albert A. Michelson and chemist Edward W. Morley, who were working out of what is now known as Case Western Reserve University in Cleveland (go Browns!), postulated that light, which was supposedly moving through this mysterious "ether," should display two different speeds when measured at two different spots on a spinning Earth: the speed into or against the ether and the speed out of or with the ether. Don't forget, Earth is spinning on its axis at 1,000 mph so that should alter the speed of light, right? Wrong. Michelson and Morley discovered, to their initial horror, that their postulate was wrong—the scientific method at its best. In other words, no matter the point of measurement, they found that the speed of light is always the same whether measured into or from a spinning planet.

Consider this: if you are at the airport on one of those moving walkways, late for your plane and cursing as you go (which will be heard), and if you run on the walkway, your overall speed will be faster—your running speed plus the moving walkway speed. Similarly, if your ten-year-old son, ignoring your protestations, is running the wrong way on that moving walkway, his overall speed will be slower—his speed minus the walkway speed. If you had a police radar gun, you could take measurements and prove all these speeds.

But light doesn't behave this way. The Michelson-Morley experiment revealed that, whether the Earth is spinning into a light source or spinning away from a light source, the speed of that light remains unchanged. How could that be?

The Michelson-Morley experiment started to put the kibosh on the existence of "ether," since light apparently was not traveling in some mysterious medium, and also further set into motion Einstein's realization that light behaves as both a wave and a particle. It additionally set into motion the weirdest thing of all: the speed of light is the speed of light based on your frame of reference. What exactly does this mean?

Suppose you're standing on the back of a train that is not moving and you turn on a flashlight and point it toward the front of the locomotive. The light emitted from that flashlight leaves you at the speed of light, or 186,282 miles per second. Now suppose the train starts moving—and I mean moving really, really fast, at half the speed of light, at a smooth 93,141 miles per second—and you once again turn on a flashlight at the rear of the train toward the engine car. You'd like to think, based on our airport moving walkway experience, that the flashlight light would travel toward the front of the training clicking along at 186,282 (the speed of light) plus 93,141 (the speed of the train), or 279,423, miles per second (the speed of light plus the speed of the train). And if you thought that who'd blame you. But that doesn't happen. The light moves away from you toward the front of the train at *only* the speed of light, unaffected by the speed of the train.

Now, suppose you have a friend standing on a train station platform. As you pass that train platform, your train chugging along at a blistering 93,141 miles per second—half the speed of light—and you turn on the flashlight. Once again, the light leaves your flashlight toward the front of the train at the speed of light, but the thing is this, your stationary friend on the train station platform also measures the same speed for the light leaving your flashlight. Even though you're moving at half the speed of light and your friend on the platform is not moving at all, both your measurements of the speed of light emitted from that flashlight relative to yourself are exactly the same. No matter what the reference point is, light is always moving at the speed of light, no more, no less. How is this possible? Time—that is how. Time becomes the variable.

Light always moves at the speed of light, but time, on the other hand, is relative. Time flows differently depending upon the state of motion of the observer, meaning that events that appear simultaneous from two points moving at different speeds are in fact not simultaneous. They just appear simultaneous. Recall our apple and Timex watch thought experiment where gravity slows down time, which is termed "time dilation." Time is relative, and Einstein gave us that theory of relativity in his third 1905 *annus mirabilis* paper. Time is relative; the speed of light is not. "Strange things did happen here / No stranger would it be."

The other oh so tiny postulate from Einstein's last 1905 paper on relativity is a pretty innocuous equation: $E = mc^2$, where energy (E) equals mass (m) times the speed of light (c) squared. But what does that mean? What is $E = mc^2$ really saying?

For starters, it's saying that energy is related to mass and that the two have been interchangeable in the universe since the Big Bang, since the Planck epoch, when pure energy created mass; simply stated, "energy" and "mass" are different forms of the same thing. For you game enthusiasts out there, it is sort of like a "zero-sum game" in game theory, where one participant's gain (mass) is exactly balanced by the losses of the other participant (energy). At the moment of the Big Bang, energy was converted neatly into mass exactly and that is what $E = mc^2$ states. It also means that there is a huge, positively humongous, gargantuan amount of energy stored in a single atom…if it were to be split.

Albert Einstein did not work on the Manhattan Project as some folks incorrectly assume, but he nevertheless played a pivotal role in relation to it. The 1939 Einstein-Szilárd letter, written by Hungarian American physicist Leó Szilárd and signed by Einstein, was a warning and an appeal to then U.S. President Roosevelt to develop an atomic bomb. Most of the German and Eastern European physicists who knew anything about harvesting nuclear energy, like Einstein, like Szilárd, had already fled the continent due to the rise of Adolf Hitler. After all, most of those physicists were Jews. They also knew that their colleagues who had *not* fled, who were not Jews, were capable of building an atomic bomb for Hitler. The race was on to harvest $E = mc^2$ with the Brits' program called Tubes Alloy, the Germans' effort named the Uranium Club, and the Americans' endeavor the Manhattan Project. The boys and girls of the Manhattan Project won that race.

Let's now take care of a minor housekeeping chore. If you have the idea that the Big Bang was a huge explosion, you would be wrong. To the contrary, it was just the opposite. An explosion, like a bomb, including a nuclear bomb, is when mass such as dynamite or uranium-235 is busted up and releases stored energy. The Big Bang was just the opposite. It was energy making mass— not mass making energy. It was called the "big bang" in a moment of a derision by a non-believer in the theory, Fred Hoyle who we'll visit with later on, and to his subsequent horror, his quip stuck.

What else does $E = mc^2$ tell us? It reveals that there is a lot of energy stored in mass, especially at the teensy-weensy level of the atom. Recall from earlier that strong nuclear force is the attraction between protons and neutrons in the nucleus. Which is why Einstein's equation established the basis for the creation of the atomic bomb, as well as nuclear power plants. A great deal of energy is stored in *them there* atoms, and if you split an atom—especially if you split a gazillion atoms all at the exact same moment—well, you get Hiroshima. If you control that splitting as a manageable, sustainable fission, rather than blowing the entire uranium-235 wad all at once, you get a nuclear power plant.

If you and me are made of protein, fat, carbohydrate and nucleic acid molecules, and if molecules are made of atoms, and if atoms made of protons, neutrons and electrons, and if protons, neutrons and electrons are made of quarks, leptons and bosons, then what are quarks, leptons and bosons made of, which is where we now turn.

8 A STRINGY WORLD

The original folks interested in harvesting atomic energy, like Leó Szilárd on a theoretical level and Enrico Fermi on an experimental level, wanted to create nuclear reactors to produce electricity for our homes and businesses, not make bombs. That was the pre-World War II emphasis, harvesting nuclear energy to run the lights in your home. But underneath that theoretical nuclear reactor to supply near endless energy—not to mention the winds of war blowing around the world in the late 1930s—was the same theoretical basis to construct a nuclear bomb, the type that could end things.

Exactly how subatomic particles and gravity formed from pure energy during the Planck epoch, which as you'll recall lasted only 10^{-43} seconds, and the subsequent moments is not known. Physicists have no definitive idea of how energy formed mass. More precisely, they have a number of ideas, none of which is currently provable and might never be provable. Provable or not, these ideas are backed up by theories, albeit ones that are nearly impossible to understand, where extra dimensions to the observable universe need to be made up like boys drawing football plays in the backlot dirt. With doing a few ends around and running the football up the middle, I'll draw a few energy converting to mass plays on this field we currently are sharing.

The most popular one is "string theory," which, in the simplest terms, posits that the initial particles generated during the Planck epoch—energy making mass—were one-dimensional points called "strings" that contained some type of mass and charge corresponding to the strings' vibration. Those one-dimensional strings at first glance appear to have had no internal structure—after all they were merely one-dimensional energy strings that moved in time and space. However, according to string theory, if we were to look deeper into a vibrating one-dimensional string, we'd see that that such a string could fold in on itself, thus creating dimension. So, a folded one-dimensional energy string could become something other than a single dimension—it could become two-dimensional and then three-dimensional mass.

Further, if the vibrating string folded in one particular direction, it would form a subatomic electron particle called a lepton; if it vibrated and folded in another direction, it would morph into a photon of light; and if it vibrated and folded in yet another direction, it would be quark, a subatomic particle that gives rise to protons and neutrons. It is, as the brilliant songwriters Brian

Wilson and Mike Love of the Beach Boys taught us, all about the vibrations: "I'm pickin' up good vibrations / She's giving me the excitations."

How those strings vibrated and how they folded and how they interacted and combined with each other is, as the fictional theoretical physicist Sheldon Cooper taught us on *The Big Bang Theory*, called a "condensation." A condensation is what took one-dimensional nearly massless energy strings into two-dimensions and then into three-dimensional string condensates with mass.

Part of string theory posits that those vibrating folding string condensates produced not only mass but also space, time, and the four fundamental forces of the universe—electromagnetism, strong nuclear force, weak nuclear force, and the most elusive of all, gravity. The problem with string theory is, although it is good at explaining how string condensates achieved mass, and good at explaining three of the four fundamental forces of the universe, it is weak at explaining gravity and weaker still at elucidating the creation of space and time.

A competing hypothesis to string theory that attempts to explain gravity in a more complete way is "loop quantum gravity" (LQG). Loop quantum gravity is less concerned about trying to explain how energy formed mass at the moment of the Planck epoch and instead looks at space and time as discrete units to which certain properties can be assigned. Because loop quantum gravity has the limited goal of only trying to understand the quantum theory of gravity, whereas string theory attempts to explain everything from energy to mass conversion and the inherent gravity in mass, it is a lesser-known theory. It has less notoriety. Even in *The Big Bang Theory*, all the scientists—Sheldon Cooper, Leonard Hofstadter, Raj Koothrappali, Howard Wolowitz, Barry Kripke—are "stringy," with the lone "loopy" holdout being Leslie Winkle, an experimental physicist played by Sara Gilbert.

Why was loop quantum gravity theory even proposed? Quite simply, it came about to fill in the gaps of string theory. The problem with string theory—other than the fact it needs ten dimensions to make the math work, that is, six dimensions beyond three-dimensional space plus the fourth dimension of time—is that, as mentioned, it does not explain space or time in quanta. It explains energy and mass in quanta but not space and time, and it is also a little weak on gravity.

According to loop quantum gravity, space is made up of loops of gravitational fields called spin networks, and these spin networks are sort of like blocks connected to each other—a continuous flow of space and time like the interconnected blocks in your grandmother's quilt placed next to the fireplace. We can't see these connected units of quilted space, but loop quantum gravity believes that the universe is nothing more than exactly that.

If you consider a glass of water as a model for the universe, then each water molecule, according to loop quantum gravity, would be part of—for our purposes—a liquid quilt in a spin network of continuous geometry repeating throughout the universe of that glass of water. Those ethereal blocks of space, like water molecules in a glass, just sort of move together, connected with this gentle, flowing motion throughout the universe, an ebb and flow that offers an explanation for gravity. Why does it offer an explanation of gravity? Because the quilt blocks of space are connected and move together, a tug on one end of a quilt will exert a tug at the other end of the quilt, albeit less of a tug the farther away it is.

In the physics world, the "stringy" physicists and the "loopy" physicists don't get along. Each group pushes its own agenda—sort of like the U.S. Congress—stringy and loopy physicists don't even go to the same scientific meetings. Returning to *The Big Bang Theory* sitcom, the conflict between stringy people and loopy people is demonstrated in episode two of the second season, "The Codpiece Topology." When experimental physicist Leonard Hofstadter starts dating loop quantum gravity believer Leslie Winkle, conflict arises when she doesn't get along with Leonard's roommate, string theory adherent Sheldon Cooper. To sabotage the relationship, Sheldon picks a fight with Leslie.

The following dialogue from the show doesn't just present the antagonism between stringy and loopy physicists, it also demonstrates how science, like religion, is sometimes taken on faith. Now, just because some science is taken on faith doesn't mean it opens the door for religious folk to claim their own set of "facts." Facts are facts. Faith is faith. And science is not always exact. But it tries. As for the TV argument:

> **Sheldon**: I will graciously overlook the fact that she is an arrogant sub-par scientist, who actually believes loop quantum gravity better unites quantum mechanics with general relativity than does string theory.
> **Leslie**: Hang on a second. Loop quantum gravity clearly offers more testable predictions than string theory …
> **Sheldon**: … Matter clearly consists of tiny strings.
> **Leslie**: [*Looking at Leonard*] Are you going to let him talk to me like that?
> **Leonard**: Okay, well, there is a lot of merit to both theories.
> **Leslie**: No there isn't, only loop quantum gravity calculates the entropy of black holes … You agree with me, right …
> **Leonard**: Sorry Leslie, I guess I prefer my space stringy not loopy.
> **Leslie**: Well, I'm glad I found out the truth about you before this went any further.
> **Leonard**: Truth, what truth? We're talking about untested hypotheses.
> **Leslie**: … Tell me Leonard, how would we raise the children?
> **Leonard**: I guess we let them wait until they're old enough and let them choose their own theory.
> **Leslie**: We can't let them choose, Leonard, they're children. [*Storms off.*]

There you have it: theoretical physicists, unable to conclusively prove their theories, partially take them on faith. One of several caveats, though, is that scientists attempt to prove or disprove their theories. The exchange above pokes fun at religion for such things with the line "how would we raise the children?" Indoctrinate them—dogma. That's how.

But all hope is not lost for friendship between the stringy and the loopy. There are some young stringy folks who are trying to incorporate loopy ideas into a unified theory.

You might be wondering what those ten dimensions of string theory are. Sounds like we're back to *The Twilight Zone* and Rod Serling's very first line in the entire series: "There is a fifth dimension beyond that which is known to man. It is a dimension as vast as space and as timeless as infinity. It is the middle ground between light and shadow, between science and superstition, and it lies between the pit of man's fears and the summit of his knowledge. This is the dimension of imagination. It is an area which we call the Twilight Zone."

As for the ten twilight zone dimensions of string theory, dimensions one, two, and three are your basic three-dimensional space on an X-Y-Z Cartesian coordinate system—what you and I see—and the fourth dimension is time. And that is where most of us get off this train. String theorists continues the ride on the train with the fifth dimension—which is not Serling's above description, nor is it the popular American R&B band The 5th Dimension that released "Aquarius/Let the Sunshine In" in 1969. Instead, the fifth and sixth dimensions in string theory are universes slightly different than ours that could have formed given the exact same initial conditions at the moment of the Planck epoch—except if the one-dimensional strings decided to fold differently than they did. The seventh dimension is if those initial conditions at the Planck epoch were slightly different, the eighth takes the three previous dimensions and runs them all out to infinity, the ninth explores how the laws of physics would vary in each possible universe, and the tenth dimension is a summation or point at which everything possible, everything imaginable, is covered. The purpose of the six extra dimensions of string theory is to appease the math gods. Our current universe went in a certain direction, but it could have gone in other directions, and those other "roads diverged in a (celestial) wood" need to be considered for consistency's sake. And to appease the math gods.

Now let's move on from string condensates to subatomic particles. Recall that this string condensates eventually gave rise to the three fundamental particles—quarks, leptons, bosons—but there wasn't just one of each flavor. Simply put, current subatomic theory states there were six types of quarks that give us protons and neutrons, six types of leptons that gave us electrons, and twelve types of gauge bosons that gave us photons and also gave us the four fundamental forces including possibly the graviton boson involved in, it is in the name, gravity. The quarks and leptons that give rise to protons, neutrons and electrons are collectively called fermions, after Enrico Fermi who in 1942 is credited with building the first nuclear reactor in the world at the University of Chicago, Pile-1. Fermions also include composite particles of quarks and leptons. The properties of these subatomic particles are called flavors and go by some interesting names. Two of the quarks, for example, are known as charmed and strange, while some of the leptons are termed electrons, muons, neutrinos, and tau. The twelve gauge bosons have special names, too, like gluon and meson and the most famous boson, the all-familiar photon.

Hadrons are an overlap, like the middle bit of a Venn diagram. A hadron is either a fermion or a boson, but it always has one fundamental feature: the hadron makes up the strong nuclear force, whereas the other fermions and bosons do not. And of course, the mother of them all, of all the fermions and all the bosons, is called the Higgs boson.

Underneath all those initial one-dimensional vibrating string condensates is the most basic vibrating string of all, a vibrating string so rudimentary that it had yet to decide which pathway it wanted to go down, had yet to decide whether or not to fold into a quark, or a lepton, or a gauge boson, and that particle is known as the Higgs boson, otherwise known as the "God particle." It is named after the brilliant Peter Higgs, of Edinburgh, Scotland, who first theorized in 1964 that such a fundamental string particle must exist. Interestingly, his initial theory was rejected by the physics world—much like how many physicists initially rejected Einstein's general theory of relativity.

Higgs, along with other researchers who were working independently, notably the Belgians François Englert and Robert Brout, continued to refine and define the Higgs boson after Higgs's initial postulate. The question soon became: Would that elusive God particle ever be discovered?

On July 4, 2012, the Higgs boson was finally identified in a fleeting moment of physics ecstasy at the CERN laboratory in Switzerland. CERN is an acronym for Conseil Européen pour la Recherche Nucléaire, or, in English, the European Organization for Nuclear Research. The discovery of the Higgs boson involved racing protons—not photons but protons—along the 16.7-mile circular particle accelerator of the Large Hadron Collider, where collisions among protons, busting to smithereens those protons, produced among other elemental particles, the Higgs boson. What exactly happened? The protons collided with each other busting up and producing for a gazillionth of a second those various one-dimensional strings, one of which was the ultimate God particle string.

Higgs and Englert were awarded the 2013 Nobel Prize in Physics for their theoretical prediction of the existence of the Higgs boson. (Brout had died just two years earlier, and sadly fourteen months before the CERN collider identified the Higgs boson.)

CERN's Large Hadron Collider located near Geneva in a semirural area that sits on, or more accurately *is under*, the border between France and Switzerland, is the world's largest and most powerful particle accelerator and particle collider. For that matter, since it is technically one machine, it is also the largest machine in the world. The location for the CERN collider is also curious. It is nearby the town of Saint-Genis-Pouilly, France, which is believed to have once been the location of a Roman temple dedicated to Apollo, who was the son of Zeus, the king of the gods, when Zeus bedded his consort—not his wife, Hera, but his consort—the goddess Leto.

Hera, also known as Juno, was Zeus's wife and oddly his sister too. While it appears Hera was not that bothered by incest, married to her brother, she was plumb pissed-off that her husband had bedded the beautiful Leto. Take-home lesson according to Hera is that incest is acceptable, but cheating is not. Leto went into hiding on the Greek island of Kos, giving birth to the twins Apollo and Artemis. Apollo is the god of prophecy, truth, healing, the sun, plague, and music, while Artemis is the goddess of wild animals, the moon, hunting, and chastity. Their mother, Leto, is the goddess of modesty and motherhood. Apparently Greek and Roman gods could have given a wit about incest and infidelity.

According to legend, that particular temple in Saint-Genis-Pouilly dedicated to Apollo was believed by the Romans to be a gateway to hell. Indeed, a puzzled choice of location for the Large

Hadron Collider. Add to that curious location, the comment made by perhaps the most famous theoretical scientist of our day, Stephen Hawking, who lived to see the God particle discovered in 2013. Since all Higgs bosons, or God particles, were theoretically consumed during the Planck epoch at just 10^{-43} seconds into this universe's existence, Hawking mused that physicists' creation of a God particle within the Large Hadron Collider could theoretically trigger a catastrophic vacuum decay, a reverse Big Bang, like a chain reaction, that would spread throughout the entire universe, causing the collapse of space and time. Sounds like one of Kurt Vonnegut's "ice-nine" doomsday stories to me.

Stephen Hawking was born on January 8, 1942, in Oxford, England—on the 300th anniversary of Galileo's death, as it were—and spent his childhood attending various preparatory schools where, despite his mediocre grades, he earned the nickname "Einstein." He read his undergraduate physics studies at University College Oxford and his graduate work in mathematics and cosmology at the University of Cambridge.

At age twenty-one, while at Oxford, Hawking developed clumsiness, and as the symptom palette advanced, he was eventually diagnosed with amyotrophic lateral sclerosis (ALS), also known as Lou Gehrig's disease after the famous New York Yankees slugger who also was afflicted with it. Due to advancing ALS, Gehrig was forced to retire from baseball in that famous tearjerker of a speech, at home plate in Yankee Stadium, on July 4, 1939. It was 1963 when Hawking was diagnosed, and neurologists gave him two years to live. Five years tops. But he defied the odds and lived fifty-five more years, dying on March 14, 2018, age seventy-six, which also happened to be Albert Einstein's 139th birthday.

What are the odds? There must be some intersection here, like the Bermuda Triangle. Hawking was born on Galileo's death date and died on Einstein's birth date. Weird. Aliens I tell you, the whole lot of them.

Although Hawking is best known for his books, such as *A Brief History of Time* (1988), which explores the beginning of time and answers question like "Can time run backward?," perhaps his greatest contribution was his work in cosmology, exploring black holes in the universe, which not only provided information in itself but also provided evidence for the Big Bang singularity as the origin of the universe. That is, Hawking showed a black hole is a microcosmic singularity, a sort of small version of the Big Bang. Another of Hawking's stellar contributions was his study of the "event horizon" of a black hole, that point where everything, including light, gets sucked into a black hole, a black hole that originated from the collapse of a certain type of huge star.

Why was Hawking's work important? Einstein's general theory of relativity—saying that time is relative, space bends, things like that—could not be reconciled with the quantum mechanics theory of the 1930s regarding what happens on the subatomic level. Macrocosmology just didn't jive with subatomic quantum field theory; the celestial macroworld and the atomic microworld seemed to behave with different sets of physical laws from each other, let alone different sets of laws from Newtonian mechanics. Even Einstein spent the later years of his life attempting to reconcile the two worlds, to no avail. In the macroworld, a good deal of space out there in the universe has to be empty, completely void; but in the microworld, space cannot be truly empty. Quantum theory says something must be going on out there in empty space, something must be in empty space—but what is out there cannot be seen or measured.

It has been known for a long time that the universe must contain more than the visible matter and visible energy that cosmologists can "see." There must exist such weird things as anti-matter (also called dark matter) and the equally strange anti-energy (also known as dark energy) that, when paired with their opposites, cancel each other out so as to satisfy the premise that all the mass and all the energy of the universe was created during the Big Bang. Sort of like an accountant's ledger.

In studying the event horizon of a black hole, it was assumed that black holes could only get larger, not smaller, because black holes suck everything in. Yet black holes can come into existence and then not only get "smaller" but apparently go out of existence altogether. How can that be? Where do the extremely large amounts of matter that have been crushed into an infinitely small space at the center of a black hole go? Do they go where donut holes go—in a separate box for sale?

Although it has yet to be proven, a proposal of Hawking's that physicists widely accept is that, at the event horizon of a black hole, it is possible for two paired particles (matter plus anti-matter) or two paired forms of energy (energy plus anti-energy) to experience a situation where, rather than both getting pulled into the black hole, just one particle or one energy could be sucked in, leaving its anti-matter or anti-energy partner stranded, shot out into the universe via what is termed "Hawking radiation." This quantum explanation of what could theoretically happen at a black hole event horizon is also used to explain what could have happened on a larger scale during the Big Bang—an explanation that allows for one possible solution to reconciling general relativity with quantum mechanics. How does it do this?

In a nutshell: empty space is not empty. There just happens to be stuff out there we can't see or measure. Simply stated, stuff got shot out of the initial singularity, the Big Bang, that cannot be seen, that is, anti-matter, and that cannot be measured, anti-energy. "Strange things did happen here / No stranger would it be."

One quick editorial note before we circle back round to Vonnegut and whatever "ice-nine" is: Stephen Hawking was, up until his death, one of the most recognizable humans on planet Earth. Not just one of the most recognized scientists, but humans, full stop. Being in a wheelchair and speaking with a computer-generated voice machine likely played heavily into Hawking being seared into our individual and collective memories. His work with matter and anti-matter, energy and anti-energy, black holes, event horizons, and Hawking radiation was more than just noteworthy. Practically every award that could be bestowed upon a scientist was bestowed on that great man, except one. Hawking never received a Nobel. What were they thinking?

Let's stop to think for a moment of all these Nobels gone awry. A 1926 Nobel went to Johannes Fibiger, who demonstrated a certain worm infection caused cancer. Turns out it doesn't, though. The exceptionally gifted Enrico Fermi, who built the first nuclear reactor, Chicago Pile-1, in 1942 and also worked on the Manhattan Project during World War II, received a Nobel in 1938 for demonstrating new radioactive elements could be produced by neutron bombardment. The catch was, he didn't produce new elements—he had only split uranium into other previously known elements, thinking the beta decay was from fusion when really it was from fission. The 1906 Nobel in Medicine went to two bitter adversaries, the Italian Camillo Golgi and the Spaniard Santiago Ramón y Cajal, for their work with the nervous system. What is odd is that they had two competing views: Golgi thought nerves were continuous, and Cajal thought nerves had synapses. Golgi was wrong, Cajal was right, but nevertheless they were forced to share the Nobel prize, perhaps because the Nobel committee couldn't decide which was what.

It is okay to be wrong in science. As we know, science advances from our mistakes as well as our successes. But how Hawking was skipped over for a Nobel is beyond me. The problem with the folks on the Nobel's science panel, as I see it, and as opposed to the literature Nobel committee, is that they all too often look for a blockbuster discovery which should not be the dealbreaker to be awarded a Nobel. What about body of work? Consider what the Nobel counterparts on the literature committee do: they look at a person's entire body of work rather than one single creation. While the body of work a scientist produces may not be flashy and headline worthy, but if all of it has significantly advanced science—well, you do the math.

The other thing about the science side of the Nobel Prize—usually true but not always true—is that especially in physics they favor experimental physicists, who prove things, over theoretical physicists, who think things. Look at Einstein: never did an experiment. His 1921 Nobel Prize in Physics was for his theoretical explanation of the photoelectric effect, and he probably only received it because the photoelectric effect had been an observable phenomenon for decades. He did not get his Nobel for explaining the Brownian motion at the heart of entropy, and especially not for his theories on relativity, such as the speed of light is always the speed of light, but time dilation is relative to gravity and space curves. Heck, even $E = mc^2$ was not considered a Nobel-worthy equation. Go figure.

Now to return to Kurt Vonnegut. What is this "ice-nine" doomsday scenario that bears an eerily remarkable resemblance to Hawking's musing that the CERN supercollider and its successful creation of the God particle could quite possibly lead to a vacuum decay that would gobble up the entire universe? Ice-nine is the fictional crystallization of water in Vonnegut's 1963 novel *Cat's Cradle*. To understand Vonnegut's ice-nine, we first need a backstory about crystallization, to refresh those of us who slept through most of high school chemistry or ditched class entirely.

A classic example of crystallization in the real world involves a supersaturated solution of sugar. As sugar is added to a solution of water, a point is reached where if you add just one more teensy, tiny speck of sugar—wham! —it triggers the entire solution to crystallize right before your eyes. In theory, that is. Rock candy is basically crystallized sugar. Snowflakes are similar. Take a near-freezing droplet of water and as soon as two adjoining water molecules join in crystal harmony—bam! —the entire water droplet crystallizes instantly into a snowflake.

In *Cat's Cradle*, Felix Hoenikker is the inventor behind ice-nine, a special molecular configuration of water that forms ice crystals at room temperature. As the story unfolds, it transpires that if one speck of Hoenikker's ice-nine were added to the ocean, just a single special molecule, it would trigger all the connected water on Earth to crystallize. The entire planet would ice over and we'd all be extinguished into one big dirty ice-ball fossil record. Except for cockroaches, of course—they seem to survive everything. Cockroaches and Keith Richards and mothers-in-law would survive ice-nine.

If the Planck epoch was caused by a quantum fluctuation produced by all the God particles, which were all subsumed in just 10^{-43} seconds into the nascent universe, making all the other string condensates, then what Hawking was saying is that placing a single God particle at high enough temperatures back into that universe could reverse the process—sort of like a crystallization of mass into energy. There has to be a Hollywood movie somewhere in there, right? Eddie Redmayne could even reprise his role as Hawking, which he first played in the 2014 biopic *The Theory of Everything*.

That Hollywood God particle screenplay could be an updated version of *Cat's Cradle*—or better yet, a James Bond or Austin Powers flick where the nemesis is in possession of a single God particle in a vial, and he will introduce it into the universe unless he is paid a gazillion bazillion dollars. Daniel Craig would be 007, because other than Sean Connery, Craig is the best Bond. And if the producers decide to go the comedic rather than dramatic route, naturally no one but no one could be both Austin Powers and Dr. Evil except Mike Myers.

CERN's Large Hadron Collider took ten years to build at a cost of $4.5 billion. After it was completed in 2008, it went live with a $1 billion annual operating cost. The 16.7-mile collider ring is buried up to 575 feet beneath the rolling farmland under the border between France and Switzerland and is basically a series of powerful magnets and electrical boosters that accelerate particles into a mindless frenzy, going faster and faster and faster around the ring, sort of like a dog chasing its tail.

If you're wondering why the designers selected precisely 16.7 miles rather than a nice round 17 miles, well, they didn't. They decided to make it 27 kilometers long, and 16.7 miles is what we get when it's converted for us metrically challenged Americans. Why 27 kilometers for the racetrack circumference—why not 10 kilometers or 100 kilometers? A smaller track would require the protons to accelerate faster around the curve to reach the same energy threshold for meaningful collision, and a much larger racetrack would add little in terms of energy but cost so much more to build. So, 27 kilometers or 16.7 miles was just about perfect. If you're furthermore wondering which direction things like protons are accelerated in the collider, clockwise or counterclockwise, the answer is "both." How else would you get them to collide with each other?

The CERN collider ring has several "stations" along its circumference, each of which is devoted to a specific experiment. The ATLAS station detects the subatomic particles produced during the acceleration of protons—not photons but protons—along the CERN ring and their subsequent collisions. It was the ATLAS station that discovered the Higgs boson or God particle.

You might be wondering why a few protons busting each other into smithereens didn't blow up the area surrounding Saint-Genis-Pouilly in one glorious display of atomic fireworks. It's simple really: first, not enough atomic elements were busted at once, and second, it was busting subatomic particles (weak nuclear force) rather than the nuclei of protons and neutrons (strong nuclear force). Following Einstein's $E = mc^2$ equation, there wasn't enough "m" to make for much "E."

A nuclear bomb needs a gazillion fissile atoms with the nuclei all going boom at once, a gazillion atoms all splitting in a gazillionth of a second. The atom bombs dropped on Hiroshima and Nagasaki were actually judged to be inefficient nuclear bombs, as most of the fissile material did not explode. Even so, it was estimated that in each bomb approximately 2×10^{24} atoms went splitsville all at once. *Kaboom!* A nuclear bomb might split 10^{24} atoms at once, whereas the CERN collider splits maybe a few dozen protons—again, not enough "m," and therefore not much *kaboom.*

The collider at CERN was named "Hadron" after the work of the noted Russian physicist Lev B. Okun. In 1962, Okun coined the term "hadron" to replace what he considered to be the clumsy phrase "strongly interacting particles," which up until then had been used to describe subatomic fermion and boson particles that made up the strong nuclear force. Fermions, as was mentioned earlier, were named after the Italian-born American physicist Enrico Fermi who as we learned built the first nuclear reactor, Chicago Pile-1, worked on the Manhattan Project and received a Nobel for an error. Bosons of course take their moniker from Satyendra Bose of the Bose-Einstein equations and the Bose-Einstein condensates.

About seven different experiments are in process along the 16.7-mile expanse of the Hadron supercollider ring. Each station has its own particle creator and particle detector, with all seven stations sharing the same accelerator. The ATLAS experiment detected the Higgs boson; the ALICE experiment is studying events that occurred right after the Planck epoch, like the formation of quarks.

Most physicists, and especially Peter Higgs and François Englert, do not like that the Higgs boson is nicknamed the "God particle." They consider it to be unnecessary journalistic sensationalism. But for us news-hungry layfolk, that name carries a certain shock value, has a certain ring to it—so good luck getting us to quit using it. Similarly, in the world of infection, referring to necrotizing fasciitis as "flesh-eating bacteria" is also journalistic sensationalism. Necrotizing fasciitis does not eat flesh, it kills flesh, and the Higgs boson was not God's particle. And the Big Bang was not an explosion.

From CERN's initial $4.5 billion construction costs, beginning in 1998, through its annual $1 billion budget, up to July 4, 2012, when the Higgs boson was detected, an estimated $13.25 billion was spent to discover the God particle. And you're going to love this: that God particle that cost $13.25 billion to create existed for only 1.56×10^{-22} seconds before it disappeared. That is $13.25 billion spent for 10^{-22} seconds of glory. There is a joke in there, but I'm not sure exactly what it is. I suppose it would be like a wine snob showing off his never-to-be-opened bottle of 1959 Dom Perignon champagne, valued at $45,000, and accidentally dropping it, the bottle shattering, the wine spilling on the floor, and all that is left is to enjoy are the tears running down the owner's face.

Of course, the Large Hadron Collider wasn't built just to produce and detect the Higgs boson. It enables other particle physics pursuits, such as the possible detection of new quarks, insights into dark matter, and research into where exactly the seat of gravity lies.

A point of clarification is in order right about now. The initial singularity is what is believed to have existed prior to the Big Bang and was the precursor to our universe. It was pure energy, no mass. A gravitational singularity, by contrast, is something else altogether: it is a black hole, or more precisely the center of a black hole where there is nearly infinite dense mass. A black hole forms out of a collapsing star. As certain stars start to explode at the end of their life sequence, they are called supernovas, and as they then collapse upon themselves with ever increasing density and gravity, the collapse reaches a tipping point or threshold point called the Schwarzschild radius, beyond which the gravitational pull of ever increasing density sucks everything into a gravitational singularity. Nothing escapes a black hole: not mass, not light (which are photons with mass), not space, and not even time … nothing except perhaps Hawking radiation. And if theoretical Hawking radiation does escape, it escapes only at the event horizon.

The world we humans live in is so strongly three-dimensional, our brains are so strongly three-dimensional, that everything we look at has three dimensions, and everywhere we look there is space, and time always runs forward, so much so that it is nearly impossible to imagine anything as absurd as two-dimensional mass, let alone one-dimensional strings, let alone the absence of time. To put our arms and minds around one-dimensional string particles, to grasp a point in space where there is no space, and to grasp a point in time when no time exists, is mind-boggling. It is the type of stuff that keeps one up at night, alongside other unimaginable thoughts such as trying to comprehend how the universe seems to go on forever, wondering if space really curves, contemplating one's own death, thinking about if life exists elsewhere in our universe, and imagining what alien life might look like. These are things that keep a person up at night, along with why the Beatles broke up and what Kim Kardashian truly thinks about Gucci bags.

If we narrow these thoughts down to just time, for instance, we might think about how time is the ability to recognize the continual progress of events that occur in an irreversible succession. For those events to occur, there needs to be particles and there needs to be space in which those particles can exist, in which they can move, change, or have an event within three spatial dimensions. This is the classic Newtonian view of time, which says that time is a fundamental property of the universe just like space, particles, and the four fundamental forces.

Another view of time as proposed by Isaac Newton's seventeenth century contemporary Gottfried Wilhelm Leibniz (both are credited with inventing calculus, though Newton more so) and subsequently expounded upon by Immanuel Kant in the eighteenth century is that time is not actually connected to space or the events that move through space. Instead, time is an intellectual construct that humans like to and need to use to sequence events. Humans are the only life forms (at least on Earth) that find it necessary to sequence events. We invented calendars and clocks and pocket watches so we could satisfy our preoccupation with the passage of time.

Of course, beneath our desire to sequence events in our lives—that is, to set the alarm clock so as not to be late for work, or to check the time on one's iPhone so as not to be late for that date with that cute guy over a cup of coffee at Starbucks—is a more fundamental driving force. Mortality. Humans and only humans measure the moments of their lives and because we are the only creatures truly aware of our own mortality, that our lives are finite, and we know that, time becomes precious. And, ultimately, because time is precious, and because we cannot stop it, like sand in an hourglass, we measure time.

In the Leibniz-Kantian world, time is neither real nor measurable nor widely accepted. This is probably the view of time of all life forms on Earth except humans. Your garden-variety house fly has about twenty-eight days to live. That Leibniz-Kantian fly cannot contemplate its lifespan, but the twenty-eight days is nevertheless real. The corollary of this statement is that time exists regardless if a human is present to say it exists or not—which makes me more of a proponent of Newton's view of time than Leibniz and Kant's. Time exists whether us puny humans say so or not. And this is where our story now turns: the science that helped to define how old our universe is, which we could otherwise call the beginning of time.

9 THE AGE OF THE UNIVERSE

Edwin Hubble was born in 1889 in Marshfield, Missouri, and grew up in Wheaton, Illinois. In his youth, although a good student, Hubble apparently excelled at sports over academics—my kindred spirit—playing baseball, football, and basketball and running track. At one track and field competition in 1906, at age seventeen, Hubble won seven first-place ribbons. He would, however, eventually leave behind his youthful athletics to study mathematics and astronomy at the University of Chicago. Then he took yet another turn of course and studied law at Oxford (I bet you didn't see that coming), out of deference to his father, who wanted Edwin to become a lawyer. With cosmology always lurking in the back of his mind, Hubble was able to leave the study and practice of law in the dust after the death of his father in 1913 (unlikely from an asteroid strike) and redirect his enthusiasm back to his first love, the cosmos. Hubble was yet another Renaissance man in a string of such freethinkers, dating back at least to Geber, the eighth century Persian polymath.

In 1919, when Hubble was about thirty years old, he journeyed to the Mount Wilson Observatory in California. It housed what was then the world's largest telescope, the Hooker telescope, named after John Hooker, a successful ironmaster who made his fortune in hardware, especially iron and steel pipe fabrication. When not attending to his iron concerns, Hooker's part-time hobby was gazing up at the stars. As an amateur astronomer, Hooker saw fit to supply the bulk of the funds needed to build the 100-inch telescope, which you can still peer through at the Wilson Observatory. Not surprisingly, when Edwin Hubble looked through the Hooker telescope, he forgot his law studies in the flash of a comet, the cosmos unfolding before his eyes.

In the time of the great ancient astronomers like Aristotle, Ptolemy, and Eratosthenes, the known universe had Earth at its center. Then came the upstarts of the Renaissance: Galileo Galilei, Nicolaus Copernicus, and Johannes Kepler. Their universe had the sun at the center of the universe, which meant at least our whole solar system was at the center of our universe. For the next several hundred years, our galaxy, the Milky Way, was thought to be the entire universe, one large Milky Way universe, with our Sun and solar system at the center. Then the shoe dropped. Many cosmologists began to believe that our star—the Sun—and our solar system were not much the center of anything. There were just too many stars in the Milky Way, and those stars' positions compared to our sun meant we were merely, desolately, scattered somewhere out there among the vastness of the cosmos. Our sun, as it turns out, is on the Orion Arm, a minor, forgettable spiral arm, located between two much larger arms of the whirling Milky Way.

Puny humans thinking we were all that. First, we thought Earth was the center of the universe, with the sun orbiting us. Then we accepted our Sun was at the center of our solar system, with Earth at least the preeminent planet orbiting the sun, finding solace that our solar system was the center of the universe. Then we discover our solar system is not the center of much of anything, just inconsequentially housed on a remote arm within the Milky Way. And things were about to get worse.

Along came Hubble, who dropped another big shoe, putting an end to any thought that our Milky Way galaxy was at least the center of everything in the entire universe, or more to the point, put the kibosh that our galaxy was the only galaxy. Hubble believed that some of the distant stars thought to be within the Milky Way, as seen through the Hooker telescope, were in fact not stars at all but entire other galaxies, each containing its own billions of stars. Hubble demonstrated this using the redshift effect, also called the Doppler effect, which we'll cover momentarily.

Hubble further demonstrated that those other distant galaxies were not stationary but actually moving away from the Milky Way, moving away from Hubble as he sat there peering through the Hooker telescope. This moving away of other galaxies from the Milky Way—their redshift—supported the concept that our own universe is not stationary, is not static, but is rather expanding, enlarging. Ours is an inflationary universe. Hubble not merely left us contemplating that the universe is bigger than imagined but that it is bigger than is *imaginable*. Edwin Hubble revealed that the Milky Way is no special galaxy, just one galaxy among billions more galaxies in a universe that is still growing.

The Doppler effect is not only used by astronomers to determine how fast and in what direction stars and galaxies are moving away from us or toward us, it is also used in medical ultrasound equipment to analyze blood flow. The Doppler effect is the apparent change in a wave, whether that wave is light or sound, as viewed when that wave is traveling either toward or away from an observer.

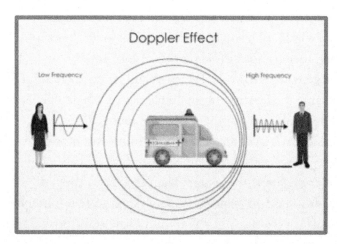

Doppler Effect

Low Frequency

High Frequency

When a wave is traveling toward an observer, the wave gets compressed, meaning the wavelength is shorter. When a wave is traveling away from an observer, the wave elongates, so the wavelength is longer. The key to understanding this is to appreciate the source of the wave needs to be moving. The classic example is sound waves. You're standing on the side of the beautiful Pacific Coast Highway as it meanders along Big Sur. As a car approaches you—likely a Ferrari 250 GTO, since it is California—the sound the car's tires and engine makes becomes a higher pitch, because the sound waves are compressed as the Ferrari is chasing after its own sound waves. Shorter sound wavelength equals higher pitch. As the Ferrari passes you and moves away along Big Sur, the sound waves are elongated, because the Ferrari can't catch up with the sound waves, so the pitch becomes lower. Longer sound wavelength equals lower pitch. It's the same as with an ambulance: its siren is a higher pitch as it is moving towards you in your rear view mirror, and a lower pitch as it zooms past you, careening down the road trying to avoid disinterested drivers refusing to pull over.

You might be wondering what the observer has to do with shorter, higher-frequency sound waves on approach, and longer, lower-frequency sound waves on recession. Well, nothing other than being the observer. It is merely that on approach more sound waves are being emitted from a position of approaching closeness because the source has speed, has velocity, as does the sound (or light) and overlapping waves increase the frequency, which increases the pitch. On recession, the sound waves are emitted from the source from a position of increasing distance, meaning the waves spread out, lessening the frequency at which the waves reach you, reducing the pitch.

This phenomenon is called the Doppler effect, named after the Austrian physicist Christian Doppler, who first discovered that the shift in the wavelength or frequency of light varies depending upon the speed of the source and the location of the observer—so Doppler's original description was discussing visible light, but the principle also holds true for sound. It is really no different than your wife yelling at you as she walks toward you—likely because you deserved it—a high-pitched sound that continues as she stands there yelling at you, and then muffles to a lower pitch as she turns about-face, storming out, still yelling at you.

Doppler, born in 1803 in Salzburg, was a gifted student. He studied mathematics and physics at the Imperial and Royal Polytechnic Institute, now known as the Vienna University of Technology. In 1841, he accepted a professorship at Prague Polytechnic in what was Czechoslovakia, where he subsequently developed and published his seminal work, *On the Colored Light of the Binary Stars and Some Other Stars of the Heavens*. In this treatise, Doppler states that the observed frequency of a wave depends on the relative speed of the source and the location of the observer. The theory was based on his observation and subsequent explanation of why binary stars have a particular color. (Those binary stars were probably other galaxies flying away from his point of observation). In his honor, this apparent shift in a wave—whether sound or light—is named after him.

Medical ultrasound, also called sonography, works by producing high-frequency sound waves we humans cannot hear, which are used to image such things as a fetus in utero, a breast mass, an abdominal mass, and any number of medical maladies that need imaging. Ultrasound is not the Doppler effect—but we'll get to that momentarily. Ultrasound is based on the concept that harmless high-frequency sound waves emitted by a transducer will bounce differentially off a static image of varying densities and send those bounced waves back to a receiver, which then images the results on a monitor. In other words, bone, cartilage, muscle, tendons, fat, and blood vessels all have different densities and will bounce ultrasound waves in different ways from the transducer back to the receiver, thus painting a picture of internal structures.

Demonstrating this effect is quite easy: just say a few words while standing in front of a wall made of brick and then one made of drywall, and you'll hear how the sound bounces off the walls differently. X-rays do not work in the same way as ultrasound—actually, the opposite. An X-ray passes *through* a structure—no bouncing here—and passes variably to the other side depending on the density of the structure, and then an image is constructed on that opposite side depending upon that variableness. The only thing ultrasound and X-ray share in common is that variableness of tissues—the former based on variably bounced off waves, the latter based on waves that variably pass through.

Then why did I mention Doppler in relation to ultrasound earlier? While your standard ultrasound is not the Doppler effect, there is in fact something called a Doppler ultrasound. This process takes the bounced ultrasound image just mentioned and combines it with the Doppler compression or elongation that occurs from the "movement" of the thing that is being bombarded with the ultrasound waves. In short, something has to be moving for the Doppler part to be useful. A Doppler ultrasound is most commonly used to image the movement of blood in vessels, and in so doing it can image such things as heart flow, blood vessels, and brain vessels, and it can also, of course, show the movement of a fetus. As the sound waves are bounced off that *moving* target, and those bounced sound waves move either toward or away from the ultrasound detector, an image of density as well as movement is constructed with a Doppler ultrasound.

But let's go back to light waves for a second, since that is where Hubble first observed the Doppler shift of distant galaxies, and which is what Doppler was postulating, too. If a light source is traveling toward an observer— say, a person peering through the Hooker telescope—the wavelengths are shortened to the blue end of the visible spectrum and the light appears blue. If the light source is traveling away from the observer, away from the Hooker telescope, the wavelengths are lengthened and appear red.

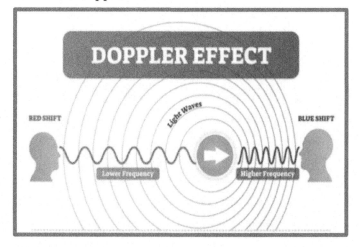

Hubble observed that what were believed to be binary star systems located in what was then the only known galaxy, the Milky Way, were in fact entire other galaxies that were redshifting—meaning they were moving away from the observer, away from Edwin Hubble sitting at the Hooker telescope, away from the Milky Way.

To summarize, as starlight travels away from the observer, it is elongated or redshifted light that is reaching the observer. As starlight travels toward the observer, it is compressed or blueshifted light that arrives to the observer's eye. Hubble not only appreciated that what we thought were other stars in the universe are indeed other galaxies, he also determined the universe is currently expanding, currently redshifting, meaning all stars are moving away from each other. No matter where Hubble looked in the celestial night, most of the stars and especially those galaxies were all redshifting.

Or are they? Is there anything moving *toward* us?

A handful of stars—hopefully dwarf-sized ones—that are possibly in our own galaxy are moving toward our sun, toward us. These include Barnard's Star, the fourth-nearest star to our sun. Not the sort of star I'd wish to meet—give me movie stars like Drew Barrymore or Jennifer Lawrence or Charlize Theron any day; those are my kind of stars. Cosmologists can tell those actual stardust-kind of stars like Barnard's Star are moving against the stream because they are blueshifted and on a trajectory somewhat toward our solar system. Don't worry, though—we have a million or so years to figure out what to do about Barnard's Star based on its current speed. Besides, it is believed the Kuiper Belt, with its trans-Neptunian objects, not to mention the Oort Cloud even farther out beyond Pluto, have enough ice chunks to absorb most of the fueling hydrogen and helium coming our way from those lost, aberrant, vagabond, and—keyword—dwarf stars.

Other examples that show not all stars in the universe are moving away from each other is the WISE galaxy, so luminous and so large that several nearby smaller galaxies are moving toward WISE, which then devours them. And when I say, "devours them," I don't mean WISE "eats" these smaller galaxies but rather, it absorbs them as part of the gang. There might be a few star collisions in such a merger of two galaxies but mostly they're just blended in. Some scientists believe that as the Milky Way formed, it similarly swallowed a dozen or so smaller galaxies during its infancy. For all we know, the Milky Way might be consuming a galaxy as I sit here and type, because at its edge—wherever the heck that is—current technology cannot quite parse whether an observed grouping of stars is part of the Milky Way or about to *become* part of the Milky Way.

The Andromeda galaxy has a halo around it, thought to be the result of it consuming, but not completely absorbing (at least not yet), an even larger galaxy two billion years ago. To make matters worse, Andromeda is expanding within itself and it is moving both away from and toward us. Its observable blueshift (moving towards us) outweighs its observable redshift—basically meaning we're doomed. One day, in about 4.5 billion years—which at least leaves enough time to finish *War and Peace* and vacuum the bedrooms—Andromeda and the Milky Way will collide. While this collision is incontrovertibly definite, it's possible we're not actually doomed. The stars in both galaxies are so far apart that star-on-star collisions are unlikely. It would be more like a mixture of two galaxies, like a gin and tonic. And it won't happen for more than 4 billion years from now.

If some poor cosmologist is sitting at their telescope in the future and realizes that the universe has stopped redshifting, stopped expanding, and has begun blueshifting, begun crunching—well, if that occurs, it will be billions of years from now and we'll still have billions of years to prepare before our current universe collapses into another singularity. It's one thing to send a nuclear warhead into space to blow up an asteroid heading toward Earth. It is quite another thing to stop a collapsing universe. To stop that we'd need the Hulk, first appearance Marvel Comics, *The Incredible Hulk* #1, May 1962, a Stan Lee and Jack Kirby creation.

Hubble's discovery of the redshifting of stars and galaxies was proof that the universe was expanding. Using mathematical models and working backward in time, he was then able to determine that the universe began about 13.8 billion years ago.

Edwin Hubble died unexpectedly in September 1953 at age sixty-three from a cerebral blood clot. No funeral was held and the whereabouts of his remains, if buried, was never revealed. He did not win a Nobel before he died, but rumor has it the Nobel committee was actually going to award him the Nobel in Physics the very year he passed away.

To review: The Big Bang, which formed our universe from an initial singularity was not a *bang!* at all, not an explosion like a bomb going off, but rather the creation of mass and a rapid expansion of space and time, all from pure energy. In fact, not only was there not a bang or an explosion, which usually gives off heat, there was an immediate relative cooling of the universe within microseconds, a cool down of the heat that was the initial singularity. This cooling is actually what allowed for the creation of all subatomic particles. If it was too hot those elemental particles would not have combined. It allowed pure energy to form the one-dimensional boson God particles, which then formed those vibrating one-dimensional strings.

"You'll have to speak up, Adam. My ears are still ringing from that Big Bang."

The descriptive term "Big Bang" was in fact a derisive joke made by the English astronomer Fred Hoyle during a 1949 BBC Radio broadcast. Hoyle did not believe in an expanding universe model, nor did he believe in an initial singularity. He instead believed in a steady state universe that was and always has been there. To Hoyle's probable dismay, his sarcastic "Big Bang" quip over the radio—made in an attempt to marginalize and deride the theory—actually had the opposite intended effect: it stuck in minds of the media and only served to advance the Big Bang theory. Just like how the media would later latch onto the "God particle" for the Higgs boson and "flesh-eating bacteria" for necrotizing fasciitis, the media latched onto the "Big Bang" to explain the origin and expansion of the universe from an initial singularity. I guess the media does that.

Fred Hoyle was one of those guys who had moments of brilliance coupled with more moments of being on the wrong side of history. Perhaps his most important contribution to understanding our universe was his singularly insightful theory on where all the ninety-four elements came from. We know the Big Bang made every last drop of the hydrogen and much of the helium and a sprinkle of the lithium, but what about carbon, what about oxygen, what about all the rest? Hoyle postulated—correctly—that through explosion of certain stars, through supernovas, there could be enough oomph to force hydrogen and helium and lithium to combine and recombine into ever enlarging atoms. Termed the "stellar nucleosynthesis theory," it was quite insightful for a man who was on the wrong side of history in other related cosmological areas. We'll get more into stellar nucleosynthesis a bit later.

In addition to deriding the Big Bang theory and denouncing an expanding universe, Hoyle was among those folks who rejected the abiogenesis theory of life on Earth, a theory that posits that, given the presence of the right chemicals on Earth four billion years ago—an ideal primordial sea—proteins and RNA and DNA and bacteria could arise.

Hoyle was having none of that. Rather than life beginning on Earth, he was of the opinion that life on Earth came from somewhere else in the universe, specifically from viral particles arriving on Earth via comets, a concept that has existed since the Greeks and is dubbed "panspermia." Panspermia was not a widely accepted theory by Hoyle's fellow scientists, its only other main supporter having been the Sri Lankan–born British mathematician Chandra Wickramasinghe. Of course, if anyone pressed Hoyle or Wickramasinghe to explain *how* life begin elsewhere—if indeed life did not begin on Earth, if it began somewhere else in the vastness of the universe and came to Earth via comets and asteroids—they were terribly silent on that point.

The fact that Hoyle single-handedly produced the stellar nucleosynthesis theory was perhaps Nobel worthy. That he did not receive a Nobel was largely due to him also backing discredited and fringe theories with too much zeal, the Nobel committee obviously taking into consideration his controversial posturing when denying him a Nobel. To Hoyle's credit, when it was pointed out he was a rather disagreeable chap, he would quip "it is better to be interesting and wrong than boring and right."

Speaking of interesting, in Ridley Scott's films *Prometheus* (2012) and *Alien: Covenant* (2017), which are prequels to the four earlier *Alien* movies starring the irresistibly irresistible Sigourney Weaver—*Alien* (1979), *Aliens* (1986), *Alien 3* (1992), and *Alien: Resurrection* (1997)—the basic story behind human life on Earth is that it arrived via an advanced human-esque alien. A single alien that looked very human and incidentally had a handsome Roman nose. He arrived on Earth, drank some potion, fell into a cascading waterfall, and presto, earthly humans arose from his DNA. That would sort of constitute panspermia à la Hollywood style.

The Russian biologist Alexandr Oparin and English scientist J. B. S. Haldane (John Burdon Sanderson) more convincingly posited in 1920 the abiogenesis theory, whereby life on Earth began on Earth. The Oparin-Haldane hypothesis states that in the primordial seas of Earth billions of years ago, given the right chemicals under the right sort of conditions, simple compounds could form more complex carbon-based polymers like proteins, fats, carbohydrates, and nucleic acids, and eventually bacteria. This is certainly different from "spontaneous generation," another disproven theory along with panspermia, which holds that entire organisms, from flies to humans, sprang forth from the dust with a snap of the fingers. Presumably God's fingers. No intermediaries. Whereas panspermia says earth life began elsewhere in the universe, and spontaneous generation claimed life on earth sprang in complete form from dust, abiogenesis offers a scientific explanation.

Comparatively, the abiogenesis theory posits that slowly, doggedly, more assuredly, from a few organic compounds combining and recombining over a billion years, bacteria eventually evolved, and from those primordial bacteria, all animals, all plants, and even viruses eventually evolved. These theories will be fleshed out later in this narrative, when we explore the world of Louis Pasteur and his germ theory.

The other major theory of how life on Earth came to be, besides abiogenesis, spontaneous generation, and cosmic panspermia seeding, is of course the book of Genesis. The Hebrew Bible (1:26–27) holds that God made man in his own image:

> Let us make mankind in our image, in our likeness, so that they may rule over the fish in the sea and the birds in the sky, over the livestock and all the wild animals, and over all the creatures that move along the ground.
> So, God created mankind in his own image, in the image of God he created them; male and female he created them.

The panspermia theory championed by Hoyle, who himself was an atheist, contains within it the idea that somewhere way out there in the cold vastness of the universe there is a higher power that designed life and sent it to Earth on meteors. Ultimately, Hoyle's disdain for the concept that life on Earth began on Earth overrode his atheism—and is so scathing that it bears quoting: "The notion that not only the biopolymer but the operating program of a living cell could be arrived at by chance in a primordial organic soup here on the Earth is evidently nonsense of a high order."

But "chance" is exactly what happened—a chemical evolution giving rise to an organic life evolution. When people think of evolution—those who believe in it, at least—they think of biological evolution, you know, apes evolving into humans, something less complex becoming more complex. (Although it can be argued that some humans are less evolved than the apes. Just saying.) But the same holds true for chemical evolution: simple compounds becoming more complex ones. Given the right sort of world, if enough things can happen, they will happen. In the end, the prevailing theory among scientists as to how life on Earth originated is abiogenesis: the origins of life arising from nonlife on Earth. An ideal primordial sea teeming with simple non-living organic compounds, the right atmosphere and sunlight, the right sort of planet located within a habitable zone of the right sort of sun—that's all you need for bacterial life to arise, that is, to acquire the capacity to replicate.

The Oparin-Haldane theory of the 1920s, stating that life could arise from organic compounds, was the first step in the abiogenesis theory. The next step came from the talented biochemist Robert Shapiro, who proposed that life did not need to come from a spontaneous emergence of RNA. Rather, he said, life arose from simple proteins followed by, and sort of along with, simple RNA, and, given enough time, the RNA started to act as a template to synthesize proteins. It's a proverbial "chicken or egg" scenario—which came first, the protein polymer or the RNA polymer? Well, the most widely accepted answer is a "protein first" model rather than an "RNA first" one, just as Shapiro proposed in the 1980s.

In the 1950s, Stanley Miller and Harold Urey at the University of Chicago performed the now famous Miller-Urey experiment. The two American biochemists recreated in a laboratory setting the conditions of Earth four billion years ago in order to test the chemical origin of life as per the Oparin-Haldane hypothesis. This experiment synthesized complex organic compounds from simple organic compounds, which themselves were synthesized from even simpler inorganic compounds. Miller and Urey demonstrated that a few amino acids, the building blocks of proteins, could be produced by electrically charging a vial containing chemicals such as water (H_2O), methane (CH_4), ammonia (NH_3), and hydrogen gas (H_2), all of which were abundant in Earth's early kitchen, cooking a primordial ocean soup sprinkled with atmospheric spices.

If you've forgotten what the main difference between organic compounds and inorganic compounds is, it's that organic compounds always, always contain atoms of carbon as the central player, usually attached to hydrogen or oxygen atoms. That is not to say all compounds with carbon are organic, some are inorganic, such as the greenhouse gas carbon dioxide (CO_2) and the highly lethal byproduct of fuel combustion carbon monoxide (CO). The poisonous chemical cyanide (CN^-) is inorganic, too. But the combination of the three basic atoms carbon plus hydrogen plus oxygen (CHO), where carbon is the ringleader, is pretty much the fundamental structure of organic compounds of life.

Interestingly, sealed glass vials from the 1950s Miller-Urey experiment were reanalyzed in 2007 with improved chemical analysis, revealing that in fact some twenty different amino acids were produced by their experiment. Amino acids, as we know, are the basic building block of proteins, and proteins are the basic structure for enzymes, and enzymes are the stuff that synthesizes other proteins as well as constructs fats and carbs. It's not long before you arrive at a fully functioning cardiovascular system and firing brain cells.

The steps in between look something like this: these simple organic molecules formed slightly more complex organic molecules, creating ever enlarging organic structures. We call these structures simple protein droplets, which evolved into amino acids, amino acids being the forerunners of complex proteins and protein enzymes. The more complex organic carbon-based protein droplets then cleaved, producing daughter molecules.

At some point in the organic chemical evolution on Earth, nucleic acids, which are different organic compounds from protein amino acids, combined to form rudimentary ribose polymers, putting us on the road to RNA. Exactly how, when, and where ribose formed these lattice polymers is currently unknown. As it turns out, ribose with nucleic acid as its core is even more difficult to form in primordial sea experiments than proteins. However, arriving at RNA through ribose and arriving at RNA through a sequence of reactions that *skips* ribose are both plausible options, so we're still on a good course here, whatever the case may be.

How do we know proteins came first, then RNA, which acts as a template to build more protein?

This issue is exactly where things got most contentious in Hoyle's day, perhaps leading to his staunch contempt of the abiogenesis theory. Predictably, the difficulty in imagining how ribose nucleic acids could form on Earth, and then could polymerize to form RNA, gave ammunition to those who supported variations of Hoyle's panspermia theory, which, as you'll recall, is the idea that RNA arose elsewhere in the universe and arrived on Earth fully prebuilt via comets. This argument is based on suggesting that RNA is just too complex to have formed randomly. Maybe simple proteins could form randomly, and surely fats and carbs could form in this way. But RNA? Preposterous, they screamed. Proponents of what was called the "RNA world hypothesis," a panspermia variant, even pointed to the Murchison meteorite, which hit Earth in 1969, as evidence that biotic life in the universe arrived on Earth via comets. (If you are wondering what the difference is between a comet and a meteor, don't fret. They are all called comets, and only that part of a comet that enters the Earth's atmosphere is termed a "meteor" or "meteor shower," and if the meteor hits the ground intact, then it is termed a "meteorite:" comet in space = meteor in atmosphere = meteorite on some farmer's cow pasture).

On September 28, 1969, a bright fireball entered Earth's atmosphere and broke up into three fragments before striking the ground near Murchison, Australia, where the meteorite further shattered into multiple pieces as well as created quite the tremor. The total collected meteorite debris, both large pieces and small pieces that made it through Earth's atmosphere without being vaporized, had a combined weight of 100 kilograms (or 220 pounds, for the metrically challenged among us).

Discovered on the surface of and within the Murchison meteorite pieces, in addition to small amounts of inorganic acids, rocks, and many other things, were organic chemical hydrocarbons, such as alcohols and most notably amino acids. This gave the panspermia and RNA world hypothesis believers some ammunition—which was odd, because amino acids are not the nucleic acids of RNA, a point that seemed to escape them. Nevertheless, they contended that RNA indeed formed elsewhere in the universe and arrived on Earth via comet debris. Their joy was short lived, however.

First off, several types of meteors have hit Earth, and meteors containing organic compounds are not unusual. It merely points to the conditions that might have, and probably did, exist 4.6 billion years ago, elsewhere in our solar system besides Earth, when planets, moons, asteroids, and comets where forming, creating rudimentary organic compounds. Furthermore, as we have explored, studies have demonstrated that not only could the conditions of early Earth, especially in the oceans billions of years ago, produce amino acids, which possessed the ability to polymerize into proteins (the Miller-Urey experiment), but that the early Earth conditions also favored the manufacture of simple nucleic acid ribose, which could then polymerize into RNA, the template to code for proteins—something *not* found on meteorites. Comets and meteors might have watery ice and silicon rock and ammonia and even some simple amino acids. But they don't seem to have nucleic acids, and in particular they are devoid of any RNA template for making bucketloads of proteins.

Another point that appears to be lost on the panspermia folks is this: all the meteors in our solar system are pretty much comets living out there beyond Neptune, in the Kuiper Belt or Oort Cloud, which means all those asteroids and comets and meteors did not form elsewhere in the universe but have been right here in our own backyard solar system for 4.6 billion years. Another way of stating it is that all meteors that have struck Earth didn't come from other star systems after traipsing about the Milky Way. Meteors don't travel across the entire expanse of our galaxy and beyond just to strike Earth. They're homegrown. Since comets and asteroids don't come from other solar systems, how could they have provided RNA life out there in the Kuiper Belt? They couldn't. Earth RNA life started on Earth.

However, having said that—having just stated that all asteroids and comets in our solar system originated in our solar system— that might not be entirely true. In October 2017, an asteroid passed through our celestial backyard, and based on its trajectory, and how it somewhat ignored the gravitational pleadings of our sun, it may have actually originated in another star system, traveling billions of years to eventually pass near our own star system. The object—named 'Oumuamua—was a really strangely shaped asteroid, not chunky and sphere-like as one might expect of an asteroid from the asteroid belt, but more spear-like. Based on its cosmological shenanigans, 'Oumuamua could have originated within the Oort Cloud but more likely came from somewhere much farther away, another star system. 'Oumuamua is Hawaiian and means "scout" or "messenger," based on the romantic notion that it was sent to our solar system on purpose.

Rather than coming from another star system as a natural phenomenon, perhaps that asteroid was thrown by Ares, the Greek god of war, or tossed our way by Thor—not just a fictional Marvel Comics superhero but the real Norse god Thor from the mythological world of Asgard. We can't be certain whether that spear-shaped asteroid was the Asgardian Thor making himself known to us Earthlings, but we do know for sure that the Marvel Thor first appeared to us in *Journey into Mystery* #83 in August 1962, thanks to Stan Lee, Jack Kirby, and Larry Lieber.

To appreciate that life on Earth had its genesis on Earth, one must understand that basic chemical reactions occur because they *can* occur. Given the right attraction between atoms, proximity, valence electrons, and favorable conditions, you get some chemical magic. I cannot stress this enough. It is the basis of how organic compounds were first synthesized on Earth: they synthesized because they *could* synthesize.

If you put two elements near each other that have an attraction for each other, that James Clerk Maxwell electromagnetism thingamajig, they'll bond—sort of like two people with a bottle of wine, with the wine acting as the catalyst or an enzyme. From atom-to-atom interaction, more complex chemical reactions start to occur, with the atoms forming chemical bonds and creating molecules because they can, and molecules near each other merging to form polymers because they can. Atoms find atoms and molecules find molecules. It really is that simple, especially if energy— such as sunlight—is added to the equation, because energy can drive a chemical bond that otherwise might not occur. When a catalyst is thrown into the mix to position the binding, even more magic occurs. Chemical reactions occur because they can occur, proteins formed in the primordial seas because they could form, RNA arose because it could arise.

Sometimes to facilitate compounds growing in length—otherwise known as a polymer—a catalyst, a third different compound, is involved. That bottle of wine helping two people become amorous would be a catalyst. At the molecular level, that wine helps molecules find each other and join in some type of intimate interaction. That matchmaker catalyst in the chemical world might be a single compound (enzyme) or it could be an array or framework called a "lattice" that positions one lovebird molecule in such a way so that the other lovebird molecule can find it and attach.

What is the difference between an enzyme and a catalyst, since both help a chemical reaction proceed? Enzymes are always organic protein single molecules, while catalysts are usually inorganic structures that may or may not be a single molecule. Enzyme-mediated reactions go faster than catalyst-induced reactions. Enzymes might be chemically altered while they're performing their job, whereas catalysts are not changed structurally. Enzymes are very specific to a single type of chemical reaction, whereas catalysts are indiscriminate and might be involved in several different chemical reactions. Enzymes might be regulated by a higher order, like a hormone, whereas catalysts take no marching orders from anyone or anything. Finally, all enzymes are catalysts, but not all catalysts are enzymes. All thumbs are fingers but not all fingers are thumbs. We won't discuss toes.

In our cells, in all living cells including the last universal ancestor, proteins form from amino acid building blocks, usually directed from an RNA nucleic acid template. In other words, the RNA nucleic acid polymer template acts like a lattice catalyst—not an enzyme but a catalyst— providing a framework for amino acids to find each other and line up in a correctly sequenced polymer string.

We've already covered how RNA came after proteins in the evolution of life as we know it, with its penchant for coding for protein enzymes, and how DNA came next, forming the double-stranded structure to protect the genetic code. But what forced this series of steps to happen, RNA giving up its role as the carrier of the genetic code to DNA? But before we venture into that territory, we must first settle a bit more controversy. Like how did a single cell bacteria evolve into a multicellular organism.

10 FROM ONE THERE WERE MANY

Those RNA world hypothesis folks who believe RNA arrived on Earth from comets, argue, understandably, that given the chemical fact that RNA is more complex than proteins and that RNA is needed to code for proteins, how could a more complex RNA compound form first to then code for *less* complex protein compounds?

Consider this: early biological evolution on Earth can be traced to a give-and-take, tit-for-tat situation, whereby simpler unicellular organisms began to form more complex multicellular life forms. That is, larger bacteria engulfed smaller bacteria, making those prisoner bacteria the nucleus of the cell—the first step toward eukaryotic, or "enclosed nucleus," cells, a step needed for multicellular evolution. Alternatively, it could have been the other way around: a smallish, very clever sinister bacterium entered a larger dimwitted bacterium, claiming it as its home, its mansion like those along Grosse Pointe Shores, a Detroit suburb, inserting its DNA into the host bacteria's DNA, hijacking its genetic machinery and taking over command of the cell. Whether big swallowed small, or small hijacked big, the end result was the same, a former prokaryotic bacteria cell with a nucleus, known as a eukaryotic cell, no longer a bacteria. But well suited for protists, fungus, plant and especially animals. Why a nucleus became important for eukaryotic multicellularity will become evident momentarily.

Other changes occurred on the road from prokaryote to eukaryote besides one bacteria becoming the nucleus inside another bacterium, and that was some cells chose to remodel the outside of the house while others did not. Many bacteria, although not all, have both an inner cell membrane and an outer cell wall, likely because the cell wall added protection and rigidity to a single celled organism faced with the harshness of the world. And for reasons that have to do with the fact they are not very complex multicellular organisms, plant, fungus and algae also kept or evolved that cell wall integrity outside the cell membrane as their cell lines of evolution advanced into multicellularity. The cell wall and the cell membrane together are known as the cell envelope.

The same is not true for the eukaryote cells that gave rise to the animal world, a cell line that either jettisoned the cell wall or felt it was not needed, in favor of just the cell membrane. Why? The cell wall presented cell-to-cell friendship issues, that is, single cells deciding to remain connected to each other on the road to multicellularity would have found the cell wall a prohibitive hurdle precisely because that is what a cell wall does. It wards off strangers. By avoiding a cell wall, those eukaryotes that gave rise to the animal world minus the cell wall could then connect to each other and remain connected permanently. The other advantage to dispatching with a cell wall as multicellularity evolved into animals is that it allowed for better communication between cells and improved transport of stuff into and out of cells. Connectivity, communication, transport—all functions that were acquired sans a cell wall.

In addition to eukaryotes having a former bacteria as its nucleus and dispatching with the cell wall, they also possess what were likely other bacteria, captured by the cell and turned into slave organelles. In the cytoplasm of eukaryote cells are various organelles, such as ribosomes that make proteins from RNA templates, mitochondria that churn out ATP energy currency (ATP to be discussed later), and trash bin lysosomes to digest spent protein enzymes, were all likely once bacteria that were captured. In the end animal eukaryote cells are nothing more than former bacteria that removed their cell wall and incorporated other bacteria to become the nucleus, the ribosomes, the mitochondria and the lysosomes.

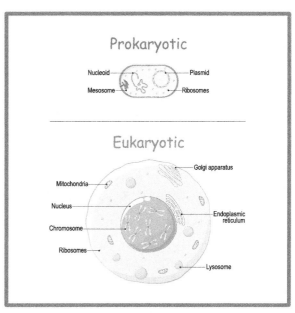

Circling back to why the nucleus evolved, whether plant or animal, and not to place a finer point on it but a nucleus was necessary for cells to "live with each other," that is, to form multicellular organisms like plants, fungi, protists, and eventually animals. A nucleus was needed for complexity, like an orchestra needs a conductor to perform Beethoven's *Symphony No. 3* (1803)—better known as *Eroica*. Without a conductor the orchestra might very well descend into cacophony. And a whole new set of rules was required. It was really no different a process than the evolution of civilization, as nomadic tribes with set rules that governed the group but with no central then hub gave way to civilizations with various forms of a central hub, a capital city, with kings and necessary administrators.

When I was in middle school biology, the taxonomy, or ordering system, of life included just two kingdoms: plant and animal. Fungus was cast into the plant kingdom and bacteria lodged within the animal kingdom. That all went out the window, just as I did when I snuck out of typing class in eighth grade to hang out with my pals. Yes—back in my day we had a class to learn how to use a typewriter. Today, a twelve-year-old can use one hand to type on a smartphone keyboard to caption a TikTok video while using the other hand to play an online game like *Minecraft* or *Fortnite*.

The taxonomy I learned back in the day with kingdom at the top of the heap has been supplanted with an even higher order: domain. Today there are two domains situated *above* kingdoms. These domains are Prokaryota, which means "no nucleus," and Eukaryota, "with nucleus." Prokaryota is further divided into two subdomains, Bacteria which might or might not have a cell wall and Archaea, which generally all have a cell wall, archaea basically being a separate division of bacteria. Some taxonomists place Archaea into its own domain, meaning there are three domains: Prokaryota, Archaea, and Eukaryota. Makes me wonder if taxonomists get into fistfights at their annual meeting over domains and kingdoms and such. Well, they're nerds so most likely not fist fights, but certainly slapping.

The Eukaryota domain includes all the rest of us that are not bacteria or archaea, and although biologists who concern themselves with the phylogenetic tree of life, termed taxonomy, believe there are upward of ten eukaryote kingdoms, for our purposes, and what is most commonly accepted, is that there are four eukaryote kingdoms: Protista, Plantae, Fungi, and Animalia.

Protists were probably the first nucleated eukaryotic unicellular life that arose from bacteria, appearing before plants, fungus, and animals. The most common protists that survive today and that precipitate human infection are the amoeba, as in amoebic dysentery or the handline-grabbing amoebas that eat your brain, and giardia, as in giardiasis, an intestinal infection. Nearly all protists have some form of locomotion, usually little hairlike tails called cilia or flagella to whip them around.

Driving toward complexity—multicellularity—required more than just a nucleus. Each cell in a multicellular organism would need to fend for itself, would need to produce its own energy, and accomplishing that feat took some more hijacking, as alluded to above. Simple eukaryotic cells captured other bacteria and turned them into a mitochondria organelle. Each cell has mitochondria, which are little slick power plants that have the sole purpose of converting glucose into adenosine triphosphate (ATP), the energy currency of multicellular life. Although cells in a multicellular organism share a bunch of stuff with each other, like hormones and enzymes and folded notes in class, they don't share their food, they don't share their ATP. Cells are rather selfish that way.

We'll get to chemical evolution in a moment, but I feel compelled to take the bacteria-to-multicellular-life pathway a few more clicks down the road.

Once bacteria evolved into cells with a nucleus (eukaryotic!), more possibilities for diversity became available simply because the cell was, for lack of a better word, *smarter*.

The first multicellular life was most likely in the plant arena or possibly among algae about 2.5 billion years ago. Multicellular animals likely arrived on the scene about 700 million years ago. Plants and algae evolved in response to their environment, they were able to reproduce asexually (where is the fun in that?), and they learned to use the sunlight plus water and the available carbon dioxide to make energy through photosynthesis. Photosynthesis basically is water (H_2O) and carbon dioxide (CO_2) forming into carbohydrates (CHO), with photons of light powering the reaction. For several billion years, Earth had bacteria and then it had plants, algae, and fungus, and these organisms grew massive. Why? Because they could. Multicellularity and large size favored an economy of resources; that is to say, cells working together were more efficient than when left to their own devices. Other than sharing that ATP thingy. The cells of multicellular life share nearly everything but not their ATP.

Animal life, well, that is another story altogether. Animal life really didn't get going until less than a billion years ago, around 700 million years back. The environment just wasn't suitable for us oxygen-breathers for the first 3.8 billion years of Earth's existence. Why did life decide animals were needed? Why didn't Earth stop at bacteria and plants and algae and slime molds and call it a day? On a cellular level, plant cells and animal cells are quite similar. But on a macrolevel, we're hardly similar at all. That fern in the corner that needs watering *looks* nothing like your poodle curled up next to that fern, yet on a *cellular level*, fern and pooch cells are very similar.

Two main distinguishing features sent eukaryotic cells down the animal path: the need for mobility coupled with the decision to use oxygen rather than sunlight to produce energy currency. Those were two key enterprises that led to animal life. Think of it this way: if you're a plant, for the most part you have to sit and wait for food sources to come your way, which is probably why plants developed roots. And as a plant, when nutrients do come your way, you're basically required to make your own food, to use sunlight to horsepower water and carbon dioxide into carbohydrates. But your still rooted.

However, if you become mobile, even if you're a lowly jellyfish, you can go out and *find* dinner, rather than while-away your time waiting for dinner to saunter by or to be forced to make your food from scratch. So, the benefits of roaming about looking for drive-thru ready-made take-out food, especially when food sources might be scarce, was quite appealing, sending some simple multicellular life forms down the animal pathway. But there was one drawback to becoming more multicellular and more mobile: the sun.

Multicellularity coupled with mobility sort of nixed using sunlight—photosynthesis—as an energy source. As a result, mobile organisms had to learn to harvest stored energy from other sources. Plants make carbs (that CHO stuff), so the photosynthesis-deprived multicellular animals started eating those plants for their stored energy. Which then entailed developing a pathway to extract that stored energy. More on that in a moment.

In a nutshell, it was environmental conditions—or more precisely, the *absence* of certain environmental conditions, namely food—that forced the evolution of mobility; and being mobile is one thing that distinguishes animals from plants. Food was becoming scarce in the neighborhood, so simple multicellular organisms deciding to abandon photosynthesis and look for drive-thru ready-made take-out, had to uproot and go looking for a fast-food joint. Uprooting meant abandoning photosynthesis, wandering and learning to derive energy currency—ATP—by eating plants that had decided to go down the stationary path.

We've sort of jumped ahead of ourselves. Having gone from less complex proteins and less complex RNA to more complex proteins and more complex RNA—chemical evolution—we then entered the world of cellularity—biologic evolution. But we need to circle back to the beginning of chemical evolution, since it predates those primordial protein and RNA polymers that gave us us. To understand this, we return to the Big Bang.

During the first few moments of the Big Bang, those one-dimensional strings made subatomic quarks, leptons, and gauge bosons, which then gave rise to the four main subatomic particles: protons, neutrons, electrons, and photons. Soon after the Big Bang, within minutes, protons and neutrons gave rise to nuclei without electrons. It was just too, too hot for a nuclei to grab an electron. It took about 380,000 years for things to cool down enough for those nuclei to capture electrons into orbit giving rise to hydrogen (H), helium (He), and lithium (Li), which are the first three elements on the periodic table. So, there we are in the first few hundred thousand years of the universe, hanging out with mostly hydrogen, some helium and a titch of lithium—and oddly, it was also around this time that photons of light appeared.

The reason no photons, no light, existed for the first few hundred thousand years or so into the early universe is the same reason why nuclei had not captured electrons, it was too hot. There was so much pressure and heat such that those gauge bosons destined to become photons could not yet form photons. For the time being, those bosons were too preoccupied reacting with unstable hydrogen, helium, and lithium nuclei—and those nuclei were unstable because they lacked much needed electrons in their outer orbital shells to transform them into solid atoms. In the words of Dr. Evil speaking to Mini Me from the film *Austin Powers: The Spy Who Shagged Me* 1999, nuclei needed electrons to "complete me." (I occasionally watch that flick mostly because it stars the stunningly beautiful and sexy Heather Graham as Felicity Shagwell. The talented Mike Meyers' parody of the James Bond 007 movies in the Austin Powers series is spot-on, although I do enjoy many, not all, of the Bond movies.) Once the heat and pressure of the universe quelled down to a dull roar, those nuclei were able to grab electrons, subsequently allowing those gauge bosons foreordained to become photons to become precisely that. And on that day, there was light, or for the creationists, the first day (Genesis 1:3–5).

Where, then, did all the other elements on the periodic table come from if not from the moment of the Big Bang, like, for example, beryllium (Be), which is next after lithium? And what about carbon (C) the ringleader of organic life, oxygen (O) along with hydrogen and carbon the basis of organic life, or silicon (Si), the central player in rocks, or silver (Ag), found in your earrings, or gold (Au), the main part of your wedding ring, or uranium (U), which gives us nuclear power? The remaining ninety or so naturally occurring elements that we are aware of formed naturally (up to uranium, possibly plutonium), and perhaps others not present on Earth but present in the vastness of the cosmos, from one of two processes previously discussed when we were knocking down Fred Hoyle's panspermia theory. Either the other elements were created within the cosmic gas from a process termed "nucleosynthesis" or more often and more likely they arose when special stars exploded, termed "stellar nucleosynthesis," that theory developed by Fred Hoyle.

The story goes that after the first 100 million years of the newly forming, rapidly expanding baby universe—composed of those cosmic clouds of mostly hydrogen and helium gases with a little lithium sprinkled in, and photons lighting up the cosmos in a brilliance—enough cooling had taken place, and enough gravitational influence had begun to take hold, that the process of star formation began (even to this day, new stars are forming while others blink out of existence). When some of those new stars burned up in a supernova explosion, they made many of the elements you find on the periodic table. It was the energy released by a supernova explosion that drove nuclear fusion, atoms combing to create large atoms.

As more and more gases condensed into stars, galaxy formation then began, with stars usually congregating around a black hole at the center of a galaxy. Black holes do that. As we know, our own Milky Way started its formation 13.6 billion years ago, early in the universe timeline, as a classic spiral-shaped galaxy apparently churning about a black hole or several black holes at the center. Our neck of the woods—that is, our solar system with our sun and the planets and whatnot—didn't take shape until roughly 4.6 billion years ago, forming somewhere along a remote outpost of the Orion Arm of the Milky Way. Soon after, about 57 million years later, or 4.543 billion years ago, Earth formed.

Orion Constellation

Before becoming an astronomer, James worked in the fashion industry.

The Orion Arm is named after the Orion constellation, which is named after the hunter of Greek mythology. Orion is one of the most prominent grouping of stars of the celestial night, and it is home to Betelgeuse, a star we've mentioned at its end sequence, itching to become a supernova. The most notable feature of Orion is his belt, made up of three bright stars.

Our sun is composed almost entirely of condensed hydrogen and helium gases, which is what most stars are, formed by grabbing available hydrogen and helium and spewing off most other debris. The core of the sun, which formed from the gravitational collapse of mostly hydrogen gas, is where hydrogen atoms fuse into helium atoms, a process that also emits heat and photons. It is the high temperatures and the high gravitational pressure that drives the fusion reaction—not fission, which is splitting atoms, but fusion, which is making larger atoms. The debris not captured by the sun when it first formed became planets, moons, asteroids, comets, and even dear old Pluto, too.

So, again, where did the elements after lithium up to uranium, possibly plutonium and beyond, come from if not from the Big Bang? Not from Amazon.com, that's for sure. Some were from cosmic fusion, which happened out there in the hot, soupy, gaseous cosmos, if and when there was abundant heat and sufficient crushing pressure to force their synthesis from hydrogen.

The second more common source of nearly all the naturally occurring elements on the periodic table arose and still arise from Hoyle's theory: supernova star explosions as well as neutron star mergers (something Hoyle was unaware existed). In fact, star explosions and star mergers are, by far, the major source of the elements found in nature.

It's worth stating that uranium, with its whopping ninety-two protons plus somewhere around ninety-two neutrons in the nucleus and ninety-two electrons in orbit, is believed to be the heaviest element found on Earth. Plutonium—which everyone has heard of, as in plutonium nuclear bomb—does not apparently occur naturally on Earth, although some traces might exist. Plutonium is a byproduct of making nuclear bombs and especially a byproduct of nuclear reactors, where some fusion takes place. But at least in our neck of the universe, whatever star or stars exploded to make all the elements that Earth ended up grabbing, plutonium was not one of them. Nor was neptunium Np, which is element 93, coming after uranium and before plutonium. It too is likely a byproduct of nuclear reactors.

The fan-favorite elements, the ones we are most familiar with, including carbon, nitrogen, and oxygen—the smaller elements on the periodic table—mostly come from supernovas. It is thought the larger elements beyond zinc (atomic number 30) were birthed mostly by merging neutron stars, and perhaps some supernovas. If a supernova is an exploding hydrogen star, what is a neutron star that merges? A neutron star is a star that is quite dense and smallish and does not have at its core hydrogen atoms like our sun—hydrogen being one proton and one electron but no neutron. A neutron star instead has at its core *only* neutrons. If you think those neutron stars formed from collapsing neutron gases, you would be wrong. Neutron stars form when electrons and photons are crushed together—the primordial leptons and gauge bosons. When stars like these merge—stars with neutrons at their core—we end up with some of the heavier elements.

The condensed *Reader's Digest* version of where the elements came from, although not hard and fast rules, goes as follows:

Hydrogen came from the Big Bang.
Helium and lithium arose from the Big Bang as well as from exploding stars.
The next two elements on the periodic table, beryllium and boron, came from nucleosynthesis in the gaseous cosmos, where high temperatures and pressure drove their fusion synthesis from the first three elements.
After boron at atomic number 5, the smaller elements beginning with carbon (atomic number 6), then nitrogen, oxygen, fluorine, neon, sodium, magnesium, aluminum, silicon,

phosphorus, sulfur, chlorine, potassium, calcium, titanium, iron, cobalt, nickel, copper, and zinc (atomic number 30), came mostly from hydrogen star supernova explosions.

After zinc, the larger elements like arsenic (atomic number 33), then bromine, krypton (not the Superman kind), molybdenum, technetium, silver, iodine, barium, tungsten, platinum, gold, mercury, lead, uranium (atomic number 92), and possibly plutonium (atomic number 94) mostly came from neutron star mergers, with some supernova explosions thrown in for good measure.

We have all these elements floating about, but what about the first molecules, especially all important water H_2O, molecular oxygen O_2, and carbon dioxide CO_2, where did they come from? First they likely formed in the local soupy cloud, bonding into molecules the way Linus Pauling would have wanted them to form. Next, the local star, our sun, whatever hydrogen and some helium that was not grabbed or wanted by the sun, all those elements and a few molecules swimming about in the cosmic soup were free to be grabbed by the soon to be formed planets, moons, asteroids, comets and other solar schmutz, each object taking up residence in whatever orbit best suited its composition, its Newtonian force, and its local Einsteinian gravitational influence.

For example, the cores of Mercury and Venus are mostly molten iron, while the cores of Earth and Mars are mostly molten iron and nickel. Consequently, you can imagine that iron and nickel was plentiful near where Earth and Mars decided to form, while only iron was plentiful and desirable where Mercury and Venus evolved. The presence of other elements that make up a planet, up to at least uranium, is directly related to what was in the neighborhood and suitable during that planet's formation. Anything not grabbed by a planet usually became a moon, and if not a moon an asteroid or a comet, and if neither of those, then just schmutz.

Earth obviously grabbed a lot of very important elements including elements from every station along the periodic table, especially iron and nickel but also oxygen, hydrogen, nitrogen, silicon, magnesium, calcium, aluminum, sodium, potassium, chlorine, and of course carbon. And those are just some of the ninety-two elements. Elements—better described as atoms—are fundamental substances that cannot be further reduced, unless you're inclined to build nuclear bombs or nuclear power stations or particle accelerators. An element, which is identified by the number of protons in its nucleus and always equals the number of electrons in orbital shells—electrons give atoms their charm, their personality—cannot be further reduced without the element losing is identity, losing its essence. Neutrons usually equal the number of protons (except an atom of hydrogen that has no neutron), but atoms can have more than their fair share of neutrons, which is where radioactivity comes in.

I mention "orbital shells" when speaking of electrons rather than just "orbit" for good reason. Electrons are, like a lot of subatomic phenomenon right out of Katniss Everdeen's reframe: "Strange things did happen here / No stranger would it be." If I remember correctly when I was a mere lad fumbling my way through high school chemistry, I was led to believe electrons orbit their nuclei like Earth orbits the Sun, in a defined, easily observable orbit. That is not true. These orbits are called orbital shells or electron shells partly because if you were to attempt to figure out the exact location of an electron around its nuclei at a precise given moment, you'll discover it is everywhere in that orbit all at once.

It is sort of the classic Heisenberg uncertainty principle, named after Werner Heisenberg, who described it in 1927, the location and the speed of a subatomic object, like an electron, cannot both be measured exactly, at the same time. The concept of exact position and exact velocity have no meaning at the atomic level.

The other galactically impossible concept to get one's arms around, is when an electron is bumped into a higher orbital shell, when it absorbs energy, like when it is hit by a photon—photosynthesis comes to mind—that electron goes from the original ground state orbit to the higher orbit without visiting the space in between. One moment it is in its ground state, the next it is in its elevated state. When the electron drops back down into its ground state, it releases that exact puff of energy and also does not pass through or visit the space in between orbital shells. Crazy stuff.

Returning to our local cosmic soupy solar system, different from elements are compounds and molecules, which are two or more atoms stuck together in a chemical bond, the charisma of electrons forcing that bond, and which can be broken back down into their constituent atoms. We have an idea where Earth got all its elements, but what about some of the basic compounds, like water, like breathable molecular oxygen, which is two oxygen molecules bound together to form O_2?

While forming, Earth grabbed, in addition to tons of elements, a bucketload of molecules floating about in the local cosmos, importantly the aforementioned molecular oxygen O_2 we breathe, the carbon dioxide (CO_2) and nitrogen gas (N_2) our atmosphere is made of, and fortunately an ocean of water (H_2O), the most invaluable chemical needed for life. All those essential compounds, like all the elements, were created in that cosmic soup by nucleosynthesis or by exploding stars. It is fascinating to think that especially the water on Earth was not made on Earth—like a shirt you thought was made in America but was only *put together* in America, all the while its constituent parts, its fabric, were actually fabricated overseas. Water was made out there in the cosmos. All the water we have on Earth was mostly there at the beginning, grabbed early on. The same is probably true of the atmospheric gases, especially carbon dioxide and nitrogen gas. Our breathable oxygen on the other hand, well, that is another story altogether, which we'll get to in a bit. But for now, it's enough to know that a good deal of our breathable oxygen was not cosmic soup cuisine, it was made on Earth.

As for the rest of the compounds and molecules on Earth, things like granite and dirt and even salt, these formed on Earth or in Earth's crust after our planet had grabbed all the ingredients it needed—formed because they could form from available elements already seized from the cosmos, forced to bond here on Earth from proximity, from high temperatures, from crushing pressure, and from gravitational pressure driving the reactions.

If we look at Jupiter, Saturn, Uranus, and Neptune—the four gas giants—to see what elements existed in their neck of the woods upon their formation, we'll find that the cores of

Jupiter and Saturn are believed to be hydrogen and helium, condensed but still gaseous, like the sun itself, alas just not *so* condensed, not *so* hot, as to trigger nuclear fusion and become baby suns themselves. In fact, you could argue that Jupiter and Saturn had the potential to be tiny stars that just didn't make it. The core of Uranus is thought to be frozen water, methane, and ammonia, and the core of Neptune is believed to be liquid iron.

All four are called "gas planets" because, despite having some type of core, they're mostly gas. (The fact that William Herschel in 1781 named Uranus as he did, the butt of jokes, all the while not knowing it was a "gas" planet, well, the irony is just too rich.) Another way of looking at Jupiter, Saturn, Uranus, and Neptune is that they have no definable surface upon which you could stroll. Where the gas *atmosphere* gases give way to the gas *planet* surface itself is sort of relative. It is not like on Earth where our gas atmosphere ends and our solid crust surface begins, an easily definable transition.

It was also fortuitous, but essential for life, that Earth is positioned where it is, since it was able to snatch basically all the elements, as well as an atmosphere and water, whereas the other seven planets (eight, if you include Pluto) were only able to grab some of the elements and some of the molecules.

For instance, Earth is positioned close enough to the sun to enjoy the benefit of a couple of the heavier elements, like iron and nickel, which sunk to and formed the molten core of our planet. The lighter Earth elements, like silicon and carbon, rose out from the core toward Earth's outer crust. And when things really coalesced enough, when Earth's gravitational pull was sufficient enough, our planet also was able to grab solar gases, most importantly the previously discussed breathable oxygen, carbon dioxide, nitrogen gas, and the best of all, water.

It bears mentioning that although Earth might have grabbed a huge share of breathable oxygen from the cosmic soup, for the most part, a great deal of Earth's molecular oxygen O_2 was not acquired at that point in time. Rather, breathable oxygen populated our atmosphere during the first few billion years of Earth's existence as a waste product from photosynthetic bacteria and then photosynthetic plants. The oxygen in our lungs is mostly homegrown.

Once Earth possessed ninety-two different elements and perhaps a handful of essential molecules, how and why and in what manner did other compounds form—you know, like dirt and rocks and diamonds, a girl's best friend? There are believed to be some 3,000 to 5,000 naturally occurring compounds on Earth, and they were not grabbed by Earth from the cosmic clouds because they didn't exist in the solar gases—like most of Earth's breathable oxygen, they were homegrown on our planet as it began to take shape.

Let's start with one of the most common minerals on Earth: quartz. Quartz is made from silicon (Si) and oxygen (O) to arrive at silicon dioxide (SiO_2). Then there's feldspar, which is basically quartz but with something like aluminum plus sodium $NaAlSi_3O_8$, or aluminum plus potassium $KAlSi_3O_8$ added to the silicon dioxide, giving chemical formulas that need not be explained further. When you smush more quartz with feldspar and a few other metals like sodium Na, iron Fe or magnesium Mg you get granite, which usually has at least aluminum, potassium and sodium in the mix with the silicon dioxide, a chemical equation I'll spare the both of us, and which is then used to make your kitchen counter. How did such smushing occur? Heat and pressure. Early Earth, with its gravity and pressure and heat and churning and leaching and separation by density and layering out and various other processes, gave rise to a vast array of inorganic compounds. It's all the stuff that makes Chemistry 101 oh so—what's the word—lifeless (devoid of organic life).

Even the once cherished and now commonplace table salt formed in this way as Earth itself was forming, and in fact salt is still forming today by leaching out from more complex compounds. Progenitors of salt deposits, that is more complex compounds that would eventually leach out salt, occurred from ancient dried-up sea beds, volcanic activity and crustal activity. Those compounds when exposed to the weathering effects of ocean and other water sources, with some acids added to the mix, leach out the salt.

As for diamonds, chemically they are not much different than the charcoal you use to cook a steak on the grill or the graphite in the #2 pencil you needed to take those dreaded multiple-choice exams. Making diamonds just took more pressure, heat, and time to force carbon atoms into a diamond lattice—diamond is just a purer and more tightly packed cousin of graphite, which means it's translucent rather than opaque, diamonds rarer rather than common graphite, expensive rather than not.

Besides molecular oxygen, which enables breathing, the other big molecule indispensable to life, not just on Earth but perhaps throughout the universe, is, as was mentioned previously, repetitively, water. There are life forms that can live without O_2, but it is unlikely there are any life forms that can arise without H_2O in the mix. I've already motioned that most of Earth's water came from "somewhere out there" in the cosmos and grabbed by baby earth, but it's actually more complicated than that (most things are!). Scientists have several theories of where all our water came from, which is to say, no one theory is entirely convincing or, more accurately, there is no sole origin for Earth's water but rather several origins. Certainly, water in its solid form, in its ice form, existed when the solar system was forming. Why else are comets and parts of asteroids and even Pluto not much more than dirty ice balls orbiting the sun?

But water on Earth—there is just so much of it that it makes one pause for a moment to contemplate its origins. One theory posits that, as our planet was taking shape, it grabbed its fair share—perhaps more than its fair share—of the ice water that was floating about in the cosmic soup. Once Earth formed an atmosphere, it was able to prevent that ice water from evaporating back out into the void. An atmosphere, or more precisely the lack of a sufficient atmosphere, is why Mercury, Venus, and Mars lost any water they might have initially grabbed. But early Earth, with its atmosphere of carbon dioxide and nitrogen gas molecules, as well as methane and sulfur dioxide produced by volcanos, was able to hold on to its precious water.

In this scenario, it is important to appreciate as ocean and lake and surface water evaporates into the atmosphere, it is in its vapor state and as a gas cannot escape the other atmospheric gases into the great void of space. Rather, as part of the water cycle, the vapor state of water, such as in clouds in the sky returns to earth in the form of rain and snow. Our atmosphere sequesters and then returns our precious water back to us.

A second theory posits that a great deal of Earth's water arrived via comets and asteroids after our atmosphere had already formed. The watery ice pellets of the ice-ball asteroids and ice comets that Earth was continuously pelted with during the early days of the solar system formation were retained by the already in place greenhouse canopy that is the atmosphere.

The full answer of water's origins on Earth is likely a mix of both theories: water has been here all along, our planet grabbing water from the cosmic soup and banking it in the crust and eventually the oceans, and then, once Earth fully took shape and acquired an atmosphere, additional water arrived via speeding cosmic ice balls. In fact, Earth has so much water—an embarrassment of riches, thankfully—that the water load really can only be explained by both theories. The only prerequisite for either is the presence of an atmosphere canopy able to retain the water once it arrived at our doorstep.

While water as we know it has been around for a long time, Earth's early atmosphere was not yet suitable for us oxygen-breathing creatures. Although the crust contained a fair amount of both elemental and molecular oxygen, the atmosphere did not. It is believed that Earth's original atmosphere was composed of mostly vapors of water, carbon dioxide and sulfur dioxide, with maybe a splattering each of ammonia, carbon monoxide and methane—the last gas of which might come out of your dog curled up on his bed, that doggy methane a constant reminder of why our early atmosphere was not suitable to breathe. Although, there is the added advantage of a dog passing gas that gives women someone to blame when they are the guilty party, because as we know, females don't pass gas.

The main takeaway point that we need to cover here is that these original atmospheric gases— carbon dioxide, sulfur dioxide, ammonia, carbon monoxide, methane—are known as "reducing agents," because they are more than happy to give up an electron and "reduce" themselves into a more stable state. Such gases would be useless in that Krebs cycle we were tortured with in high school biology, the citric acid cycle that makes ATP, the energy currency all life runs off of. Krebs needs an oxidizing agent, something that grabs electrons, not gives them up like a reducing agent, and that role is played by none other than oxygen (O_2). Grabbing electrons is what oxygen likes to do, which is what makes that Krebs cycle go around and around and around, an oxygen-rich, oxidizing atmosphere suitable for us O_2 breathers who would come later in Earth's history. Later, much later, we'll have a more thorough discussion on how the Krebs cycle works. But panic not—I'll keep it streamlined so I can understand it myself. There will be no pop quiz,

"I've devised a new paternity test, or as I call it, a pop quiz."

and I don't mean a paternity test pop quiz.

The first life forms on Earth, bacteria and archaea, didn't need oxygen. Instead, they mostly used or eventually used photons of light as an energy source—that good old photosynthesis—and the reducing gases in the atmosphere gladly gave up their electrons to help power these photosynthetic energy reactions. When you and I think of photosynthesis, we think of the chlorophyll in plants that absorbs the photon, elevating an electron inside the chlorophyll that then fuels the making of organic compounds based on the carbohydrates' CHO structure. But that's the plant world. The initial photosynthetic life forms on Earth were bacteria.

Photosynthetic bacteria—cyanobacteria—do not have chlorophyll, but they did and still do have other pigments that acquire the photon energy to elevate an electron, which, same as for plants, drives the synthesis of carbon dioxide and water into carbohydrates with breathable oxygen as a waste product. Imagine that: oxygen is a waste product in the photosynthesis world of both photosynthetic bacteria and photosynthetic plants. The chemical equation in its simplest form, understanding that it is elevated electrons from photons of energy horse-powering the reaction is:

$$CO_2 + H_2O = CHO + O_2$$

Cyanobacteria, also called blue-green algae, were the first photosynthetic bacteria on Earth—they're still with us—to derive energy through this type of photosynthesis, just like plants later would. These early energy reactions are what helped lead to the polymerization of the basic organic building block of carbon-hydrogen-oxygen (CHO), which itself went on to polymerize into longer, more complex organic polymers. Fascinatingly and fortuitously for us oxygen-breathing life forms, the waste product of photosynthesis was and is that breathable oxygen. Today, a great deal of our renewable oxygen on Earth comes from the continual output of cyanobacteria and its cousins, collectively termed phytoplankton, in the oceans, as well as the bountiful plants and especially trees on land.

Those original ocean plankton—a catchall phrase for cyanobacteria, archaea, algae, mindless creatures that have no means of locomotion—floated along in the seas and oceans of Earth, soaking up sunlight like sunbathers on Coronado Beach, San Diego, plankton deriving their energy from a reduction atmosphere (gases that gave up electrons), and produced copious amounts of oxygen by-product that then populated the atmosphere. Basically, atmospheric oxygen arrived on the scene for us oxygen-breathing life forms later on, only after some the early multicellular descendants, like the sea sponge, tested out the situation first, using oxidation chemical reactions rather than reduction chemical reactions to derive energy.

Although I'll delve deeper into it a tad later on, the very first bacteria on earth did not come equipped with a photosynthetic apparatus in place. That would have been too much to ask for. That came later. The very first bacteria, sometimes termed Stage 1 bacteria, would absorb nearby chemicals into their cytoplasm, compounds of which would then undergo fermentation, that fermentation would supply elevated electrons for the bacteria to harness but also resulted in waste byproducts building up inside the bacteria. Not a very good system.

You're likely wondering why Earth switched over from an initial reducing atmosphere, where photosynthesis was the only means of deriving energy, to an oxidizing atmosphere that still allows for photosynthesis but where oxygen is used to accept an electron along the pathway to deriving energy? It's rather simple, really. Photosynthesis had pumped so much oxygen into the atmosphere it forced it to become an oxidizing atmosphere, it had no choice. An oxidizing world still allowed for photosynthesis, but it also had one crucial outcome, an oxygen-rich atmosphere favored multicellularity, favored evolution. I can't stress this enough. In order for single cell organisms to evolve into multicellular creatures, they needed to abandon photosynthesis and develop another apparatus to derive energy.

Reduction reactions using photosynthesis can't make life forms that are able to progress much beyond unicellular or very simple multicellular configurations— that is, organisms that are complex. This is because photosynthesis, whether in bacteria or plants, needs to occur in the cell, and the sunlight that powers photosynthesis simply cannot penetrate into a multicellular creature. The mechanics of photosynthesis demand the apparatus be on the surface of the organism where photons strike, like cyanobacteria or the leaves of plants. Can you imagine us humans being photosynthetic, like a tree, needing to lay out in the sun like a solar panel for all our energy needs? Oddly though, people do precisely that, spend gobs of money, travel great distances, lay out in the sun on some

"Yech...I hate when they put that sauce on."

beautiful sandy beach soaking up UV light that could be harnessed to drive photosynthesis within our skin cells.

It's possible that a few billion years ago some early multicellular creatures tried to evolve into complexity using photosynthesis—but life would have chosen them for extinction.

There is also not any easy way for the reducing atmospheric gases—especially carbon dioxide—to diffuse into complex multicellular structures to offer up their electrons. Although it is conceivable that some early multicellular life forms tried to use carbon dioxide, less likely methane, sulfur dioxide, carbon monoxide or ammonia, internally to power reduction reactions, it too would have been a set up for extinction.

By its very nature, a reducing atmosphere only favored unicellular and terribly simple multicellular organisms that enable the reduction and energy conversion to occur within the cells that sunlight can penetrate; that is to say, a reducing atmosphere favored photosynthesis. A new way to acquire energy and transport that energy into the deepest recesses of a multicellular body needed to come along in order for complex organisms to evolve. The answer: molecular breathable oxygen (O_2). If you can figure out a way to get oxygen into every single cell, you can evolve a Krebs cycle to produce ATP energy currency. That is, rather than multicellular creatures attempting to carry on the photosynthesis of plants, why not just eat those plants and harvest their stored energy. Which is where we now briefly and apologetically turn for a moment, the dreaded Krebs cycle, the heart of what is termed cellular respiration, using oxygen to extract energy from glucose.

ATP just doesn't accept the electrons outright from glucose which is your most basic carb Krebs likes; there are linked reactions—the "cycle" part of Krebs—that eventually create ATP currency, reactions I believe life chose in order to make us miserable in science class. ATP is linked to the Krebs cycle we were all tortured with in 3rd period high school biology class, but it can be (over)simplified to make tasting it more like sugar than vinegar.

We know oxygen likes to accept electrons but if coaxed the other direction, will pass those electrons along, that transferring produces stored energy states. As discussed earlier, when it comes to producing ATP energy currency in a multicellular creature, each cell must fend for itself. That means each cell needs a personal supply of oxygen, so what did complex life eventually turn to deliver oxygen to every single cell: hemoglobin. Of course, hemoglobin which can bind both oxygen making it a red hue, or carbon dioxide a blue hue, is a rather complex 3-dimensional structure very susceptible to injury. It can't very well tumble on its own in the blood stream like hormones or antibodies. It needs its own transport vehicle to carry out its duty, sort of like a mailman needs his mail truck to deliver the mail. That transport vehicle is the red blood cell, a highly specialized cell that early in evolution when it accepted the duty of chauffeuring oxygen around, jettisoned its nucleus so as to make more room for hemoglobin to pile in. As red blood cells work their way into capillaries, the hemoglobin releases its oxygen which diffuses out of the red blood cell, briefly into the serum, and then is taken up by cells starving for oxygen. The oxygen traverses into the mitochondria and if glucose (pyruvate) has also meandered its way into the mitochondria, all you need is the Krebs cycle to extract the energy of glucose and convert it into ATP using oxygen O_2 as the patsy to eventually accept the lower state electrons.

The Krebs cycle simplified, which is the only way to understand it without tearing your hair out, is what follows. Glucose converted to pyruvate enters the mitochondria where it is driven by an enzyme gives up its electrons to a compound called oxaloacetate producing citrate. Then over a series of much smaller steps – apparently about ten steps – that high energy citrate with its elevated electrons is broken back down to low energy oxaloacetate, each step requiring an electron carrier called NADH to accept the electrons becoming NAD, which then passes those electrons on to ADP converting it to ATP.

At the end of it all a bunch of NAD would normally be built up jamming the cycle; that doesn't happen because our good friend oxygen O_2 strips the now low energy state electrons from NAD—oxygen loves electrons which is key to the Krebs cycle, accepting low energy state electrons—letting NAD return to NADH.

What happens to the oxygen that accepts those final low energy state electrons? That oxygen desiring a more favorable outer valence binds with carbon from the digested carb glucose to form carbon dioxide CO_2. And as oxygen enters the cell via diffusion from red blood cell hemoglobin, carbon dioxide leaves the cell by diffusion into that same red blood cell hemoglobin; in the lungs the hemoglobin then forfeits the carbon dioxide with exhaling and grabs more oxygen with inhaling.

As for definitions of the above acronyms, for those of you who are curious: ATP adenosine triphosphate, ADP adenosine diphosphate, NAD nicotinamide adenine dinucleotide, and NADH is *reduced* nicotinamide adenine dinucleotide.

The Krebs cycle is a very efficient way to grab the energy from glucose, you know, from the foods we eat, the carbon-hydrogen-oxygen CHO organic compounds from plants and animals that form our diet. The entire machinery is quite elegant and efficient, and get this, whereas as oxygen is the waste product of photosynthesis, the air we breathe, the waste product of Krebs is the carbon dioxide we exhale, the base component plants (and photosynthetic bacteria) desire to make carbohydrates. The Krebs is merely the reverse of the photosynthesis chemical reaction:

$$\text{Photosynthesis: } CO_2 + H_2O = CHO \text{ (stored energy)} + O_2$$

$$\text{Krebs: } CHO \text{ (stored energy)} + O_2 = CO_2 + H_2O + ATP \text{ (energy currency)}$$

It really is an efficient bit of machinery, the same red blood cell hemoglobin that evolved in order to bring inhaled oxygen deep into the interior of multicellular life is the same red blood cell hemoglobin that removes carbon dioxide waste from the cells and delivers it to the lungs to be exhaled. The oxygen waves to the carbon dioxide as they exchange in and out of cell mitochondria, and the wave again as they exchange in and out of the lung alveoli, ships passing in calm waters. And as for the "cycle" in the Krebs cycle, the elevated citrate at the beginning of the cycle breaks back down to oxaloacetate at the end of the cycle, to be used over and over again when glucose (pyruvate) shows up. Not only is this a marvel of life, it is a marvel of science to have elucidated this.

The ATP produced by Krebs is the energy currency used to drive all cellular reactions within that cell, whether that's unzipping DNA, building proteins, fats, and carbs, tearing down those proteins, fats, and carbs, allowing an impulse to conduct along a nerve or a heart muscle to contract, transporting goodies into and out of a cell across the cell membrane transport gates, or even attaching then detaching oxygen to hemoglobin all require the expenditure of energy. Nothing proceeds for free; it all calls for energy currency, which, in biological parlance, means moving an electron from one point to the next. Oxygen does that to produce ATP and ATP does that to spend its energy. When ATP has blown its cash, it becomes ADP. It recycles itself back to ATP through the Krebs cycle, with each molecule of ATP replenishing itself over 1,000 times per day. Or so I am told. As for one molecule of glucose, it produces about 38 molecules of ATP currency. Good return on investment.

When life switched to oxygen, or more precisely oxidation reactions as an energy conversion source, it not only would need to create the cleverly evolved hemoglobin and red blood cell to transport oxygen and carbon dioxide, it would need to provide for a highway for the limousine to cruise along. Life evolved a slew of artery and vein boulevards and back-alley capillaries—complexity upon complexity—as multicellular intricacy unfolded. It only took a few billion years for simple multicellular life using sunlight and photosynthesis to evolve into complex multicellular life using cellular respiration. And it all happened on an ideal planet, which is where we now turn, to astrobiology.

11 THE SWEET SPOT

In astrobiology, the well-known concept of the "habitable zone" describes the orbital region around a star that could potentially support life. Too close to the star, it is too hot, and the right stuff isn't around. Too far from the star, it is too cold and also lacking the right stuff. A planet of a certain size, endowed with certain elements, able to make certain compounds, and in possession of water, possibly oxygen, and an atmosphere, located in the habitable zone of the right sort of star, is just what your god ordered—a planet with the right stuff to support life.

Although it's debatable what the "key ingredients" of life are, most astrobiologists agree that for it to exist elsewhere in the universe, as has been stressed before in the pages of this narrative, the *numero uno* requirement is water. Not oxygen, but water. So, just why is water so important for life here, there, everywhere?

On Earth, water exists in three states—gas, liquid, and solid—and it exists in these three states within a narrowly defined temperature range. That last bit is important: narrow temperature range, because water needs to be able to do things in all three phases in order to support life and life can only exist within a slim temperature range, the more complex the life is usually translating into the narrower that range need be. Water also dissolves everything. Meaning that water, due to its polarity, brings things into solution, which then allows for things to happen, notably chemical reactions to happen within its solution. When it comes to the chemical evolution that preceded the biological evolution on Earth, a fundamental principle is that enough things that needed to happen could happen, and that happening happened thanks to water acting as the universal solution providing the level playing field. Simply stated, most chemical reactions, and certainly the chemical reactions needed for life, must take place in solution that doesn't interfere with most of those chemical reactions, and positively nothing does the job better than water.

In chemistry class, you're taught that some things hate water—they're hydrophobic and will not dissolve in water—while other things are hydrophilic, meaning they love water. But, ultimately, the reactions needed for life to evolve mostly dissolved in water, and those reactions hesitant to dissolve in water—that were a tad hydrophobic—at the very least went into suspension in water. What's the difference between "solution" and "suspension?" A fluid mixture like stirring sugar in your coffee, the water evenly disperses the sugar throughout, the water molecules surround the sugar molecules, and this is termed a solution. If you take some really thick fatty cream and stir it into your coffee, initially the delightfully tasty creamy fat molecules become suspended (not dissolved), making for a nice cup of joe before work; this is termed a suspension. But if you let that cup of coffee sit long enough, the fatty cream, less dense than water, will settle out to the top (while the sugar will not). That's the difference between a solution and a suspension, at least for our purposes.

In Earth's early days, when life was struggling to form, there was nothing that didn't dissolve in water, and if it didn't dissolve in water, it at least became nicely suspended in water. Fast-forward 4.5 billion years and humans make things, gooey things, sticky things, oily things, that do not dissolve in water, like two-part epoxy resin, for example. These things are exquisitely hydrophobic, they hate water. They won't dissolve and more likely than not they won't suspend either. They're human made usually for that purpose. But for life essence to evolve and survive, for chemical evolution and biological evolution, water is the ticket.

Water also flows, which is essential for many obvious things besides surfing crashing waves off Malibu. For our purposes, the flow of water is essential to life on a cellular level. The fluidity of cytoplasm inside all cells is due to water, that is where most of our body water hangs out, inside cells. Humans are 60% water and most of that water is intracellular. That cytoplasm water allows, as we just reviewed, things to dissolve or suspend in it for reactions to take place, molecules and elements floating about, here and there, for every cellular process to proceed, whether it is messenger RNA coding for protein enzymes, or the goings about in the mitochondria making ATP, it happens in a water solution.

The transport of things into and out of cells also proceeds in a watery environment, which we term either active transport or passive transport across a cell membrane. Active transport means the substance may not pass across the membrane without going through a gate and without the expenditure of ATP currency. Why? Because life decided that whatever molecules or elements that requires active monitoring also needs regulating.

Passive transport, on the other hand, is that semipermeable mumbo jumbo stuff we learned in biology, molecules and ions that don't need to show their passport to cross the border, regulation either not needed or their need so great that regulating would be too time consuming. Oxygen and carbon dioxide freely diffusing across a cell membrane is an example of a mumbo jumbo semipermeable process, their passive diffusion warranted by their immediate need.

Finally, but very importantly, water's fluidity allows blood to flow and blood flow delivers oxygen to the cells and alleviates them of their carbon dioxide. The fact that blood also carries immune agents to fight infection—white blood cells and antibodies—was a nifty trick life evolved later along the timeline. But the red blood cell hemoglobin uber trick was the primary reason blood and blood vessels emerged, not possible without water.

Water allows a great deal of movement on so many levels. Not to put too fine a point on it, but water truly "goes with the flow," and there is no other fluid known to humans—even as a thought experiment—that can accomplish what water does on the cellular level that would support life. Ergo, life cannot exist anywhere in the universe without water. Or without Godiva chocolate, but for different reasons.

"Can you believe that people inhale the gases we expel—sick, right?"

As we've seen, oxygen was *not* essential for LUA, the last universal ancestor, to thrive. Early bacteria—cyanobacteria—instead used photosynthesis to derive energy and actually expelled molecular oxygen as a waste product: the type of oxygen needed for more complex life forms to be fashioned. Imagine that. Oxygen, which is so essential for us oxygen-breathing, end-of-the-genetic-line life forms, was and still is a waste product produced by plankton and plants. So, go ahead and hug a tree. They and plankton and their other green friends continue to replenish our atmosphere with delicious oxygen. We inhale tree waste gas.

Let's recap: Once Earth's atmosphere was populated with enough oxygen, life evolved from using photosynthesis to using cellular oxidation respiration to derive energy—more efficient machinery for more complex organisms. As life became increasingly complex, especially animal life—and I don't mean John Belushi and his friends in their college frat house in *National Lampoon's Animal House* (1978)—that life required a system that allowed for each and every cell in an organism to derive energy lickety-split so that every cell, and as a result the whole complex organism, could run smoothly. Every cell in the human body makes its own mitochondria ATP for personal use and then burns those ATP energy stores to power all cellular reactions, which, if you're a chemist is nothing more than moving electrons along.

Earth is very nicely situated in the habitable zone of our Sun and grabbed enough water to eventually give us *us*. Earth is in the sweet spot in our solar system, about 93 million miles away from the sun. Astrobiologists suggest that if we were 5% closer or 88 million miles from the sun, or further out by 15% or 107 million miles from the sun, life would not have proceeded on our planet, and least not our type of life. That is a narrow zone.

Being in the habitable zone, also called the Goldilocks zone, on its own isn't enough, however. A planet must have an atmosphere so that it can hold on to its water. Holding on to its water is assisted by what was just mentioned, too close to the sun and the water evaporates into the cosmos, too far away the planet freezes over.

Oddly, or not so oddly, there is a habitable zone on Earth itself just as there is a habitable zone within our solar system. We can't live in the oceans which makes up two-thirds of earth's surface, nor can we live much higher than 10,000 feet above sea level for sustained periods. Yes, La Rinconada, Peru is a town at 16,700 feet that boasts permanent inhabitants whose physiology must become highly adapted for such diminished oxygen, but that is an exception. Leadville, Colorado at 10,152 feet is a more reasonable upper end of the inhabitable zone on earth.

There are also habitable zones within a galaxy. Our solar system itself is ideally located on the remote Orion Arm within the Milky Way, far from the galactic center and the black hole at that center. If we had been too close to the galactic center, there'd have been too much influence of gravity, such that enough of the things that needed to happen—chemical evolution, biologic evolution—for organic life to form would not have happened. In other words, just like there is a sweet spot habitable zone on Earth, a sweet spot habitable zone within our solar system, there is also sweet spot habitable zones within our galaxy, which, for other off-Earth worlds desiring to create life would need in order to evolve life. Being too close to the galactic center not only involves way too much gravity, but the burden of radiation would likewise be too great to survive. The types of stars found at galaxies' centers are not ideal for life-sustaining solar systems for this reason. On the flip side, if you go too far out toward the edge of a galaxy, the orbit of the star you're hitched to as it moves around the galactic center might be too erratic for life to arise.

Life happened in our solar system and on Earth because it could happen. Sweet spot in the solar system, sweet spot in the galaxy, and us puny humans within the sweet spot of the planet.

When you think about it, all those protons, neutrons, electrons, and photons that formed as a result of the Big Bang in the first few seconds, years, millennia, or mega-annums into the nascent universe, all those larger atoms that next formed and continue to form via nucleosynthesis in the soupy cosmos or via exploding hydrogen stars or via merging neutron stars, traveled across the expanse of the universe, over distance and time, to eventually assemble you and me, a series of habitable location evolution steps. Maybe not all people were formed in this way, but definitely you and me.

Given this narrative of sweet spots and habitable zones and the right sort of stuff and enough things that needed to happen, that could happen, that did happen—what then is this boggle about which came first, proteins or RNA?

If you were to compare complex proteins and complex RNA, it is indeed troubling to imagine that RNA, which templates those proteins, came first. How on earth could RNA know what it should template? It would be like Henry Ford building the automobile assembly line, but the automobile had not been invented yet. The error, of course, occurs when one assumes that complex proteins and complex RNA arrived on the scene at the dawn as chemical evolution merged into biological evolution. Wipe that out of your mind. Instead, think about those very simple proteins that were around at the dawn of evolutionary chemistry on Earth. Similarly, think about the very simple RNA sequences also present at the kickoff of organic polymers. If you limit yourself to imagining simple amino acid proteins dancing about while similarly simple but inherently smarter ribonucleic acid polymers are also sauntering around, one can imagine they sort of evolved *together*, both feeding off each other and feeding each other, a seesaw back and forth, polka dance of ever more complex protein polymerization in synchrony with ever more complex RNA polymerization. Proteins and RNA mostly arose together and mostly became more complex together, although proteins arrived on the scene first.

One proof lies in the research performed by folks like James Ferris and his group at the Rensselaer Polytechnic Institute in Troy, New York, in the 1990s. Ferris has shown that something as simple as the clay in Earth's crust—clay being a very soft mineral from the silicate group (SiO_4)—could have provided a perfect crystal lattice catalyst billions of years ago, which then acted as a framework upon which simple RNA polymerized from nucleic acid building blocks, alongside the polymerization of simple proteins from that RNA. Think of it this way: you have your clay lattice that lined up a few nucleic acid RNA into a short polymer, which in turn lined up a few amino acid proteins. Eventually, the clay lattice was no longer needed once the RNA became self-aware, became intelligent.

The thing is this: protein enzymes are good at catalyzing chemical reactions, such as fastening structural proteins—collagen, keratin, elastin, muscle, tendon—and constructing functional proteins—enzymes, antibodies, hemoglobin, hormones—as well as enzymatically helping to build the two other components: fats and carbohydrates. Protein enzymes are even good at building nucleic acid polymers like RNA and eventually DNA.

RNA is good at sequencing or coding for those protein enzymes over and over and over again by the bucketload as well as remembering the sequence. But RNA is not much good for anything else. In chemical evolution, prior to the dawn of DNA life, RNA became the potential code that would be able to mass-produce oodles of protein enzymes, functional proteins and structural proteins. It engaged in these various chemical activities in a non-cellular world. Simple RNA coded for simple proteins because it could. They became more complex in lockstep together.

You're perhaps wondering why, if proteins came first, was there even a need for RNA? Why didn't proteins develop more and more complexity on their own? Well, it just wouldn't happen. The more complex proteins became, the lesser the chance of their component amino acids randomly assembling into complexity . Take insulin for example, it has two protein chains, A with 21 amino acids exactly sequenced and B with 30 amino acids exactly sequenced, each chain using five different amino acids, and not a single amino acid can be out of place. The chance of insulin self-assembling anywhere, let alone inside a cell, was not going to happen. A template would be needed to assemble the amino acids exactly sequenced by the bucket load and to remember that exact sequence.

As the road to life evolved, something had to remember the gazillion sequences the twenty or so amino acids there are that, in convoluted sequences, ended up making all our functional and structural proteins. If life wanted to burst forth in all its glory from the primordial soup, it could not very well expect complex protein enzymes to self-assemble on their own. That something that would remember all the sequences was RNA.

Let's jump to DNA for a moment and then circle back as to why RNA gave way to DNA as the eventual genetic code bearer. In the nucleus of a cell (or in the nuclear region, for bacteria), unzipped DNA codes for a messenger: our old friend messenger RNA. That mRNA then leaves the nucleus and enters the cytoplasm, where it codes for protein enzymes. Those protein enzymes then catalyze the mass-production of structural and functional proteins and helps with carbohydrates and fat synthesis, storage and breakdown. On days you tip the scale way too much, you wish those darn protein enzymes that build fat storage would take a hike.

We have an idea where and how protein and RNA evolved, they evolved together, but what about DNA? Why as life was knocking on the door, was it necessary to move from RNA to DNA? The answer to this question is where we're heading now. But first we need to figure out why life chose to evolve a "cell." What is it about a "cell" that is so gosh darn important for life which is where we now turn.

How exactly did RNA and DNA come to arrive inside a cell membrane? Or, more precisely, why did RNA create the first semi-cellular structure, which would eventually evolve into our ancestral bacteria, LUA, the first true life on Earth? And why and how did RNA give way to DNA?

Let's put it this way: RNA was looking for a place to hide from the harsh realities of the world. Its nucleic acid code was vulnerable, exposed—not tightly wound into a double helix like its descendant DNA would become—and so RNA was seeking "shelter from the storm." Over time, like on a cold night when you wrap yourself up in a blanket, those RNA strands that were smart enough or lucky enough to tuck themselves inside the fold of some nearby fat polymer membrane survived the cold, gloomy nights of chemical evolution. Those RNA strands unable to find a blanket of fat to wrap themselves in did not survive. So even in the pre-cellular, pre-life world, self-replicating RNA molecules underwent natural selection in Darwinian terms. That fold of fat cloaking RNA acted much like a woman's shawl on a cold afternoon—"Come in, she said / I'll give you shelter from the storm," as Bob Dylan put in his 1975 song "Shelter from the Storm" on the *Blood on the Tracks* album, albeit Dylan was probably not thinking about RNA and fat. Despite its good intentions, a little cape-like fold of fat proved to be insufficient to fully protect RNA.

With the passage of more time in the primordial soupy oceans, the folded fatty cape eventually and completely polymerized around the enclosed RNA, forming a spherical structure like a space suit, the forerunner of a cell. Instead of a shawl to keep the RNA protected from the elements, on the road to life RNA chose a cellular membrane—a heavy-duty raincoat topped off with sou'wester and Wellingtons—to help the code weather the storm. Another way of looking at it is that the fat shawl was like Superman's cape partially covering the RNA, and the completely enclosed RNA was like Iron Man's suit. Superman first appearance was in June 1938 in *Action Comics* #1, thanks to his creators Jerry Siegel and Joe Shuster. Iron Mans first appearance came many years later, in March 1963, in Marvel Comic's *Tales of Suspense* #39, created by Stan Lee (who else), Larry Lieber, Don Heck, and Jack Kirby.

The point is, chemical evolution allowed for RNA's code to eventually become completely ensconced within a fatty membrane, a precursor cell structure—and there we have our next step on the road to RNA becoming the first primordial harbinger of life on Earth. Although cellular, these strands were not much more than very simple self-replicating RNA molecules enclosed within a simpler, still fatty membrane. They were not life, and most specifically they were not Stage

1 bacteria, the first life on Earth that might have had a brief stint with RNA but very quickly transformed into Stage 2 bacteria, or DNA bacteria.

Why, then, did those ensconced RNA droplets that were able to code for themselves as well as code for proteins not remain content being themselves? Why did RNA, and if Stage 1 RNA bacteria, find it necessary to evolve into DNA bacteria? Afterall, the RNA had successfully acquired the sanctity of an enclosed home. Shouldn't that have been enough? Well, no. We've already touched on why RNA converted into DNA, but some finer details still need to be added to the picture. For those of you who might be thinking that these RNA constructs were viruses, they were not. Viruses came after bacteria—another point to be fleshed out later on. These constructs were either non-living RNA cellular droplets or very crude, original Stage 1 RNA bacteria.

RNA was the original code template for itself—self-replicating—and for coding protein enzymes, but cell division was not yet clearly part of the picture, a fundamental principle of cellular life. Nor is RNA double stranded, meaning the base pairs (our old friends adenine A, guanine G, cytosine C and thymine T) were yet to pair. Since RNA could not base pair to form a protective double helix, by its very single-stranded nature it was, as we know, very vulnerable despite being ensconced. This is especially true if other chemical reactions like fermentation were starting to occur inside these pseudo-cells, the byproducts would damage the RNA.

DNA, on the other hand, would evolve to protect the all-important nucleic acid sequence, that would become the genetic code, in its spiraling double helix. The key changes are that RNA allowed DNA to first evolve, and then DNA agreed to unzip as needed so as to form RNA, which then would continue with its preferred job of coding functional and structural proteins. It was a match made in heaven…or in the primordial seas.

At some point on the road to the first true bacteria on Earth, the later Stage 1 or the Stage 2 bacteria, things switched from RNA to DNA by evolving these two key changes, specifically by removing the hydroxyl group from "ribose" to make the sugar "deoxyribose" which allowed for the tightest of tight helical twirls, and substituting the nucleic acid "thymine" for "uracil," which paired nicely with "adenine" giving rise to the double helix part of the equation. Those two chemical evolutionary changes allowed the RNA sequence to form double-stranded, tightly woven DNA protecting the code. Voilà, you have your first true primitive bacteria with one new favorable wrinkle, cell division.

Somewhere in that goop, when DNA learned to unzip to form RNA, it also unzipped to code templates of itself. In modern times we term this cell mitosis but back then a few billion years ago it was binary fission. To you and me mitosis and binary fission appear to be the same, a cell duplicating its DNA and spitting in half, with each half grabbing half the cell membrane (it would be later in the evolution of bacteria that they also acquired a cell wall). But binary fission is a much simpler process, first the DNA separates and as it codes its mirror image the nucleus or nuclear region also separates. In effect you have one cell with two nuclei, two brains, a double-headed monster that then separate from each other into two cells. It is a clean one-step two-step process.

In mitosis the DNA separation occurs at the last moment through these spindles that produces two cells, each with a nuclei. None of that double-headed rigmarole seen in binary fission.

In a world where life was beginning to desire stability and willing to expend energy for a small amount of un-molestable DNA over molestable RNA, base pairing and a helical arrangement offered the needed vault for the all-important genetic code. Life needed fidelity. But in that same world where you need to protect DNA, you still also need a bucketload of RNA transcribed from the unzipped DNA to make all those protein enzymes that then make functional and structural proteins, fats and carbs. Life needed quantity. That's why life kept both: the highly energy-intensive fidelity of DNA and the less energy-intensive quantity of RNA.

Once all these necessities were in place, the first bacteria, the first life—which by definition means "able to reproduce"—appeared on Earth. It was a mere nine billion years or so from the Big Bang and about one billion years after earth first formed. Evolutionary biologists refer to this little guy bacteria bubbling into existence as the Big Birth, providing symmetry to the Big Bang newspaper headline grabbing moniker.

Through one-dimensional string condensates to quarks, leptons, and gauge bosons, to protons, neutrons, electrons, and photons, to the gravitational collapse of hydrogen, helium, and lithium into stars, to star explosions, star mergers, and the nucleosynthesis of at least ninety-four elements and several molecules, water especially, to the local gravitational collapse of our bit of the solar system, to an beautifully sized planet within the habitable zone in an properly located solar system with an aptly churning sun, to a place endowed with all the stuff needed for the chemical evolution of proteins and RNA into a biological evolution of cells and DNA—to arrive at the Stage 1 of life, 3.8 billion years ago, a cellular bacteria what was little more than a bag of proteins and some genetic material. Whether the very first, the very, very first bacteria on earth used RNA or DNA is not known—it's not as though bacteria leave fossil records—but what is known is that if RNA bacteria arrived, it fairly straightaway switched to from molestable RNA to un-molestable DNA bacteria.

These Stage 1 fellows were tiny bacteria, much smaller than their modern day descendants, if you can imagine that. They were even smaller than the next stage of evolution 3.5 billion years ago, Stage 2 bacteria, affectionately known as the *last universal ancestor* or LUA, sometimes dubbed LUCA for *last universal common ancestor*. All organism on Earth, all bacteria and all plants, animals, fungus and protists now living have a common descent, and that would be LUA, a DNA bacteria with all the right attributes, which is why we hold that now extinct single-celled ancestor in such high esteem.

Stage 3 of life was another major leap, the cyanobacteria, wiggling and wriggling its way into existence three billion years ago, about 500 million years after LUA arrived on the scene. These little guys learned to harvest sunlight as an energy source through photosynthesis. In its simplest definition for our purposes, photosynthesis is a photon of light bumping up an electron to a higher energy state that then propels the synthesis of energy currency, in modern parlance, the production of ATP energy currency.

If you're wondering what Stage 1 and Stage 2 bacteria "ate" for dinner, for their fuel, since they lacked photosynthesis to produce sunlight energy, the answer is in that bottle of wine you polished off last night. Whatever geometrically produced organic molecules that were floating about, we call them carbs today, were absorbed by those pre-cyanobacteria organisms, the energy stored in those carbs were then harvested through fermentation. The breakdown of a more complex sugar into a simpler one—glycolysis—without oxygen is termed fermentation and as fermentation proceeds, high energy electrons are transferred. In order for fermentation to have worked, the appearance of ATP as the energy currency had to have evolved simultaneously with the very first bacteria. What does that mean? Fermentation inside bacteria was only able to proceed with the near simultaneous evolution of ATP. Another way of looking at it is that ATP predates photosynthesis, making itself available to chemosynthesis, which also predates photosynthesis. Deriving energy from chemicals is termed chemosynthesis—as compared to photosynthesis—and the first life on earth accomplished that nifty trick.

The accolades for bacteria don't even end there. We've discussed previously that in order for multicellularity to pass through the next door of evolution, it needed to switch from photosynthesis or a reduction reaction that drove energy currency to an oxidation reaction where oxygen accepts the final electrons. But it wasn't the multicellular organism that accomplished that nifty trick, too. It was bacteria. Although bacteria back then either survived on chemosynthesis or photosynthesis, they put in place the first crude Krebs cycle, using oxygen. This step had to have been accomplished on a single cell bacteria level before multicellularity could unfold. Bacteria did it all. Wow, huh? Smart little buggers. Smarter than us.

And speaking of wow, it is high time we turn our attention to Darwin and evolution.

12 YOU REMIND ME OF MYSELF—A WHILE AGO

Charles Darwin was born in 1809 in Shrewsbury, in England's Shropshire county, along the River Severn, historically a farming community notably involved in the wool trade and beer brewing. Shropshire is located east of the Welsh border and roughly 140 miles from London. Both Darwin's paternal grandfather, Erasmus Darwin, and maternal grandfather, Josiah Wedgwood, were abolitionists and members of the Unitarian Church—the Unitarians believing that God was a single entity and not the Trinity put forth by other Christian doctrine. The Trinity of course is God the Father, God the Son (Jesus), and God the Holy Spirit; this last hypostasis, the Holy Spirit, is God in the active form.

Darwin's father, Robert, was a physician and his mother, Susannah, was a member of the crystal glassware Wedgwood family. Charles Darwin was, in fact, a Wedgwood both through his mother and through his wife, Emma. To put it in blunt terms, Charles married his first cousin. Can you believe that, the father of evolution—a theory that promotes genetic mixing—marrying his cousin seems like a plotline designed to dangerously tempt fate, no? Was he totally clueless about consanguinity and inbreeding and kids with two heads? Was it even legal? Actually, it was in England in the early nineteenth century.

Today in the United States, current legislation has mixed opinions. About twenty states allow first cousin marriage, twenty-four states do not, and about five states allow it so long as both parties are over fifty-five years of age, presumably beyond the ability to have children. Some states allow first cousin marrying provided one of the couple is infertile.

Charles Darwin apprenticed as a doctor with his father, tending to the poor of Shropshire before entering medical school at the University of Edinburgh, where he found the lectures boring and surgery unnerving. Not exactly a promising start to a medical career. Darwin was not your blood-and-guts kind of fellow. Being a surgeon is not for everyone, but even forgoing the slicing and dicing, being a medical doctor is not for everyone, either. While at college, Darwin became involved in the Plinian Society, a club devoted to the natural sciences and, as such, a club that held a dim view of religious orthodoxy.

The Plinian Society likely took its name from Pliny the Elder, a Roman scholar and writer born in 23 AD who was one of the original naturalists. His most important work was the multivolume *Natural History* published 77 AD, a thirty-seven opus encyclopedia of all that was natural: from mathematics and astronomy through geography, agriculture, and mining and on through physiology, zoology, botany, and pharmacology. Not only was the work itself a compendium, it set the structure of all future encyclopedias. Just like the *Encyclopedia Britannica* we had growing up. I still have it, a twenty-four volume beautifully bound reference marvel, our edition published in 1968 (the first *Encyclopedia Britannica* published in 1768 in Scotland), now made sadly obsolete thanks to the internet and the boys and girls at Google. As for Pliny the Elder, apparently, he died as a result of the infamous eruption of Mount Vesuvius on August 25, 79 AD, which destroyed the ancient Roman city of Pompeii. In his honor, a certain type of volcanic eruption, marked by high plumes of ash debris into the stratosphere, is termed a "Plinian eruption."

So annoyed was Darwin's father by the young lad's disdain for religion—and by his standoffish attitude toward medicine—that he sent Charles to Christ's College at the University of Cambridge to study divinity, supposedly placing him on the road to becoming a county parson. Of course, many parsons in Britain at that time were well educated in the natural sciences, especially botany, geology, entomology, and ornithology, a point apparently lost on Darwin's father. If Darwin Senior had hoped that Darwin Junior would soon forget about the natural sciences and devote himself to God, he was wrong. Being sent to Christ's College didn't end up being much of a punishment.

Things often do not unfold the way fathers want them to unfold. Aren't sons put on Earth to disappoint their fathers? I certainly disappointed my father at various times. In my family, I was the youngest of five children with a father who, overwhelmed by the household commotion, distanced himself from his brood—I'm not sure he even knew my name. I have two daughters, so on that front I don't know a wit about sons disappointing their fathers. Eventually I rose above the chaos of adolescence, ultimately establishing a bond with father, mostly through our common ground in medicine.

The approach that many parsons of Darwin's day took was to explain natural history without marginalizing the existence of God. Many nineteenth-century clerics were devout Christians and also used natural science discoveries not only for an educational benefit, for the sake of advancing the body of knowledge, but actually as an argument for "divine design" or "intelligent design" in nature. The basic premise of many naturalist parsons, as they merged divinity with science, was the theological argument that the world was just too complex to have been left to chance. God must exist. All those awe-inspiring things in the preceding pages of this narrative that could happen and did happen—atomic evolution, chemical evolution, biological evolution—would understandably give a person a moment of pause, and give ammunition, albeit unproven ammunition, to those who believe God created everything.

Prior to Darwinism, an earlier theory of evolution was Lamarckism, proposed in 1801 by the French biologist Jean-Baptiste Lamarck. He posited the concept that an organism can acquire an adaptive change during its own lifetime and then pass that change on to the next generation. Through use or disuse of a biological aspect, Lamarck proposed, a change can occur in one generation and appear in the next. This, of course, is not true; genetic change must happen at the level of DNA and DNA is loath to accept change. Lamarck also posited that organisms are driven to ever greater complexity, which is pretty much true. To his credit, Lamarck was a core proponent of the idea that biological changes occur through natural processes and not by divine intervention. Where he went astray is positing that adaptive change can happen in one generation. It takes eons for an adaptive change to work itself into the DNA fabric of a species.

There is, however, some change at the genetic level that can happen faster than eons. This type of faster change has to do with epigenetics, which is to be distinguished from inheritable genetic traits. Inheritable genetic traits are the genetic code; they definitely do not change overnight. Epigenetics, on the other hand, refers to the ability to turn on or off certain inheritable genetic traits—sort of like light switches along the chromosome that allows specific genes to be turned on, that is, to be expressed, like blue eyes, or to be turned off, not expressed, like green eyes. These epigenetic switches can be injured. For example, an injured "UV light" epigenetic switch that is designed for tumor suppression through DNA repair can fail, which then might allow a damaged pigmented melanocyte DNA in the skin to become a deadly melanoma, that is, a bad type of skin cancer.

Perhaps the most important aspect of epigenetics is cell line differentiation. Every single cell of an organism possesses the complete complement of DNA for that organism—humans, for example, have twenty-three pairs of DNA chromosomes, for a total of forty-six chromosomes, and we have a gazillion genes, also called alleles, along each chromosome. Each and every cell in your body carries your genetic blueprint. (Well, almost every cell; red blood cells jettison their nucleus so as to house more hemoglobin.) Yet a nerve cell with all forty-six chromosomes becomes a nerve cell, and a fat cell with all forty-six chromosomes becomes a fat cell, and a heart-muscle cell becomes a heart-muscle cell, and so on. How? How does a single cell with the entire genetic blueprint for that organism know what it was "born to be"?

The answer is that certain sets of epigenetic switches along the chromosome control which *alleles* are to be expressed and which are suppressed. As a fertilized egg goes down the embryologic road, as cells differentiate to form the fetus, certain epigenetic switches come into existence for that cell line and may very well go out of existence as that cell line further differentiates, allowing yet other epigenetic switches to have their day in the sun, and so on. I suppose if a brain cell's epigenetic switches messed up and decided to code for some fat expression rather than pure brain neuron expression, it would give new meaning to the phrase "fat head."

As for the overlap between epigenetics and medicine, the manipulation of these switches might have future medical therapy potential. Imagine turning on a "tumor suppression" light switch to shut down a certain cancer or turning off another switch that leads to early coronary artery disease. Unfortunately, though, I doubt there is a switch to make your potbellied, balding husband of twenty-five years look like he did in college—lean, long haired, and ruggedly handsome. Some biological changes we just have to live with.

No one knows for sure exactly how long it takes, but—counter to what Lamarck believed—genetic changes that are durable (that is, not epigenetic expressions) can take thousands of years, if not longer, to work their way into the genetic code, into the DNA of that species. Some changes, through a genetic aberration, might occur in a mere few hundred years or so, but those changes are usually not durable: they are just as likely to revert back or vanish altogether, never to be seen nor heard from again.

It simply would not be a very durable, stable organism that allowed its genetic code to be changed easily, frequently, every few hundred years. It is critical for survival over the eons that the genetic fabric is not so easily swayed into an adaptive change. Because what if something that initially seemed like a survival trait turned out, instead, to doom the entire species into the fossil record? The longer an evolving species spends testing a new genetic mutation, the more time that species gives itself to back out, if necessary.

Consider this, humans are taller than they were hundreds of years ago, but that is not a favorable genetic change that became durable as evidence that in more modern times tall people tend to select taller mates and push the average height upward, while "short people" get mentioned in a Randy Newman song. Newman's song *Short People* released in 1977 was highly criticized as being mean-spirited, making fun of short people. That was not the point of the song. That criticism failed to understand Newman's penchant for constructing songs as a biased narrator—the opposite of what he thinks—in this case revealing the crazy stupidity of prejudice. In other words, Randy Newman was not referring to "short people," he was referring to short-sighted racists.

To provide another example of a false change in genetics over a short period of time, an urban legend is floating around out there that suggests the breasts of American women are getting bigger, and getting bigger over one generation—and that's putting aside breast augmentation, and also ignoring brassiere companies penchant to boost women's confidence by intentionally upsizing bras. Younger women today appear to have naturally bigger boobs than mother. Has there truly been an increase in female breast size in America, or, more to the point, has there been a change in the genetic fabric for female breast size? The answer is a resounding "no."

Female breasts might be bigger in the US and in other parts of the world simply because women, on average, weigh more than they did a few generations ago. The average weight of an adult American woman between 1960 and 2010 went from 140 pounds to 165 pounds (and for men 165 pounds to 195 pounds). Since female breasts are composed of a good deal of fat tissue, as opposed to true glandular breast tissue, girls with a few pounds on—their freshmen 10—have bigger boobs. It is that simple. Additionally, birth control pills introduced in the 1960s with their estrogen and especially progesterone formulation can make breasts fuller, especially during certain times of the month, similar to a menstruating woman's breast is fuller from mid-cycle trough the period.

There is also the suggestion that some pollutants and food additives in our environment, ones that mimic estrogen, might be stimulating female breast tissue to grow. But these factors—

plumper women, the pill, and estrogen mimicking compounds—are not genetic changes. They're behavioral and environmental changes. Nothing at the level of DNA.

Lamarck and later Darwin relied on similar evidence as the other to arrive at their theories of evolution, such as vestigial structures and artificial selection. Despite using the same evidence, they interpreted it differently. Examples of vestigial structures in humans include our appendix, which was once needed for digestion; our wisdom teeth, once needed for grinding plants; and especially our coccyx bone, a vestigial tail. As for artificial selection, the domestication of the wolf into about 340 different breeds of dog comes to mind.

Artificial selection, or selective breeding, as witnessed in animal and plant breeding, is not a "survival of the fittest" genetic adaptation adjustment in the DNA code as much as a forced mechanism of reproduction. That is to say, half the DNA complement comes from one parent and the other half comes from the other parent, which gives rise to a selective result, not a change in DNA. It's the same reassortment of DNA that occurs with humans, but it's sped up in the dog world and the cattle world and the corn world and any world other than the human one—hopefully—because in those worlds someone is in charge of pushing the DNA reassortment through the selection of specific traits coupled with rapid breeding turnover.

The main problem with Lamarckism—although there are several—is that genetic inheritance due to a favorable genetic change that was pushed by the environment (that is, by natural selection, not by selective breeding) cannot possibly occur in a single generation. No plant, no animal, can acquire a trait during its lifetime that is totally novel that can then become a part of its inheritance, its phenotype, which today we know is its DNA genetic code.

Natural selection is the ability of a species to pass through the next door of evolution, to survive over the long haul—eons—due to a favored phenotype, a favored genetic adaptation to the environment. Just take your pet dog for instance, descended from the now extinct gray wolf ("grey" if you're British), the wolf was domesticated about 15,000 years ago and eventually, perhaps during antiquity, man started selectively breeding dogs. And certainly, over the past several hundred years that selective breeding went astonishingly motley, the American Kennel Club now recognizing 195 dog breeds. And not a single one of them could survive in the wild on their own, which is precisely the point, not that I'm complaining. It is merely to illustrate how selective breeding is incompatible with "survival of the fittest." We took the stately gray wolf whose evolutionary tree made it fit for the wild, a human-friendly subset of the gray wolf started to hang around human campsites 50,000, perhaps 100,000 years ago likely thinking what could possibly go wrong hanging out with these humans? We domesticated the wild gray wolf into the family pet, forced selective breeding on it into 195 dog breeds, and we now make the once proud family poodle wear Halloween costumes posted on Facebook.

"How would you feel if your owner dressed you up in a silly costume, then posted it online?"

156

The other huge problem with Lamarckism is its sinister adoption by Adolf Hitler's National Socialist Party, which held a belief that a "super race" of blond-haired, blue-eyed Aryans could be bred overnight. Lamarckism really dovetailed with the racial program of Hitler—who, ironically, was neither blond haired nor blue eyed but rather dark haired and even darker eyed. To make such a eugenics program work, to create a so-called super race based on hair and eye color, especially since both are recessive traits, you would have to kill nearly other every person on Earth who did not match the desired phenotype—the Holocaust was a start—and you'd even need to eliminate your own people who did not meet muster, all the while only allowing your favored phenotypes to do the heels-to-Jesus procreation.

Such selective breeding fails to take into account that natural breeding, natural selection, has the intended goal of producing favored phenotypes, favored mixing of phenotypes, favored genetic adaptations pushed by the environment, to produce the most durable species. Selective breeding cannot, does not, and will not take into consideration what nature demands there are in the race for life. Subsequently, selective breeding among humans would unleash horrors that would have been prevented by naturally selected phenotypes.

Hitler might have wanted a blond-haired, blue-eyed Aryan race, but imagine his surprise if, in pushing selective breeding, he ended up with a "favored race" with two heads, one head with blond hair and blue eyes and the other blue hair and blond eyes.

Darwin, upon completing his Bachelor of Arts at Christ's College and through a recommendation by John Stevens Henslow, a British priest and naturalist specialized in botany and geology, accompanied Captain Robert FitzRoy on that pivotal second voyage of the HMS *Beagle*. Darwin Senior was disinclined to allow his son to go to sea, but his relative, the physician Erasmus Wedgwood, who was held in high esteem by Charles' father, convinced dad to let the poor lad go.

The first voyage of the HMS *Beagle* was 1826-1830 captained by Pringle Stokes, sailed from Plymouth, England. Somewhere tooling around Tierra del Fuego Stokes went mad, shot himself, and soon after died. The next in line was advanced to captain, sailed the *Beagle* to Montevideo, Uruguay where higher ranking Rear Admiral Sir Robert Otway, commander of the HMS *Ganges*, placed his trusted Flag Lieutenant Robert FitzRoy as captain of the *Beagle*, who, only twenty-three years old, then eventually piloted the ship back to Plymouth. Whatever was accomplished on that voyage need not concern us here.

The second voyage of the *Beagle* with Darwin aboard left Plymouth, England, in December 1831 and did not return until 1836, five years later. The voyage was charted to circumnavigate Earth, first heading to South America, rounding the tip at the convoluted Strait of Magellan (apparently preferring that sight-seeing route to the more treacherous Drake Passage further south), and then sailing north along the Chilean coastline to the all-important Galápagos Islands off the coast of Ecuador, a volcanic archipelago that formed millions of years ago. Ecuador derives its name from the Spanish word for "equator."

There on those somewhat barren, lava-formed Galápagos Islands, Darwin began to cement his theories of evolution, based on his observations of the differences between finches from one island to the next. These finches are today affectionately known as Darwin's finches.

This is the long and short of it: Perhaps two million years ago, the single ancestor to Darwin's fifteen distinct finch species arrived on the newly formed Galápagos archipelago, which back then was pretty much one large island off the nearby coast of Ecuador. With rising sea levels after the last ice age, plus continued volcanic activity

along the archipelago, mixed in with some tectonic plate activity, that single large Galápagos island became an archipelago of dozens of individually isolated islands.

As a result of that geological process, that one species of finch became individually isolated on dozens of separate islands, as the distance between those islands was too far for the finches to fly. Over the last 20,000 years or so, that single species of finch, now separated by time and distance on thirteen islands, speciated along slightly different evolutionary pathways, each path determined by the demands of the specific island. One of the most noticeable genetic changes observed by Darwin was beak shape, adaptations the finches acquired unique to the varying insects encountered island to island. When the islands separated out, so did the insects separate out, undergoing their own bug evolution, forcing the finch to evolve too.

A finch, by the way, is a passerine bird, and a passerine is a bird that both perches and is a songbird. Besides the finch, other common passerine birds include the thrush, sparrow, swallow, crow, starling, wren, warbler, bluebird, cardinal, oriole, and one of my favorites because of its song, the meadowlark.

In 1839, three years after completing the nearly five-year voyage to the Galápagos, Darwin published his epic first book, *The Voyage of the Beagle*, a landmark book which features such novel evolutionary concepts as "common descent" and "natural selection." Common descent implies that species that appear similar yet distinct, diverged at one time in the distant path from a common ancestor. Natural selection is the ability of a species to survive over the long haul from a favored phenotype, a favored genetic adaptation, to pass through the next door of evolution. That random attribute then keeps popping up eventually leading to a heritable trait within that population, becoming part of the DNA fabric of that species. Of course, science folks like Darwin and Lamarck knew nothing of DNA, but they imagined some operating system was at work.

Charles Darwin's theories on evolution continued to evolve after publication of *The Voyage of the Beagle*, in no small part due to his professional relationship with the leading geologist of the day, Charles Lyell. Lyell, who was a lawyer and naturalist, had propounded his uniformitarianism geology theory in his epic 1833 book *Principles of Geology*. Central to Lyell's thesis was the foundation that the Earth evolved from past forces that were still in operation and that these forces were slow-moving, acting over very long periods of time. In example, the Earth according to Lyell could not possibly have been created at a moment in time, 4,000 BC as ascribed by Biblical scholars, but rather at a much, much earlier time. It is said that Charles Lyell freed geology (and science) from the "well-meaning but incorrect dispensations of Moses."

This raises another important point about science. Geology evolution and biology evolution occurred around the same time, during the nineteenth century, and this was by no means an accident. Both evolutions were described independently and also in tandem with one another. In other words, Darwin and Lyell could see natural changes that called for extensive timelines in their own disciplines; as a result, each theory reinforced the other. It was precisely not by accident that Lyell and Darwin arrived at their theories around the same time. Fields of science often makes parallel leaps together. When "standing on the shoulders of those who came before," those shoulders might be in the same discipline or in a complementary discipline.

Darwin spent the next twenty years reviewing articles, studying the evolution of other species, such as barnacles—that's right, the crustaceous barnacles that are related to crabs, the ones that stick to the bottom of boats—and he also spent a good deal of time being ill.

The exact nature of Darwin's ailments, which plagued him much of his adult life, remains mysterious. Some forensic medicine types have attributed his vague symptoms, mostly intestinal, to having acquired some "bug" while on the half-decade HMS *Beagle* voyage, but others point out that his infirm constitution began in college, before he boarded the *Beagle*. Possible diseases the great man might have labored under include Crohn's disease, ulcerative colitis, lactose intolerance, Chagas disease (a tick-borne disease he would've picked up from his travels), chronic fatigue syndrome, and a few psychiatric diseases thrown in for good measure, such as autism spectrum disorder, panic disorder, and obsessive-compulsive disorder.

Darwin continued to delay finishing, and more importantly publishing, his opus *On the Origin of Species*—its full title being *On the Origin of Species by Means of Natural Selection, or the Preservation of Favoured Races in the Struggle for Life*—until prompted to do so by the convergence of several motivating events to be discussed in a moment. For twenty years, he wrote, revised, pondered, wrote, revised, pondered—never pulling the publishing trigger.

Alongside Charles Lyell, Thomas Henry Huxley, an English biologist and renowned comparative anatomy expert, strongly supported Darwin's theories. The pair knew of Darwin's theories, and likely discussed them over tea and crumpets or read excerpts from his letters and manuscripts. The precipitating event that ended up forcing Darwin to publish his opus was when Lyell received an unpublished paper written by Alfred Russel Wallace, perhaps forwarded to Darwin by either Huxley or Lyell. This paper presented Wallace's independently arrived at and astoundingly similar theories on biological evolution.

Wallace developed his theory while performing fieldwork along the Amazon basin in South America as well as within the Malay Archipelago of Borneo and Java in the South China and Sulu Seas. Wallace's theories were especially fortified after he observed evolutionary changes along what was called the faunal divide, later renamed the Wallace Line in his honor. Just like the Galápagos Islands, the Malay Archipelago was at one time one large landmass. But as geological processes unfolded, as the ice age ended and the seas rose, a "faunal divide" occurred whereby some species were caught above that imaginary line along the Malay archipelago, species inhabiting Borneo and Sumatra, while members of the same species were trapped below that line, like along the Celebes and on New Guinea. As we learned from Darwin's finches, once divided by time and distance, a species will evolve differentially depending upon the demands of its particular environment, with natural selection favoring certain phenotypes.

Northwest of the Wallace Line live the Asiatic animal species, and southeast of the divide are the ancestrally related but evolutionarily divergent Australasian animal species. Some of these species on either side of the faunal line include birds and large mammal species that once, millions of years ago, were DNA coded the same. But through the pressures of time and distance, with rising oceans and tectonic plate movement, those species evolved differentially. Wallace appreciated this fact and the treatise he wrote said as much.

On June 18, 1858, Darwin received Wallace's unpublished manuscript to read, and he found contained within it such phrases as "natural selection"—most likely causing Darwin alarm from the top of his head to the tips of his toes. I'm sure his heart sank into his stomach. I'm sure whatever near life-long intestinal ailment he suffered flared up. Huxley suggested to both Darwin and Wallace that they present their theories simultaneously. Darwin immediately forwarded his tome (despite the fact it was to his mind incomplete) to the one man who could conclusively settle a bet in the world of naturalists: the geologist Sir Charles Lyell. Darwin then disappeared off the radar, as strep scarlet fever was ravaging his English village, including afflicting his daughter, who had strep throat. Today in the antibiotic world a bout of strep throat is a minor inconvenience; back in Darwin's day it could just as easily end a life as recover.

Lyell along with Joseph Dalton Hooker, the father of geographical botany, decided the best course of action was a joint presentation before the Linnean Society of London on July 1, 1858—a presentation designed both to establish biological evolution and to settle who was the father of evolution, Darwin or Wallace. The Linnean Society of London was a membership only natural science club named after the great eighteenth-century Swedish botanist and father of modern taxonomy Carl Linnaeus. Taxonomy as you might recall and might like to also forget uses the binomial nomenclature of naming species in Latin into "genus" and "species," whether be it plant or animal. Examples of binomial nomenclature to refresh your repressed memory include the bacteria *Staphylococcus aureus* or the flower *Rosa madame* also known as the Peace Rose, a favorite of gardens and bees, or *Canis lupus*, our friend the dog, or *Homo sapiens* which is us, or at least some of us. Carl Linnaeus, the Swede, and this taxonomy scheme was actually more widely received, enjoyed and celebrated in Britain—home to most of the then world naturalists—than in his home nation of Sweden. And which is why the Brits honored him with The Linnean Society being located in London. The Swedes were more interested in chemistry stuff, an area where they simply excelled.

When July 1, 1858, arrived, everyone who was anyone within the Linnean Society waited in excited, anticipated breath, crammed into the society's headquarters Burlington House, to hear Darwin and Wallace present their theories. Actually, the attendance was not that impressive at all since it was a members-only affair, and the uninvited outside world was also not quite ready to accept that evolution mumbo jumbo anyways. It *should've* been the debate of the century, but it got little notice other than among a rarified group of British naturalists. Additionally, there were two small hiccups: Wallace was still in the Malay Archipelago and Darwin was not in attendance either, his baby son having died three days earlier from scarlet fever. Later along this road we are travelling, we'll cover strep throat and the three syndromes associated with strep: rheumatic fever, scarlet fever and St. Vitus dance.

So, instead, without Darwin and without Wallace in attendance, the secretary of the Linnean Society first presented the two men's theories based on a synopsis each provided and then read an introductory letter from Lyell, an excerpt from Darwin's book, a letter from Darwin to famed Harvard botanist Asa Gray dated 1857 that laid out some of his theories, and parts of Wallace's manuscript were read. Although Wallace had a delightfully convincing manuscript, given the several decades Darwin spent formulating and refining his theories, not to mention The Voyage of the Beagle published eighteen years earlier in 1839, taken as a whole Darwin's theories were unsurprisingly more defined and fleshed-out than Wallace's, it was clear to most everyone that Darwin had arrived at biological evolution first. As the Linnean members shuffled out of Burlington House into the muggy London night, it was clear they favored Darwin over Wallace. The following year, Darwin published *On the Origin of Species*, and the sheer volume of it, with its meticulous research and detailed documentation, cemented Darwin's title as the "father of evolution."

One of the reasons Darwin's theories of evolution were slow to be embraced even by the science-minded folk of his era, the naturalists, let alone the pedestrian man and woman strolling along the Thames, was the paucity of proof. Yes, his book was huge, but there wasn't actually a great deal of evidence in the dirt, in the fossil record. Here was a theory that proposed species evolved to be more and more complex over millions and millions of years, whether plant or animal, and yet there was little in the fossil record to back it up. The fossil record held such a dearth of clues because, quite simply, up until that time, there just weren't that many fossil hunters (today we call them paleontologists) looking in the dirt for bones. To make matters worse, of the few fossil hunters who did exist, they weren't quite sure where to look. You can't just grab a shovel and go out into your backyard, start digging, and expect to find the evidence you're looking for.

"You remind me of myself—you know—a while ago."

The most common reason British commoners—heck, all commoners—had a difficult time believing in Darwin's evolution was quite simple. To a man, especially to the even more advanced woman, believing humans evolved from the humble ape was a tall order.

Even though Darwin's friend Charles Lyell—one of the fathers of geology, although it should be mentioned that James Hutton, for his work a century earlier, is the true father of geology—had given him the geological timeline Darwin needed for the millions upon millions of years for evolution to unfold, even Hutton's and Lyell's theses were mostly worked out on paper with a dearth of evidence in the dirt, nothing physical, to prove either geological or biological evolution. At least, nothing that had been discovered *yet*.

The other problem with Darwinian evolution was the existence of the opposing concept of "dilution theory." Naturalists argued that, if by chance a species acquired a new favorable trait, a sudden mutation, it would be diluted or washed out during mating based on the strength of the ordinary traits of that species, would thus override the novel trait in the next generation of offspring, no matter how advantageous it may be.

It doesn't take a rocket scientist to imagine that that's how reproduction works. Even our dimwitted Neanderthal relatives probably had some sense that their children bore remarkable resemblance to both parents. And if the child didn't look like dad, well mom had some 'splaining to do. Humans didn't know how this was—until DNA was discovered—but we have known for a long time, since prehistory, that our children look like a blend of each parent. So, the naysayers of Darwinism argued that a favorable trait would be washed out by the tried and true ordinary everyday traits of both parents, traits that were presumably durable, traits that had fidelity, and traits that were not about to be corrupted by some upstart novel genetic mutation.

How could a Darwinian acquired trait overpower a preexisting trait? The answer would come in the form of an unassuming Augustinian monk by the name of Gregor Johann Mendel, who was experimenting away in the pastoral abbey of St. Thomas's in the remote village of Brno, Moravia, modern-day Czech Republic. Indeed, the answer came in the form of Mendel's soon-to-be famous peapod experiment. Gregor Mendel was born in 1822 in Silesia, which was then part of the Austrian Empire, born into a German-speaking family of farmers. A gifted student with little family money, he struggled to pay for his education at the University of Olomouc, receiving help from his younger married sister.

Gregor went on to become a friar, not necessarily due to any profound religious convictions but because it would allow him to continue his science studies at the university level without having to pay tuition. Mendel then entered the Augustinian St. Thomas's Abbey to become a priest, a place where he could be left alone to pursue his interests. He then took a leave from the abbey to conduct further studies at the University of Vienna, which included studying under physics professor Christian Doppler, he of the famed Doppler effect which we discussed at length in an earlier section, that red-shifting blue-shifting light and that high-frequency low-frequency sound mumbo jumbo.

After finishing this next step of his education, Mendel returned to St. Thomas's, where he selected seven traits of the peapod plant—seed shape, seed-coat tint, pod shape, pod color, flower color, flower location, and plant height—by which to study genetics. From 1856 through 1863, he chronicled a whopping 28,000 peapod plants, from which he posited three laws. Pod color would turn out to be the hero of this story.

The *law of segregation* states that every organism contains two alleles for each genetic trait, one from mommy and one from daddy. When that organism makes an egg or a sperm, the resulting cell (gamete) results in a germ cell that has only half of the complement, a process termed meiosis. Recall mitosis is cell division with each cell have the full DNA complement; meiosis is cell division where each germ cell gamete has half the complement of DNA. These mature male or female germ cells (half complement) then are able to unite with another of the opposite sex in sexual reproduction to form a fertilized egg or zygote (full complement). You'll love this: gamete is from the Greek *gamein* and means "to take a wife." As for zygote it is also from the Greek *zygotos* from root word *yeug* which means "to yoke, to join."

In example, human DNA has 46 chromosomes, so female eggs and male sperm each have 23 chromosomes. Every cell in your body with a DNA has the full complement except your germ cells—egg and sperm—if you have any left. And as was mentioned earlier but deserves re-mentioning, epigenetic switches along a cell's DNA constantly reminds that cell what it was born to be, such that, in example, when a skin cell divides the two daughter cells know they're skin cells. Any hope of those epithelial skin cells desiring to become highly vaunted brain cells—to be in the immortal words of Marlon Brando's Terry Malloy from the wonderful, Academy Award winning movie *On the Waterfront* 1954, a "contender"—is simply out of the question.

In Mendel's experiment—without knowing anything about DNA and cell division and all that rigmarole—he proposed that pea pods had two colors, green and yellow, and that when their flower parts did their thing—the female pistil making the egg, the male stamen the pollen—green and yellow were separated out into half-complements for each gamete. Individual plant female egg and male pollen gametes each possessing half the complement, either a green allele or a yellow allele, and that is, boys and girls the law of segregation.

The *law of independent assortment* states that the separation of alleles is generally passed independently from other alleles even if they seem to be closely related; pea pod color for instance had nothing to do with pea pod shape. In humans consider this overly simplified example, hair color and hair pattern assort differently and are not tied to each other. A person can have brown curly hair or brown straight hair, the hair color is inherited separately from hair pattern.

The *law of dominance*, which is what this discussion is really all about, proposes that a trait can be dominant or recessive. In Mendel's pea pod experiment, he determined that green pod color ruled over yellow pea pod color. When pea pod egg and pollen with either green or yellow combined in plant fertilization (in humans we call it the fastest sperm reaching the female egg, the ovum), the resultant fertilized egg called the zygote, would have 25% green-green, 50% green-yellow (green dominant), and 25 % yellow-yellow; the result would be 75% green pea pods and 25% yellow.

So how does a random mutation enter into the DNA fabric of an organism? If the acquired trait was not a favorable mutation, it would die out, being outnumbered by the fidelity of both the dominant and even the recessive traits. But if it was a favorable trait, whatever the percentage of the species who enjoyed the benefit of that mutation, would live, and live long enough to see their mutation survive while their fellow pals would enter into the fossil record. In other words, if a random mutation offers survival, it will over thousands and thousands of years, permeate long enough to become part of the DNA fabric of that species. A Darwin finch that mutated a different beak shape was probably laughed at, but when that finch was able to eat bugs the other finches could not, that laughter soon trickled off into moans of hunger, and over enough time those non-adapting finches would fall from their passerine perch, never to sing or laugh again.

Darwin's theories and Mendel's experiments were eventually combined in what is called "the modern synthesis." It took quite a few decades and probably some more visible proof dug up by fossil hunters for evolution to become established. What is odd is that Darwin and Mendel knew of each other but apparently did not know of each other's theories, or if they knew they both failed miserably to combine them. Eventually other folks pieced it all together, with the final stamp of approval coming from, of all people, Julian Huxley—the grandson of Thomas Huxley who was known as "Darwin's Bulldog" for how voraciously he defended Darwin's theories of evolution. In a beautiful unfolding of science, the grandson of Darwin's most ardent supporter put forth the modern synthesis of evolution in his aptly titled book *Evolution: The Modern Synthesis*, published in 1942. And yes, Aldous Huxley, author of the wonderful dystopian novel *Brave New World* (1932) is Julian's younger brother, also the Bulldog's grandson. That Huxley family was quite the intellectual dynasty. At the risk of being a bit on the nose, the Huxley alleles for intelligence were most certainly dominant genetic traits.

After the publication of *On the Origin of Species* in 1859, what followed were a series of public debates—for commoners, not just academics—between Darwin's Bulldog Thomas Huxley who was pitted against Bible creationism proponent Bishop Samuel Wilberforce. Darwin was no-good at public speaking and certainly shied away from anything remotely resembling a public debate. So Huxley carried the torch of evolution by debating Wilberforce. It is unclear to history how the early debates unfolded, until that is, the legendary debate on June 30, 1860.

The setting was the beautiful campus of Oxford University in the scholarly University Museum. You can imagine the scene, all these professors dressed in their long black robes, aristocrats dressed in fancy duds, the air thick with suspense, the night warm, and the beer warmer. As the debate unfolded no notes were taken, so there is no actual verbatim document as to the veracity of what was said and by whom. This story I am about to tell comes to me not through reading books or using Google, but as relayed to me by my college physics professor at Antioch, Oliver Loud, who was friends with Linus Pauling who had heard the story from Aldous who in turn had heard it from family lore.

The air dripping with suspense; everybody who was anybody wiggled their way into the Oxford Museum, lining the great hall, standing shoulder-to-shoulder along the gallery above, hanging from the rafters. Bishop Wilberforce went first attempting to eviscerate Darwin's theories with his opinionated loquacious tongue, lacking any proof of his statements. But this night Bishop Wilberforce decided to get clever and go off script. That would be his undoing…and I'm paraphrasing since no minutes were recorded.

> **Wilberforce**: Charles Darwin would like us to assume that our grandfathers are descendants of apes. Are we to presume that our grandmothers are also descended from apes?

The audience chuckled. Someone sitting near Thomas Huxley heard Huxley whisper under his breath, "God hath delivered him unto me." This was oddly ironic or poetic because Huxley was an agnostic. Huxley then got up, mounted the lectern, and said:

> **Huxley**: Bishop Wilberforce jokes about an important subject. Quite frankly, I would rather be descended from humble apes than from a man who goes around the countryside misrepresenting earnest men and women who are in search of truth.

The audience took a moment of pause to digest what Huxley had just said; the lecture hall was deafeningly silent. Slowly those in attendance began to realize that Huxley, and in turn Darwin, were merely in search of truth, that even religious people are, or should be, or were in search of truth, and that Wilberforce would much prefer to deny those religious people truth, he preferred dogma to truth, especially if it came in the form of scientific proof. The tide had turned that very warm night in June 1860 within the great halls of Oxford, Huxley's defense of Darwin's theories grew in strength and numbers while Wilberforce's diminished. Prior to that debate, public opinion was

wavering with many people making fun of Darwin, caricatures of Darwin in the newspapers looking like a half man-half ape. But after that debate, few if any made fun of Darwin.

Although Alfred Wallace, Darwin's rival for the "father of evolution" title, is not given his just due with regard to evolutionary theory, neither in the 1800s nor now, his command of the geographic distribution of animal species on Earth in the 1800s at least secured him the title of "father of biogeography"—a runner-up prize to be sure, but not a booby prize. Wallace was familiar with nearly every animal and most plants on Earth known at the time and where they were located. And, of course, the Wallace Line running through the Malay Archipelago was named after him—perhaps just another runner-up trophy, but still better than only receiving a participation trophy. It would nevertheless be an understatement to say that Alfred Russel Wallace almost became extinct in the lore of biological evolution, but most appropriately his name and contributions have been duly noted by the field of evolution.

Here at the end of this chapter, it's time to ask ourselves anew: Where do we come from? That age-old question. On the one hand, we have the book of Genesis's answer as to the origin and diversity of life on Earth. On the other hand, we have the science-based narrative provided in these pages, a tale that runs from a celestial singularity, to the seemingly random byproducts of atom synthesis, to the apparently random acts of molecular synthesis, to the biological circumstances that ultimately gave rise to humans and all other life on Earth.

I suppose you could also choose a third answer to the question: Ridley Scott's *Prometheus* answer—a version of panspermia—where engineers elsewhere in the universe built mankind and sent an emissary to populate a new planet with humans in his image. Scott's film is terribly silent on where those engineers came from, who built them in the first place. In the next prequel *Alien: Covenant*, we seem to visit the planet where those engineers were based, but it still doesn't answer the question of who built those engineers or where did they come from. I guess we'll have to wait for the next Ridley Scott installment.

When it comes to the riddle of the origins of life, you must pick your poison. Here is my poison, in condensed form, from seconds to minutes to millions then billions of years, from the moment of the Big Bang those 13.8 billion years ago to the moment man appeared on the scene:

- Time Zero: *The Big Bang* from an initial singularity 13.8 billion years ago.
- 1 second in: pure energy forms strings, which form the three main elemental particles (quarks, leptons, bosons) and the four forces of the universe (electromagnetism, strong nuclear force, weak nuclear force, gravity).
- First few seconds in: elemental particles form the three subatomic particles: protons, neutrons, electrons.
- 20 minutes in: first atoms arise from subatomic particles, mostly hydrogen and helium.
- 200,000 years in: photons of light emerge from bosons.
- 300 million years in: stars form.
- 800 million years in: Our Milky Way galaxy takes shape with at least one huge black hole at its center.

From this point onward, focusing on our little neck of the universe, things started to pick up the pace, an inflection point we'll consider from the past time perspective of "years ago" rather than "years in" because, quite frankly, it is just easier to assimilate.

- 4.6 billion years ago: our solar system forms on the Orion arm of the Milky Way, a very obscure lonely outpost.
- 4.5 billion years ago: Earth forms.
- 4 billion years ago: organic molecules develop on Earth because they could, a planet in the sweet spot habitable zone.
- 3.8 billion years ago: *The Big Birth*, bacteria with no nuclei emerge on Earth, a team effort where protein enzymes, RNA and DNA created a cellular world, "seeking shelter from the storm."
- 2.7 billion years ago: cells with nuclei emerge and take charge, the progenitor eukaryote cell, and the first step enabling multicellular life.
- 1.5 billion years ago: the eukaryotic cell lines with nuclei are complete producing four primitive offspring—animal, plant, fungus and protist; while bacteria and archaea go off continuing to do their own thing.
- 1.4 billion years ago: virus emerge on the world scene, basically a spit of genetic material with a fatty cape capsid having jettisoning their cellular bacteria structure.
- 900 million years ago: true multicellular life arises in the sea, such as marine sponges, part of the animal world, yet they have no central brain.
- 600 million years ago: more complex marine life emerges, essentially organisms with a body plan, like the fish.
- 570 million years ago: the Cambrian Explosion with every major animal phyla and many plant phyla appearing in the fossil record.
- 500 million years ago: land plants emerge, mostly mosses.
- 400 million years ago: insects emerge, those annoying pests.
- 390 million years ago: cold-blooded tetrapods conquer land—fish to amphibian to reptile.

- 300 million years ago: dinosaurs emerge from the reptiles.

• 200 million years ago: warm-bloodedness evolves, allowing for a wider range of life on Earth, allowing for a wider distribution on the planet.
• 150 million years ago: primitive birds take flight with modern birds appearing 60 million years ago.
• 100 million years ago: placental mammals arise—horse, zebra, dog, elephant, rhino, rodent, camel, cattle, pig, and primate.
• 63 million years ago: Old-world monkeys evolve (macaque, baboon).
• 15 million years ago: apes arrive on the scene.
• 6 million years ago: the first hominids arise, including fan favorite Lucy 3.2 million years ago.
• 2 million years ago: *Homo habilis* appears, able to use tools but still not walking upright all the time.
• 1.6 million years ago: *Homo erectus* appears, no longer dragging his knuckles on the ground.
• 200,000 years ago: *Homo sapiens* emerge on the scene. That would be us.

Its challenging to put one's arms around a timeline that stretches billions of years, so let's look at it another way, through the lens of a 24-hour clock, beginning with the Big Bang but skipping some of the steps above to tighten the comparison.

• Midnight: *The Big Bang*. Quarks, leptons, bosons to protons, neutrons, electrons, photons to atoms, to stars, to galaxies.
• 4:10 PM: Earth forms in the Goldilocks zone grabbing 92 elements, an atmosphere and a few molecules especially water.
• 6:30 PM: *The Big Birth*. Stage 1 bacteria.
• 10:30 PM: Multicellular marine creature surfaces.
• 11:16 PM: The Cambrian Explosion produces all current animal, plant, fungus and protist phyla.
• 11:17 PM: Amphibians crawl onto land which then give rise to the reptiles.
• 11:30 PM: The age of the dinosaurs.
• 11:40 PM: Warm-bloodedness and mammals take the stage.
• 11:58 PM: The apes bound about.
• Forty seconds before midnight the first hominids evolve.
• Fifteen seconds before midnight, *Homo habilis* starts using tools.
• Thirteen seconds before midnight *Homo erectus* takes center stage.
• Two seconds before midnight the dawn of *Homo sapiens*.
• Half second before midnight our direct ancestor Cro-Magnon or the extant species *Homo sapiens sapiens* rears its matted head
• 3/100th of a second before midnight marks the dawn of recorded human history, about 5,000 years ago.

This is the scientific answer to the question "Where do we come from?

13 THE FUNNY THING ABOUT VIRUSES

That phone rang.

It was clear to me that Deborah had crossed over to the dark side of human infection, had fallen off the curve into a 911 situation. Alan told my mother-in-law he would call emergency dispatch right away and he hung up the phone. Whether a viral cold or the viral upper respiratory infection Deborah had had several weeks earlier, that initial infection had morphed into something a heck of a lot more serious. I was thinking either a viral pneumonia or bacterial pneumonia.

Colds and the flu are viral in origin, and as we all know they usually run a predictable course. A cold is significant for a few days and peters out before the week is up. The flu is worse than a cold, more aches & pains and moans & groans, but also peters out by a week, maybe a few days longer. If you are wondering what the difference between a cold and the flu is, it's really simple. Colds are pretty much confined to the upper respiratory passages, the nose, throat, and sinuses and produce little in the way of a fever or muscle aches & pains. With a cold, you don't feel particularly lousy, you don't have total body systemic symptoms. Unless of course, you're a man who does not suffer lightly anything remotely resembling a cold or the flu, whereas a woman carries on with her job and the kids despite a cold or the flu.

The flu virus is not confined to the upper respiratory passages, able to journey down the airway a titch and turn into a bronchitis, produce an exhausting cough. The flu is often associated with a low-grade or a not-so-low-grade fever, a general tiredness and weakness termed "malaise," and of course there is that total body muscle aches & pains so you can't get any rest symptoms. The flu virus accomplishes these systemic manifestations because, unlike the cold virus, it has the capacity to travel throughout the body, known as a viremia. Overall you feel lousy, that is, you feel "awful, terrible"; not the other definition of "lousy," as in "covered with lice."

Infrequently the flu virus can assault the sanctity of the lower respiratory passages, the lungs, giving rise to a viral pneumonia. About one-third of all instances of pneumonia is caused by the seasonal flu, and which is why the flu is especially harsh on the elderly, a group that already has pre-existing lung issues for no other reason than they're elderly. Add a history of cigarette smoking to the mixture and viral pneumonia can be a real problem.

A viremia is a virus that travels in the blood and is generally, in and of itself, not harmful, just annoying. It is not an "infection" in the blood since viruses need host cells to infect and replicate and that just doesn't happen in the blood. It is more or less viral particles hitching a ride in the blood to distant shores. It is sort of like a bacteremia, bacteria present the blood but not actually infecting the blood; that would be septicemia or sepsis. Viruses can't infect the blood, only travel within it; bacteria can accomplish both tasks.

The thing about a viremia is not that there are viruses in the blood, as it is they have a destination, which ends up giving rise to the systemic symptoms. A common target for many viruses—measles, German measles, chickenpox—is the skin where viral injury to skin blood vessels—not the blood but the blood vessels—causes the classic exanthem rashes. The muscle aches & pains of the flu are similarly the result of the virus targeting muscle and joint tissue. But it is not that simple, the immune system, in its attempt to annihilate the viral buggers, also contributes to the muscle aches & pains, immune mediators and immune cells joining the fray. In their noble effort to kill viruses, they kick up quite an inflammatory storm.

As for that low-grade fever often accompanying the flu, it is usually due to the immune system fighting back. Why? Most simply stated, flu viruses are heat sensitive, finding it a drag to replicate even when the body temperature bumps up a notch or two. The immune system knows this and releases pyrogens to cook those little viral muggers. The fever the immune system instigates also increases blood flow to the battlefield, allowing immune cell troops to enter the fray. In bacterial infection a fever can be due to either the immune system response to the invader— trying to cook those mothers, trying to increase blood flow—or due to the bacteria releasing its own pyrogens designed to aid the culprit's ability to escape from the initial site of infection, most notably through increasing blood flow at the focal infection.

A bacteremia is bacteria that travels in the blood and is often more than annoying—it can be a serious health concern. If the bacterial load in the blood is minor, like from a deep dental cleaning or even from taking a number two where a hemorrhoid, anal fissure or whatnot might let a few bacteria gain entry into the blood, the introduced bacteria passes through the body unnoticed. A small burden of bacteria in the blood usually does not cause harm, and the immune system mops things up.

But if the bacterial load is greater, all bets are off as to whether or not harm will come. With a larger bacterial load in the blood, the body knows it, you know it, because it is accompanied by symptoms such as a fever and chills and the shakes, with an uptick in heart rate and the beginnings of a more impressive fever. If the bacteremia targets something like a heart valve or an artificial joint, we have a real problem on our hands. If the bacteremia morphs into a septicemia, a

171

bacterial infection in the blood—that is, bacteria not just traveling within the blood but *living* in the blood, growing, dividing—well, then, that is a horse of a different color, the symptoms will become so pronounced you'll feel like you've been hit by a Mack truck. Septicemia, or sepsis, is a killer.

When one is bedeviled with a cold or the flu, the course is predictable, usually, and several days into it, you start to feel better. Sometimes the respiratory infection can linger longer, more so with the flu than a cold, since the flu virus is fussier, more tenacious, able to target the bronchus above the lungs, resulting in your classic bronchitis. But even with that, the course of flu infection is generally a steady, reasonable, albeit frustratingly slow, march toward health. It is not unusual for someone to feel some remnants of an upper respiratory infection a few weeks after the main event, but those lingering symptoms, those remnants, are almost always not active infection, the virus having long since been vanquished to the afterlife.

The protracted symptoms of a cold and especially the seasonal flu—those occasions where it lingers for a few beleaguering weeks—is often the result of the aftermath of the earlier infection, the smoldering battlefield where the immune system has won but was put through the wringer. This results in a temporary but annoyingly persistent *inflammation* state—not *infection* state—that slowly clears over a week or two. That frustratingly lingering bronchitic cough after the flu, that's not usually an infection either: it is an inflamed bronchus from a vanquished infection causing inflamed bronchial muscles to contract spasmodically, a subject to be further discussed later on.

Infection and inflammation are two different things, although they are often confused with one another. Infection implies that a pathogen is active, doing its utmost to wreak havoc as it attempts to extend its lineage, whereas inflammation associated with infection is swelling from the milieu in one's internal environment, during and after the infection has been vanquished. Here the inflammation is due to both the damage the pathogen mercilessly rendered plus the necessary disturbance the immune system kindly rendered—the two combined renderings giving rise to "the milieu." It is a battlefield with the strewn bodies of dead pathogens and fighting white cells. The immune response causes a good bit of the swelling and protracted inflammation, too, as it brings in white blood cells and antibodies to do battle.

So, it stands to reason, after the infection has run its destructive course, it is the lingering inflammation that gives rise to such annoyances as a slowly resolving dry and swollen, sore throat, that pesky ever-present bronchitis-of-a-cough from bronchial inflammation, and fullness in the sinuses, the result of failure to equilibrate air pressure. In other words, those enduring, and hardly endearing, symptoms after a cold or flu virus has been quelled are usually from *inflammation*, not active *infection*.

The problem is, many people don't understand this—and frustratingly some doctors, too—don't understand that protracted inflammation after a respiratory infection is par for the course. They exclaim they still "have the bug" when in fact they don't, and so they subsequently seek out antibiotics or prescription stuff to squelch what most likely is a recovering battleground. It is possible to develop a bacterial sinus infection after a viral cold or flu, or to develop a viral or even bacterial pneumonia after the flu, either of which require medical attention. But more times than not, the protracted symptoms are just normal resolution. The way to parse this out would be if there was actually an increase in symptoms, worsening symptoms, a period of improvement followed by a relapse of sorts, or a persistent or new fever. But by and large, the slow resolution after a cold or the flu is the norm.

The last universal ancestor bacterium, LUA, was the first cellular life on Earth with the right stuff that survived long enough and so moved on to form all life on Earth—plant, fungus, protist, animal—yet LUA is no longer with us, itself having been vanquished to the fossil record as its bacterial descendants passed through the next door of evolution.

Oddly, even viruses, which are not much more than bits and pieces of DNA or RNA strands, arose from the comparatively more complex bacteria. Bacteria came first, then viruses. You would think it would have been the other way around, since, as we learned earlier, primitive proteins and primitive RNA gave rise to more complex proteins and more complex RNA, then DNA. From that painted picture it seems logical the virus, which is not much more than strands of RNA or DNA sauntering around with a cape, would have inserted itself into our world before bacteria arrived on the stage, not after. It is believed by those who have studied this bit of viral evolution that, despite Stage 1 bacteria arising 3.8 billion years ago, LUA Stage 2 about 3.5 billion years ago, viruses arose more like 1.5 billion years back. How do virologists know this? Working backwards in time using the protein key that viruses need to break into a cell, based on speed of genetic changes of that protein key, viruses must have shed their cellular confines and decided to go rogue.

"Hey baby...You wanna go viral?"

With that history you'd think viruses arose independently of bacteria, based on the logical argument that RNA and DNA strands existed before the first bacterial cell, so why not viruses arise from those similar strands? Those bits and pieces were not viruses, because they lacked certain viral features like a capsid covering. But one of the most glaring and obvious reason viruses did not arise from floating strands of genetic material is they lacked the ability to insert themselves into a host DNA genome to replicate daughter viruses. The very essence of a virus is its ability to hijack a host cell's genetic machinery in order to extend its lineage, and it couldn't do that without harboring some similarities to the cell it eventually turned on, the bacteria. The death nail to the independent evolution of viruses—as opposed to going through the bacteria cell evolution first— is that the genetic sequence of viruses is all too similar to parts of the genetic sequence of bacteria that lived 3.5 billion years ago.

Our attention now turns to modern-day bacteria, those descendants of LUA who certainly look a little more like LUA than they look like us, despite the fact we share that common ancestor. If you go back far enough in time, the bacteria that cause human infection are indeed our cousins. What can you do? Family, right!

That phone rang.

The first time Alan called Tampa, earlier that New Year's Eve, I wasn't sure what type of upper respiratory infection Deborah had previously had a few weeks earlier, a cold or the flu. I knew likely it was one of those two, but whatever it was she was experiencing now was not a relapse with that same virus, nor was it a chronic smoldering cold that she couldn't shake. Colds and the flu don't do that. Usually. They're generally one-and-done affairs, like the one-night stands many of us dream of having in our twenties but most of us are never lucky enough to actually experience. Listening to Alan describe his wife's condition coupled with her previous respiratory infection easily swayed me away from thinking Deborah's illness was a lingering trifle, the remnants of an extinguished virus, and pushed me more toward believing it was a new, secondary infection, likely bacterial.

The funny thing about viruses—if there is anything funny about them—and this is not always true, but it is mostly true, is that if a person with a normal immune system has a respiratory viral infection, like a cold, like the flu, they can't get another cold until multiple weeks later, at least not from the same cold virus. How the immune system works, how it launches an attack on that viral putz, and how it "remembers" that viral putz—immune memory—confers some protection in a person for at least several months. So, you can't catch the same cold virus after another cold or the same flu virus after another flu, that involves the same virus, unless something is terribly wrong with your immune system. You *can*, however, suffer another bout of cold or flu if the virus causing that new cold or that new flu is different from the one weeks prior. But generally speaking, there is only one cold virus variant or one flu virus variant running roughshod through a given community at a time, so getting two of the same cold viruses or two of the same flu viruses, back-to-back would be freakishly abnormal.

When people say a "bug is going around," meaning a cold or flu virus, it means a specific phenotype of virus is circulating in your community. This phenomenon is especially true with the seasonal flu, as each new fall/winter season brings with it a fairly specific virus phenotype, which, by the way, is what each particular year's flu vaccine is directed at, or attempts to be directed at. Of course, this year, 2020 will be hallmarked by not only the seasonal flu virus but the emergence of the novel coronavirus pandemic flu virus, the actual virus termed SARS-CoV-2 and the illness it caused **co**rona**vi**rus **d**isease, first appearance 20**19** in China, the portmanteau COVID-19.

Flu vaccines, which we will be discussed in *The Second History of Man*, are usually trivalent, meaning they include the epidemiologists' best guess of what three flu phenotypes are most likely going to strike the next flu season. Epidemiologists and virologists who concern themselves with next season's flu vaccine basically make an educated guess about how to configure that vaccine well in advance, so as to give the pharmaceuticals who build the vaccines time to manufacture the concoction.

Cold viruses are so many, so varied, and possessing varying year-to-year sub-phenotypes, that a vaccine would be discouragingly impractical. Virologists would tear their hair out trying to predict what exact phenotype of cold virus to construct a vaccine. As for getting the cold twice in the same season, for the most part, as we go about our little daily circle of life, most of us stick to the same routines—the school scene, the work scene, the grocery store scene—and only occasionally deviate from our pattern, usually on weekends. But unless we board a plane and fly to another community enjoying the wrath of a different cold virus, odds are it should be one cold, and one flu, per season. In other words, if a cold bug is out and about, it's typically just one phenotype. That means if you get a cold from your snotty-nosed kid who brought it home from kindergarten, you're likely not going to get that same cold virus that year as it makes the rounds. And most certainly not within a few weeks. This was my reasoning when first confronted with Deborah's supposed relapse; I knew it wasn't that. It was a horse of a different color.

Bacteria are different sorts of fellas. A person can have a bacterial infection of a specific species, and between the immune system and prescribed antibiotics, beat it into submission. Yet if every last remnant of that bacterial pathogen is not eradicated, and sometimes *even if* every last vagrant is annihilated, that rogue bacteria can reemerge to cause another infection, and cause another infection in the not-too-distant future, and with a vengeance. All things being equal, a virus is fortunately incapable of relapsing very quickly and seldom does so; bacteria, while still not terribly likely to relapse, are at the very least more "relapse-able."

Deborah most certainly had a cold or flu a few weeks earlier. What was going on that New Year's Eve day was not a lingering viral infection, nor was it the lingering inflammation of a viral infection, and it wasn't even a new viral infection. Her new set of symptoms were too pronounced to be viral; they had to be bacterial. Which is where our story now turns: to who and how and when bacteria were discovered, even though their discoverer had no clue that they caused infection.

14 THE CELL AND THE ROMAN EMPIRE

If you can believe it, it took nearly 200 years after the discovery of bacteria in the seventeenth century for physicians to connect bacteria with human infection, a connection made in the 1860s, and designated the germ theory. This discovery is largely attributed to Louis Pasteur in Paris and Robert Koch in Germany. Men of medicine prior to the bacteria/infection connection was made, as it turned out, preferred to dogmatically stick with Galen's miasmas—that is, until a series of discoveries and events in the mid-to-latter half of the 1800s forced them to reconsider. But first, we need to dial back about two hundred years prior the emergence of the germ theory which is where our story now turns.

In 1675, Antonie van Leeuwenhoek became the first human to observe single-celled microorganisms when he viewed protozoans and bacteria, squinting and peering through a sort of crude, dimly lit microscope. He called these microscopic cellular oddities "animalcules," from the Latin, which means "little animals."

Van Leeuwenhoek was born in 1632 in Delft, in what was then called the Dutch Republic. His father, a basket weaver, apparently died when Van Leeuwenhoek was five years old. His mother came from a reasonably well-to-do family of beer brewers. For reasons that are unclear to history, his mother sent little Antonie away when he was ten to live with his uncle, an attorney. My guess is she probably had a new beau and needed to keep her prying-eyes son out of the picture. Think about it: his eyes would become the first to visualize microscopic organisms, so the little observant tike would've been able to "see things" in the Van Leeuwenhoek home without the aid of a microscope, things his mother would not have wanted him to see. The young lad went on to become a bookkeeper apprentice in a drapery shop, and from there he opened his own drapery boutique. You'd think he would have become a beer brewer or a lawyer with that family line, but such was not in the stars for Van Leeuwenhoek.

Being a purveyor of fine linen and cloth for draperies, Van Leeuwenhoek naturally became curious about the quality of the fabrics he was buying. He was not confident that what he could see with his eyes or feel with his hands was the quality he was paying his hard-earned Dutch guilders for. In order to "see" the attributes of the cloth thread character and thread count better than what a simple handheld magnifying glass could provide, Leeuwenhoek invested time in improving the quality of glass lenses. The Dutch Republic was already known as a nation skilled in the making of top-notch lenses, for eyeglasses and for telescopes, but Van Leeuwenhoek provided a significant leap forward for microscopes. He stumbled upon a heating process that greatly improved the purity of the glass. Taking molten silica glass and bending it into lenses was easy, but obtaining pure, high-quality, bent silica glass by which to refract and focus light was an order of a totally different magnitude.

Peering at linen threads through his newly improved magnifying glass ended up being not enough amusement for the inquisitive Van Leeuwenhoek—proof why mother sent him to his uncle—so Antonie began looking at other things under the microscope he had built. It was due to this piqued curiosity that Van Leeuwenhoek, when peering through his high-quality bent glass, first saw microscopic forms of life. He would go on to describe the microscopic world to the human-sized world.

Using his dimly illuminated microscope—lit by reflected candlelight, no doubt, or sunlight through a window—he described the microcosmic world of bacteria, protozoa (later to be called protists), algae, and molds. He also described red blood cells and, for reasons that escape history, sperm, presumably his own. Yet despite Van Leeuwenhoek's seminal work—double entendre intended—and the early conception of the entire field of microbiology, not many scientists were on board with the significance of his findings. To state it another way: these microscopic critters were all fine and dandy to be amused by, but their connection to human infection escaped notice.

Technically, Van Leeuwenhoek did not invent the microscope, because he did not invent the concept of magnifying things down to such an infinitesimal size. Who precisely holds that honor is unknown, but likely either Dutch eyeglass and telescope makers Hans Lippershey and Zacharias Janssen or Cornelis Drebbel, a Dutch engineer and inventor also known for building in 1620 the first functional submarine. Nevertheless, Van Leeuwenhoek perfected the microscope to such an extent that it ushered in the discipline of microbiology. Peering at fine linen was his original objective but founding the field of microbiology was apparently his destiny.

Quite simply, Van Leeuwenhoek's magnifying glass lenses were far superior to anyone else's, making his microscopes the Mercedes-Benz of the microscope world. All the other seventeenth-century microscopes were the equivalent American Motors Company 1970-78 Gremlin or the 1971-80 Ford Pinto, two poor excuses for cars to come out of Detroit. Especially the Pinto was notorious for catching fire—pathetic fuel tank location and design—with your typical rear-end fender bender. Being an astute businessman, Van Leeuwenhoek kept the secret behind his technique of purifying molten glass for as long as he could. There was after all Dutch guilders to be made.

Historically speaking, Van Leeuwenhoek is considered the father of microbiology, yet despite his published observations, coupled with evidence from other scientists elsewhere, the medical world failed miserably to make any connection between the microscopic world of bacteria and protozoans and that of human infection. Medicine held fast to Galen's miasma theory and gave little attention to microbes, leaving that microworld to scientists less interested in medicine and more interested in the wider world of the natural sciences.

With his attention firmly turned from linens to microbes, Van Leeuwenhoek presented his findings to a friend, the Dutch physician Reinier de Graaf, who immediately appreciated the advancement in microscopy and subsequently contacted the Royal Society in London, which at that time was the center of the world for medicine and science. But even London's Royal Society was not overly moved by that upstart Dutchman across the North Sea.

And to be completely honest, despite Van Leeuwenhoek title as "father of microbiology," the discovery of the *cell*—viewing bacteria, protozoans with his microscope—also does not belong to him and, in fact, it predates him. It was actually Robert Hooke, the British naturalist, architect, and polymath, who first proposed the concept of the cell being the basis of life. Born in 1635 and a contemporary of Isaac Newton—they famously did not get along—Hooke discovered around 1665, using his own crude microscope, that cork from the cork tree was composed of tiny enclosed structures he named "cells." Although the English word "cell" is derived from the Latin *celere*, meaning "to hide, conceal," that was not what Hooke was thinking of when he coined his phrase. He was rather reminded of those barren, tiny rooms that monks and nuns lived in at monasteries, also called "cells." Ten years or so after Hooke described cork cells in 1665, Van Leeuwenhoek started his pioneering microscopic observations of his sperm and what not.

As for the feud between Robert Hooke (1635-1703) and Isaac Newton (1642-1727), well, that was something else altogether. They were both polymaths and both were strange fellows, Newton no doubt the stranger of the two. Hooke was known to be an irascible sort, often prone to publicly attacking other scientists and getting in people's faces, whereas Newton was very much the retiring type, your classic hyperfunctioning autistic savant who got in no one's face. In fact, Newton made no eye contact with anyone, presumably precisely because he had high-functioning autism and eye contact was nerve wracking. I am sure whenever those two would meet, strolling about the campuses of Oxford or Cambridge, it never went well—Hooke likely descending into yelling at Newton, while Newton's pained expression would glance elsewhere, avoiding all eye contact with his tormentor and looking for a tree or shrub to hide behind.

The biggest issue between the two of them had to do with gravity. After Newton published his *Principia* in 1687, in which he defines the three laws of motion and defines the math of gravity, Hooke claimed—rather publicly—that he had given Newton the idea of gravity. That did not go over well with, well, everyone, but especially not with Newton. The science public tended to favor

Newton over Hooke at every turn, which over the course of history has resulted in the brilliance of Robert Hooke being mostly obfuscated.

There was only one known portrait of Robert Hooke, and Newton supposedly had it destroyed. When Newton became head of the Royal Society, he buried Hooke's papers in the back of some drawer. Eventually, the Hooke journals would be rediscovered, long after his death, and his genius posthumously recognized. It should also be pointed out that nowhere in his journals does Hooke mention gravity. If you're to be so bold as to claim you gave the idea of gravity to Isaac Newton—a huge idea, you might agree—your claim holds absolutely no water if you never even bothered to once jot down a few notes about it with your quill.

As for Van Leeuwenhoek, over the course of the next twenty-five years, he became the world's authority on infinitesimally tiny microbe life. Other scientists with formal training as naturalists were rather chagrined, perturbed, downright irritated that Van Leeuwenhoek had backed his way into the field of microscopy from a dusty drapery shop in Delft. He became the authority simply because no one could figure out how to reproduce his microscopes. Those who wished to contend with him had to pay the price of admission to the microscopic world by buying one of his very, very expensive handcrafted scopes. Not many in history have accomplished so much in a science specialty with so little education, whether formal or self-developed, going from being a purveyor of fine linen to a purveyor of fine and, as it turned out, not-so-fine microbes.

It is perhaps time to mention the intersection where the duty to share a medical discovery meets the desire to make bucketloads of money off that medical discovery. Van Leeuwenhoek went down the money road, whereas someone like the American virologist Jonas Salk did not. When Salk developed the first polio vaccine in 1955, he had the option to license the "recipe" and make gobs of American greenbacks. He chose not to, walking away from a lot of bank, instead giving the world his polio vaccine for free. It was vintage Salk, selfless.

To be fair to Van Leeuwenhoek, the duty, the honor, to share medical knowledge was not exactly a fleshed-out, entrenched part of medicine in the 1600s. There was no unwritten nor universally accepted code that scientific discoveries are to be shared, for free, with other science nerds. And, technically, Van Leeuwenhoek was not even a formal scientist. Some things in science, especially in medicine, are freely shared. Other things in science, like the highly secretive Microsoft and Mac operating system codes, are kept classified, unpublished, and generally mystical, and thus have made huge fortunes for some very smart people. As far as Microsoft's Bill Gates, he has ended up taking the splendid road, sharing his fortune to make the world a better place than how he found it. As for Steve Jobs of Apple, he died too young to allow his philanthropic side to blossom.

I have a special fondness for Gates and his wife, Melinda, as they spend huge chunks of their money in the health care arena, attempting to put a leash on such vicious infections as malaria and Ebola. This is not to mention Melinda's work with advancing the rights of women worldwide, a segment of society that has long suffered in a male-dominated world.

The secret to Van Leeuwenhoek's glass recipe had to do with the fact that, when starting from a blob of molten glass, rather than shape it into a lens and let it cool, he instead used a glass bead to draw out the purest glass, repeating the drawing-out process over and over. With each reheating and with each drawing off, impurities were driven away, leaving behind ever purer glass. Van Leeuwenhoek died in 1723 in his birthplace, Delft, at the age of ninety, which is quite a long life even by today's standards, let alone at the dawn of the eighteenth century.

As an aside, the famous Dutch painter Johannes Vermeer (1632-1675) was also from Delft, born the same year as Van Leeuwenhoek. They were pals. Vermeer was not a well-recognized painter in his life time, only having been rediscovered several centuries after his death as one of the better Gold Age of Dutch painters, a period spanning 1575 to 1725, their lead-off hitter of course Rembrandt van Rijn, a list that includes Frans Hals and Jan Steen. In the nineteenth century Vermeer was redeemed art critics, especially appreciating how frightfully skilled he was with light, a skill that some say the optics guy Van Leeuwenhoek might have played a role in. Everyone, or nearly everyone is familiar with Rembrandt's *The Night Watch* (1642), one of the best known paintings of all-time; yet Vermeer's *Girl with a Pearl Earring* (1665) is possibly equally recognizable to people even if the composition is not always known to be a Vermeer. It is one of my favorite paintings.

While Van Leeuwenhoek purified it, glass as a material dates back to antiquity, likely discovered by accident when early human troglodytes, sitting around a campfire, noticed sand near the fire had transformed when it became molten. Sand, being mostly silica rock, which is silicon dioxide (SiO_2), when heated and then cooled, morphs into a glassy appearance, the molecules taking on a new transparent arrangement. A proud Neanderthal probably had such a shiny, translucent trinket in his satchel, along with a piece of flint, some obsidian, and maybe an eagle feather, pulling out that piece of crude glass to peer through, for no particular reason other than pure amazement. It's really no different than contemporary folks gazing at the rainbow from a prism hanging in their backyard. Not all glass is formed from pure silicon dioxide, however. Other compounds in the admixture can include sodium oxide (Na_2O) and calcium oxide (CaO), the latter called lime.

In addition to appreciating the sand-to-glass transformation, Stone Age man was also very curious about naturally occurring obsidian, which is black volcanic glass—neither clear nor translucent but proudly shiny—formed from molten igneous rock, the main rock of Earth's crust. Apparently, stone was not the only material man used to make tools during the Stone Age: obsidian glass was turned into some pretty sharp and nifty knives to cut and stab with and spear points to stick into things—a wild boar, a bear, a gazelle, a neighbor who had the audacity to make a pass at your mate.

Beyond noticing molten sand turning into translucent glass by the campfire and finding pieces of obsidian lying around at the base of a volcano, the first true manufacturing of clear silica glass—that is, the deliberate enterprise of heating sand into glass—is believed to date to the earliest civilizations in Mesopotamia and Egypt, around 3000 BC.

Exactly how does translucent glass occur from something as solid and opaque as sand? The answer, as it turns out, is crystal clear. If you look at silicon dioxide sand in its solid state—the type of sand in the neighbor's sandbox that the cat likes to leave a little surprise—you'll see the molecules are arranged in one direction. When that silicon dioxide sand is heated into a molten state, the molecules take on a totally different orientation to each other and are pushed further apart in the lattice, making the molten silica transparent to light waves. What is important to the story of making glass is that once the molten silica cools, rather than the molecules reverting back to their solid, opaque compact configuration, the molecules instead maintain that new transparent albeit solid expanded crystal configuration. The same is sort of true of graphite, which does not allow light to pass through—but, when it is heated and compressed in Earth's crust for a few billion years, the carbon atoms reorient into a very tight, very strong tetrahedron shape that not only allows the light to pass through but is also just about the hardest mineral on our planet, next to wurtzite boron nitride. That is why diamond burrs are diamond burrs: they're generally stronger than the material they're cutting. And a few billion years for diamonds to lie deep in earth's crust before being dug up so a man can get on one knee and propose "gamete," from, as mentioned earlier, the Greek "to take a wife," is why diamonds are a girl's best friend. The diamond, not the husband, not by a long shot.

The next bit of the tale is a hard pill to swallow, even with a glass of water. Technically, glass is the liquid form of sand, although some say it exists in a mysterious Twilight Zone region between solid and liquid. As odd as it might seem to conceptualize and accept, because it defies our eyes and our sensibilities, both the clear glass you drink water out of and the water in it are liquids, the latter (water) being a true liquid and the former (glass) being a supercooled liquid or an amorphous solid, depending on your bent. That is, is the glass half full or half empty?

If you want visual proof of glass behaving as a liquid, you need look no further than a very old home with very old windows or the windows of a very old cathedral. We're talking glass a few hundred years old, like the original glass in some of the homes in Deerfield, Massachusetts, home to Deerfield Academy and one of the quaintest colonial villages from the 1600s you'll ever see. Carefully inspect the thickness of a 300-year-old glass pane at its base and at its top, and you'll see that the base flares—it is thicker. Old glass panes are thicker at the base not because they were made that way but because there has been some gravity settling. The window seems solid because the liquid silica molecules have organized themselves—not as well as crystal glass, which is more solid, but they are organized, nonetheless. Over the expanse of time and gravity, old glass will settle.

What's the difference between standard glass and crystal, besides the degree to which their molecules are organized? And the degree to which their costs are different. They're both made from silica, but crystal has some of the heavy metal lead (Pb) evenly mixed into the lattice structure, making the glass more durable, more solid feeling, and, oddly, more crystalline. What do the lead atoms do? Just like adding salt to water stabilizes the water molecules raising the boiling point, adding lead to silica makes the silica molecules more stable. In the US, if the lead content is 1 percent or higher, glass is allowed to be sold as crystal. In Europe the standards are more exacting, are higher, and the lead content must be more like 10 percent, which translates into an even more durable, more crystalline lattice.

And that is the path our narrative is now headed down: one of crystallization. The transition from Galen's miasma theory, which held that bad vapors cause human infection, to the reality that tiny microbes lie at the core of human infection was a concept that began to *crystallize* during the 1500s, even before the microscope arrived. As we have found out though, this wasn't a case of instant crystallization, from a state of miasma to one of microbes, like the instant crystallization that Kurt Vonnegut's fictional ice-nine substance experiences. Despite the invention of the microscope, a device that revealed the world of bacteria and protozoans, getting to the realization that microbes were at the heart of infection—including heartworm infection that targets pet dogs—was a long process that took some intermediate, well-intentioned but flawed, concepts to push the science world in the right direction. Are we crystal clear about that?

In 1546, Girolamo Fracastoro, an Italian physician, mathematician, and would-be poet, proposed that infections, especially epidemic infections like the Plague and the flu, were caused by tiny particles he called "spores." Now, this next bit of his theory is important, because he held that those "spores" were not alive, not like how we know today that microscopic pathogens are alive, such as fungal spores or the bacterial spores of anthrax. Fracastoro's theory postulated that these tiny spores were "chemical" spores, sort of like dust or pollen, that entered into a person and transmitted a signal that unleashed a cascade of symptoms that we'd label an infection.

What is interesting about Fracastoro's proposal, other than that it was wrong, but clearly on the right road, is that a chemical spore of sorts is precisely what does cause seasonal allergies and hay fever, conditions that share a similar symptom palette to colds and the flu. It is probably this realization—that something like pollen caused sneezing—that led Fracastoro to postulate colds and flus were also chemical-initiated events. Who could blame him; it made perfect sense. Although wrong, Fracastoro's chemical spore theory was at least a radical departure from Galen's miasma theory—a scientific move in the right direction. Fracastoro developed his theory 100 years before Antonie van Leeuwenhoek perfected the microscope and 300 years before Louis Pasteur established germ theory.

Fracastoro's theory was apparently the first theory to propose an alternative to Galen's miasma theory, or at least the first such theory that some members of the medical community actually took seriously. Galen's dogma was entrenched, a shadow cast unnecessarily long for over 1,500 years. An alternative explanation of infection to counter miasmas was an unusual foothold at the time, especially since there was little science for Fracastoro to base his postulate on.

While the "spore" theory perhaps had a foothold, it turned out it didn't have a very secure position on that cliff as the sixteenth century drew to a close and the seventeenth century dawned upon the scientific world—a world, at least in Europe, that was being ravaged by the sins of syphilis. Syphilis at that time was a new infection hitting Europeans like a blitzkrieg, but since conventional wisdom about human infection was still deeply entrenched in bad vapors, there was little that was done, or could be done, to successfully combat it. It boggles the mind to ponder that seventeenth-century physicians knew syphilis was spread by sex but still believed that the transfer was through the air or some such mishegoss.

The state of seventeenth-century medical knowledge in Europe is especially frustrating in hindsight, since other scientific disciplines were shaking off their Dark Ages shackles and entering the light of the Renaissance. The 1600s saw major strides in astronomy, with the observations of Galileo Galilei and Nicolaus Copernicus as probers of the celestial sky, and Hans Lippershey, Isaac Newton, and James Gregory refining the telescope, which helped to rearrange the cosmos from an Earth-centric to sun-centric solar system. The seventeenth century also saw advances in other

"NICE TRY, BILLY, BUT DOGMA IS NOT A PUPPY'S MOTHER!"

science disciplines, including in microbiology, with Antonie Van Leeuwenhoek's perfection of the microscope; in engineering and physics, with the invention of the steam turbine in 1629 by Giovanni Branca and Newton's publishing of the *Principia*, which contained the laws of motion and gravity; in mathematics, as Newton and Gottfried Wilhelm Leibniz invented calculus as well as a calculating machine; in mechanics, with the development of the pocket watch in 1675 by Christiaan Huygens; and, most importantly, in food sciences, with Dom Pérignon's 1670 invention of champagne. Yet the medical world lagged behind its sisters of science, holding fast to the old ways, to the old world, to the ancients, to miasmas, to dogmas.

Ancient Rome and ancient Greece, and for that matter ancient Persia and ancient Egypt, were all societies as pointed out earlier, tolerant of scientists, so it is really rather strange to consider that during the first one and a half millennium AD, human inquiry not only did not pick up speed but was entirely snuffed out. As the Roman Empire fell and the Christian Church rose to beat down heretics, humanism and science came to a crashing halt. As did advancements in medicine.

We earlier talked about the rise of the Roman world, how a guy like Galen, of Greek ancestry living in Turkey, was considered a Roman citizen. Wars do that. But while Rome was a far-reaching realm, phrases like the "Roman Empire," "Roman Republic," and "Holy Roman Empire" are bandied about as if they are various names for the same nation, when they are not.

Each name describes a different iteration of Rome. Julius Caesar was actually not an emperor of the Roman Empire, since that nomenclature had not yet been established. Caesar lived during the time of the Roman *Republic*, with its Roman Senate, and after he had squashed the enemies of Rome, there was nothing left for him to squash but the Roman Republic itself, which he gladly did. Caesar ended up appointing himself dictator for life in 45 BC while standing over the smoldering ashes of the Roman Republic, a republic he helped destroy through the Roman Civil War of 49–45 BC, sometimes called Caesar's Civil War. The Roman Republic had existed for 500 years prior to that war, surviving the expansion of Rome from a small enclave to the world's biggest power.

The Roman Republic was the first iteration of Rome, which was ruled by noblemen who called themselves the Roman Senate. The Senate would appoint a Roman king, usually from their small group of insiders, to lead the Senate, or a king from the royal lineage, since Rome had started with a single ruling king. As Rome expanded, from 500 BC onward, the Roman Senate decided to dispense with appointing just one of them as king and instead decided to collectively rule Rome themselves—not a single leader, no king, just the collective authority of a bunch of rich noblemen. That is, until Julius Caesar took over as "dictator for life" in 45 BC. How did that happen? How did Caesar eviscerate a form of governance that had endured for nearly 500 years?

Around 60 BC, Caesar and two other military buddies, Pompey and Crassus, formed the First Triumvirate in order to rule over the ever expanding Roman world, based on the argument that the Senate was too slow to react to threats and needs, especially to troubles in far-flung reaches of the realm. A smaller band of three overlords, it was believed—or those three believed— could more effectively manage the vast Roman domain. This Triumvirate essentially began the process of minimizing the power of the senators. Caesar especially gained fame and power when his legions conquered Gaul, modern-day France, a huge feather in his cap, during the Gallic Wars of 58–50 BC, which was no small undertaking. Crassus soon died, leaving Caesar and Pompey to stare over their shoulders at each other, each coveting absolute power, or so we assume. Eventually a civil war broke out that saw Caesar, who believed in a single autocratic leader—the dictator model—pitted against Pompey, who still held true to the Roman Senate model, the Roman Republic, but with a single leader.

The legions of Caesar and his general, Mark Antony, fought against the Roman Senate and its generals, Pompey, Brutus, and Scipio. I'm sure you see where this is going. Early military victories favored Pompey, but Caesar and his horde soldiered on, eventually defeating Pompey, who then fled to Egypt—where he was assassinated as soon as he stepped foot on Egyptian soil.

After Pompey's assassination, Brutus then surrendered to Caesar, who, rather than killing him, granted Brutus amnesty. Big mistake for Caesar, and an error he soon enough paid for with his life. Caesar went on to defeat Scipio, after which he appointed himself dictator for life, a dictatorship that, as I intimated, turned out to be short-lived.

"Et tu, Brute?," that famous line from William Shakespeare's 1599 play *The Tragedy of Julius Caesar*, is the lament Caesar expresses when he realizes Brutus has turned on him. Brutus was a close friend and protégé of Caesar, but he aligned himself with Pompey during their civil war. Nevertheless, his life was spared by Caesar. It is unclear why Caesar granted his onetime protégé amnesty, but maybe that is precisely why: Brutus was his protégé.

Within a year into Caesar's dictator-for-life rule, in 44 BC on March 15, a day also known as the Ides of March in the Roman calendar, a group of Roman senators attacked Caesar in a conspiracy to assassinate him. According to Plutarch, the Greek essayist, historian, and Roman citizen, in his account of Caesar's murder, written 100 years after the fact, Senators Cimber and Casca attacked Caesar on the portico of the Theater of Pompey—a rather ironic location in which to die, since Caesar had had Pompey murdered. Bad karma is a bitch. Caesar, who was a great warrior, fought back, and Cimber and Casca were no match for him. Casca cried out for help, upon which the remaining conspirators who had been hiding, including Brutus, appeared on the Portico of Pompey.

"ET TU BRUTE."

Julius Caesar initially fought off even more attackers, and fought them off convincingly, even as they piled on. But it is said when Caesar saw Brutus among those conspirators, his heart stopped. If Caesar didn't actually say, "Et tu, Brute?," if that's just Shakespeare's poetic license or the musings of Plutarch, he certainly must have *thought* it, especially after having granted Brutus mercy a year or two previously. Some historians, like Plutarch, claim that Caesar, upon catching sight of Brutus, simply laid down his sword and resigned himself to his fate.

The Roman Empire sprang out of the ashes of Caesar's dictatorship, which had risen from the cinders of the Roman Republic, which was begot of the debris of the first Roman monarchy. If the Roman Senate had thought, had hoped, that by dispensing with Julius Caesar things would return to the good old days of the Roman Republic, they were sadly mistaken. What was instead ushered in was a greatly weakened Senate and a powerful single leader—this leader bestowed with the title of "Caesar." It's all rather odd, really, that the plotters who plotted to kill Caesar, successfully assassinated him, and went through a short-lived rebirth of the Republic that was then snuffed out, ultimately favored the very dictator-for-life model Caesar had been assassinated for. And not only that, but they somehow allowed the next and subsequent supreme leaders of Rome to take on his name, calling themselves "Caesar"—morphing from a family name to become synonymous with "emperor" or "dictator." I guess Caesar's assassins felt bad for knocking him off, so as a tribute to him, they picked the title "Caesar."

To the surprise and chagrin of Julius Caesar's family, especially his third wife, Calpurnia, the bulk of his estate was left to his little-known eighteen-year-old great-nephew Octavian, the grandson of Caesar's sister Julia, but who had also been "adopted" by Caesar, whatever that means? Octavian, later to be known as Augustus Caesar, would go on to avenge Caesar's assassination. Families do that. Especially Sicilians, according to *The Godfather* (1972).

The young Octavian, along with Mark Antony and Marcus Lepidus, formed the Second Triumvirate of Rome, spelling defeat for Caesar's assassins and the Roman Senate. The Triumvirate then divided the Roman Republic among themselves into the Roman Empire, which is the first time the name the "Roman Empire" was used in place of the "Roman Republic." But the Second Triumvirate ended up fighting each other—egos do that—in another civil war of sorts to the bitter end. Triumvirates didn't seem to last very long in ancient Rome. Probably wouldn't last long anywhere, except for years ago when my wife and two daughters formed a female triumvirate to lord over me. Even our male dogs joined their cause. The two cats, well, they could've cared less. Cats do that, not care. Lepidus ended up being driven into exile, and after Octavian defeated Antony at the Battle of Actium in Greece, Octavian became the last man standing—the sole ruler of Rome, the first to take on the name Caesar, not as a surname but as a title. His great-uncle Julius Caesar would've been so proud.

As for Mark Antony, we all know how that ended. His story is yet another tragedy brought to life by Shakespeare, in 1607 … or was it Richard Burton and Elizabeth Taylor in 1963? Shakespeare's account of the story, the play *Antony and Cleopatra*, is again from the writings of Plutarch. When Antony rose to power and was selected to control the Middle East and Egypt by the Second Triumvirate, Cleopatra set her sights on the dashing young general, Antony fourteen years older born in 83 BC and the fair-maiden Cleopatra born 69 BC. Cleopatra apparently was that beautiful and that seductive, and she was also the Ptolemaic pharaoh of Egypt, born of a Greek lineage that ruled over Egypt. Not that Egyptian women were not beautiful, but Cleopatra was a Greek beauty living in Egypt. Wooed by stories of Cleopatra's beguiling and provocative ways, Antony first met Cleopatra in 41 BC after sending for her, the fell in love and subsequently made Cleopatra queen not merely over all of Egypt but, in a galactically foolhardy decision, over huge swaths of Greece—a decision that enraged Romans and Greeks and especially Octavian. Cleopatra might have been of Greek ancestry, but her realm was only supposed to be Egypt. (Romans knew that to rule over a conquered foreign land it was always a good idea to keep a local lord in power, if only as a puppet. It made for controlling the masses that much easier).

Not surprisingly, another Triumvirate civil war reared its ugly head. After Antony lost to Octavian's forces at the Battle of Actium along the western shores of the Ionian Sea, Antony and Cleopatra fled to Egypt to seek refuge. With Octavian bearing down, Antony was in retreat. For reasons that are not clear to history, Cleopatra spread the rumor that she had committed suicide, likely to throw off Octavian so she could hide. It backfired. Antony, upon hearing that Cleopatra had committed suicide, committed suicide himself, on August 1 of 30 BC, by a self-inflicted stab wound, presumably to the gut. Incidentally, in medieval Japan, a disgraced samurai would in a similar way commit *seppuku*, a self-inflicted ritual suicide by disembowelment. No thanks.

Antony did not die straightaway. In his final moments of anguish, he was made aware that Cleopatra was indeed alive, and that her suicide was a ruse of war. Antony requested and Octavian granted that he be brought to Cleopatra, who was now a prisoner, locked up in a mausoleum, of all things. It was in that mausoleum, in Cleopatra's arms, that Mark Antony died. You can't make this stuff up—but if you're William Shakespeare, you can write a romantic tragedy about it. Or, better yet, if you're Hollywood film director Joseph L. Mankiewicz, you can write the 1963 film screenplay *Cleopatra* about it. And, really, who else could have starred in that movie but Elizabeth Taylor and Richard Burton, another real-life set of star-crossed lovers?

After Antony died, Cleopatra was brought to the palace where Octavian, who was just twenty-seven years old to her thirty-nine, had stationed his victorious forces. Cleopatra attempted to move Octavian to either lust or pity, but it was in vain. Octavian was not moved. Seeing she had no other way out, Cleopatra committed suicide—for real this time. On August 12 of 30 BC, she enticed a very poisonous Egyptian cobra to bite one of her presumably perfectly shaped breasts.

With Mark Antony and Marcus Lepidus out of the picture, Octavian, the nephew of Julius Caesar, became the first true emperor of the Roman Empire. He subsequently honored his uncle by changing his name to Augustus Caesar and held command of the Roman Empire from 27 BC until his death in 14 AD. And it is from that point on that the emperors of Rome were called Caesars. Tiberius Caesar succeeded his adopted father, Augustus, ruling until 37 AD, making Tiberius the Caesar during the time of the crucifixion of Jesus of Nazareth. Pontius Pilate, the Roman prefect of Judea, was the general who presided over the trial of Jesus and his subsequent crucifixion, the exact date of which is in dispute but widely accepted to be April 3 of 33 AD.

A little unknown fact is that Cleopatra issued four children, the first, Caesarion was born in 47 BC and his father? Julius Caesar apparently, and perhaps the only biologic child of Caesar. The other three children of Cleopatra were Alexander Helios, Cleopatra Selene and Ptolemy Philadelphus, all children of Mark Antony.

Recall from a much earlier discussion that, back then, years were not dated, not counted, and calendars were way off-kilter. Named days, like the Ides of March, were necessary to provide an idea of the middle of a month. It was that Roman monk Exiguus who, many years later, around 500 AD, attempted to calculate the birth of Jesus, setting it at 0 BC or 0 AD, depending on which direction you're looking at it from. But Exiguus, as we know, was off.

Using modern calculation methods with references from historical records, it's been determined that Jesus was likely born in 4 BC, his ministry began around 27 AD, and his crucifixion was in 33 AD, although some scholars place the crucifixion anywhere between 30 and 36 AD. Additionally, Jesus was likely born in the summer of 4 BC, not on December 25, as is widely held. That summer was the season of Jesus's birth is based on the scriptures, such as Luke 2:8, which describes shepherds "living out of doors and keeping watch in the night over their flocks," which could only have been in the summer. So why, then, was the birthdate of Jesus moved from summer to December 25? Why indeed?

The early Christian Church sometime in the fourth century AD, nearly 300 years after Christ, moved Christ's birthday to December 25 to coincide with the winter solstice and the pagan Roman festival that marked the birthday of the unconquered sun. Why did Christian leaders do this? Likely to get more Roman pagan followers to associate the winter solstice with Jesus so as to eventually convert those Roman pagans to Christianity. If you're a Roman in 336 AD, and you witness the rise of Christianity all around you, and as the winter solstice arrives, you look at your pagan gods, you look at this rising religion of Jesus, the Son of God , you look at your diminishing pagan gods, you look again at the meteoric rise of Jesus and his resurrection, you probably would take a moment of pause to consider switching religions. Pretty soon that winter solstice celebration of your pagan gods is supplanted by that Jesus guy and his birthday, which you had no way of knowing was moved from summer to winter in order to accomplish precisely that—to convert you.

The Roman Empire began to crumble in the late third century AD, when it had grown so vast that Emperor Diocletian (the title "Caesar" had fallen out of fashion by then) decided to split the Roman realm into four regions—an arrangement that did not last. It was never going to last. The result of Diocletian's actions was one man left standing, General Constantine, the winner of yet another Roman civil war who became the sole ruler over the Roman Empire. In a move that angered many within the Roman world, especially Romans in Rome, in 330 AD Constantine moved the Roman capital from Rome to "his city," Constantinople, modern-day Istanbul. If Rome's hold over the expanse of Europe was already showing cracks, it's easy to imagine that the process was accelerated when Constantine moved the capital out of Europe, across the Bosporus, to Turkey. It's a move that sort of makes sense geographically—if you believe your empire is invincible—placing your capital smack-dab in the middle between Europe and Asia. Constantine also became the first Christian Roman Emperor when he adopted Christianity as the official religion of the Roman empire and formally dispensed with the old pagan gods, dealing those old gods their final death knell. Moving the capital to Constantinople also allowed the newly minted Christian Constantine to be nearer to the Holy Land.

The Western Roman Empire, based in Rome, plodded on until it started to peter out and pretty much disappear around 480 AD. It was unable to withstand the invasion of northern Germanic marauding tribes, mostly the Vandals (from which is derived vandalism), a demise that was accelerated with the rise of the Christian Church as the ultimate authority in the West. One wonders: If the capital of Rome had remained in Rome, would the strength of the empire have endured longer? Would history have been different? Would the power and authority of Rome been a tonic to balance the excesses of the Church?

The Eastern Roman Empire, based in Constantinople and also known as the Byzantine Empire, fared significantly better, surviving until around 1453 AD, at which time the Ottoman Turks came crashing down upon it. And that was that for the last vestige of the Roman Empire. It was also the end, as previously reviewed, of the Silk Road and spice trade, the Ottomans shutting down trade across their lands, which then ignited the Age of Discovery, a sea route to the Far East.

And we can't forget about the Holy Roman Empire, can we—yet another, fresher iteration. After the division of Rome by Constantine and the fall of the Western Roman Empire in 480 AD, Europe was left without any empirical Roman presence for several centuries. But on December 25 in the year 800, Pope Leo III revisited the concept of a Western Roman Empire by crowning Charlemagne as its emperor. The Pope cleverly labeled this new God-blessed empire the "Holy" Roman Empire.

Charlemagne was the great Frankish king who united warring Germanic tribes, the Franks being of Germanic origin just like the Vandals. Charlemagne's Holy Roman Empire, through a series of emperor-to-emperor successions, lasted until 1806—that's 1,000 years, which isn't too shabby. It remained undefeated until the French statesman Napoleon Bonaparte came along to conquer the regions of Prussia in Germany, Austria, and Hungary, thus dissolving the last remnants of Charlemagne's legacy, the last of the Holy Roman emperors, which was by that time known as the House of Habsburg.

The Habsburg dynasty was a great European royal family that was in power for most of the second millennium AD, controlling vast swaths of what is now modern-day Germany, Austria, and Hungary. Their reign began around the eleventh century and survived into the twentieth. So prominent was their house that a Habsburg held the throne of the Holy Roman Empire from 1438 until 1740. And their sway wasn't just limited to Europe. Habsburg Spain, which existed from 1516 until 1700, controlled most of the newly colonized Americas as well as controlled trade with the East Indies.

Habsburgs did not marry for love but rather for the acquisition of land, titles, wealth, and power, resulting in the greater balance of European nations having a Habsburg at the helm. The Habsburg clan find their origin in the eleventh century with Bishop Strasbourg and his brother-in-law Count Radbot. The name Habsburg is derived from Habichtsburg (Hawk's Castle), their ancestral home. From there the dynasty grew to become essentially the ruling monarchy of Austria by the fifteenth century. From 1438 onward, the Holy Roman Emperor was a Habsburg, until, that is, the extinction of the male line in 1740, after which the next Holy Roman Emperor was not of the true Habsburg bloodline. The next Holy ruler was Charles VII who was from of the House of Wittelsbach. It was not until twenty-five years later, in 1765, when Joseph II assumed the title, that another bloodline Habsburg was back in power, although from the House of Habsburg-Lorraine, not pure Habsburg.

It is unclear why the male lineage of the Habsburg line became extinct in 1740, but it *is* well known that the family were prolific inbreeders. The frequent consanguineous marriages perhaps resulted in a not-very-durable genetic line. The Habsburg inbreeding occurred under the mistaken belief that it would consolidate power. What it provided for instead was a bucketload of birth defects. So prominent were some of the facial defects—not to mention defects of the internal organs, like the heart—that they went by specific names, such as the "Habsburg nose" and the "Habsburg lip" and especially the "Habsburg Jaw," so prominent of a protruding lower jaw it looked like, well, Jaws played by Richard Kiel in the James Bond movie *Moonraker* 1979, with Roger Moore as 007.

Habsburg Spain (1516–1700) materialized when the House of Bourbon, a French Habsburg dynastic family, seized the country. In addition to the monarchies of Austria, Germany, Hungary, France, and Spain having Habsburg bloodlines, so did those of Italy, Portugal, England, Ireland, and Mexico.

Like all good things that must come to a bitter end, so did the Habsburg dynasty, with its various lines eventually falling like dominoes, until just its core of Austria and Hungary remained. The Austrian Habsburgs kept the Holy Roman Empire legacy going. But when Napoleon defeated Austria in 1806 during the Napoleonic Wars, he dissolved the Holy Roman Empire once and for all—an empire that had existed since the time of Charlemagne in 800 AD. Napoleon did, however, keep the Habsburg name intact so that he could adopt it—I guess he admired the family lineage, or, more precisely, admired their sweeping *power* lineage, not their *birth defect* lineage.

Napoleon, who valued a powerful family over a powerful religion, corralled the capacity of the Holy Roman Empire's religious dispensations, limiting its spiritual influence. He certainly did not dispense with the wealth and prestige of the now much diminished Habsburg dynasty, that Bonaparte preserved under his sphere. The last Habsburg to rule Austria and Hungary was Charles I of Austria, who abdicated control over Austria in 1918 to the evolving Austrian government as World War I was coming to an end, and then ceded control of Hungary to its own evolving government in 1921.

It's pretty impressive that papal authority—the mere waving of the pope's hand and maybe his signature or seal—was all that was needed in 800 AD to declare Charlemagne emperor of the resurrected Roman Empire, an empire that had been dead for 300 years. That was a lot of power vested in the pope. It's sort of like a storyline from the classic *Indiana Jones* franchise, that Steven Spielberg and George Lucas collaboration starring another great, Harrison Ford. In the first of the series, *Raiders of the Lost Ark* (1981), the "ark" is the Ark of the Covenant which we briefly visited with in the first few pages of this narrative, a wooden chest adorned with gold containing the original stone tablets bearing the Ten Commandments given by the hand of God to Moses on Mount Sinai. It is said that any army that carries the Ark of the Covenant before them would be invincible.

If a pope could possess that much power in 800 AD to create out of thin air the Holy Roman Empire, imagine what he and his fellow popes, those who came before and those after, could do to stifle science, art, music, humanism? The rise of the Church seeded the rise of the Dark Ages.

And speaking of tiny seeds of infection, we now turn to the lead up to the germ theory.

After Antonie van Leeuwenhoek introduced the microscope in the 1600s, it still took another 200 years to throw aside the cloak of Galen and begin to consider other options as to the causes of infectious diseases. More specifically, to consider that those tiny microscopic pathogens that Van Leeuwenhoek saw under his lens sat at the heart of infection. Dogma dies hard.

In the nineteenth century, with the brilliant work of Louis Pasteur and his germ theory of infection, the concept of microscopic critters causing human infection began to take hold. Several researchers had made headway prior to Pasteur, including Agostino Bassi, an Italian scientist specializing in the study of insects, a discipline known as entomology. One of my childhood chums, Marc Epstein, is an internationally recognized entomologist, having started collecting butterflies and bugs when he was in elementary school, naming the genus and species in Latin, while us less intelligent knuckleheads in third grade were still struggling with English. There are thought to be 1.3 million different species of insects on Earth, and Marc probably knows most of them. That's a lot of biodiversity and, as we will come to appreciate, that's one of the reasons why some microbes, like malaria and yellow fever, selected the insect as their vector model to spread infection.

Agostino Bassi was born in 1773 in the Lombardy region of Italy, a northern province that borders Switzerland. Bassi's father, oddly for a biologist, wanted his son to become a civil servant. The young Bassi initially obeyed his father, but eventually his desires drove him into science, much to the chagrin of his dad. It's not so dissimilar to Robert Darwin's attempt to steer his son Charles Darwin into the ministry. Like Darwin, it is fortunate that Bassi followed his own path. His research published in 1835 after twenty-five years of investigation demonstrated that a microscopic parasitic organism from a fungus family causes a particular disease in silkworms known as muscardine. This was perhaps the first time in history that an infection was linked to a microscopic living organism, a microbe, the individual cells of which could not be seen with the naked eye but only with a microscope. Oddly, it was a plant biologist who made this first microbe-infection connection and not a physician. It stood to reason, however, that if a silkworm could become infected with a microorganism, couldn't a person become infected with a microscopic pathogen in much the same way?

Between Girolamo Fracastoro's 1546 "spore" theory that held tiny nonliving chemical spores sat at the heart of infection and Bassi's 1835 research demonstrating a microscopic parasitic organism caused infection in silkworms—and thanks largely to Van Leeuwenhoek's invention of the microscope—science was able to start drilling ever deeper down into the bedrock of what truly causes human infection. Science was almost there. Almost. There was one more hurdle to clear— an offshoot of that simplistic miasma theory. It had to do with the belief that royalty, nobility, and the wealthy did not get sick with infectious illnesses nearly as often as did the poor and the common. This is an important observation to dissect, because it was a belief that supported and was supported by Galen's miasma theory, and it required some much needed disassembling.

"Worse than a cold. It's a common cold."

Simply put, royalty and the wealthy and emperors and priests did not get sick as often as did the poor, also called "the great unwashed"—a phrase often attributed to the eighteenth-century Irish English writer Edmund Burke. The did not get the "common cold." Only commoners got the common cold, he said sarcastically. Nor for that matter did generals in the field of battle get sick as often as did their lowly foot soldiers. Infection, it appeared, was not in the stars for the well-to-do, it was not their providence. Medical scholars pontificated that infection was instead the providence of the poor because, quite simply, the poor smelled funny.

Take foot soldiers, for instance. Then, just as now, many soldiers were conscripted from the poor, and as the pages of this book will explore, and in subsequent volumes, soldiers often died in droves during war—but not from pointy spears, not from the sharp edge of swords, not from zinging bullets, not from explosive devices, not from any sort of weapon. Rather, soldiers died during war in droves with so much frequency and fervor due to infection.

To give you a disheartening example, during the American Civil War of 1861–65, of the approximately 370,000 Union soldiers who died, 60 percent (or 225,000) died from disease, mostly infection, some from starvation, and others from exposure to the cold. Of the approximately 250,000 Confederate soldiers who died, 66 percent (or 165,000) died from infection, starvation, and exposure. Sobering thought. Up until the latter half of the twentieth century, which saw the advent of antibiotics, as well as an appreciation for hygiene, food, and protection from the elements, a soldier was more likely to die from an infection than a battle wound.

And what was with this infection is the providence of the poor malarkey? If a tiny indiscriminate organism caused infection, didn't it stand to reason that the wealthy should be struck down with as much zeal as the destitute? After all, a mindless microscopic critter would lack the ability to tell the difference between a rich man and a poor man. Because the "great unwashed" suffered infections at an alarmingly higher rate than the "washed," pre-mid nineteenth-century scientists held on to the belief that "miasma" or "bad air" was at the root of infection, since it was quite an observable fact, one that could scarcely go unnoticed by the olfactory apparatus, that the wretched poor, *les misérables*, lived in abject smelly squalor. Likewise, soldiers in their trenches, usually conscripted from the ranks of the poor, starving and trying to stay warm, reeking of all manner of odorous vapors, died from infection. According to Galen, smelly unwashed people died because smelly vapors permeated their world. Sometimes simplicity can be elegant, but in the case of Galen's miasmas permeating the medical landscape for 1,500 years—well, this simplicity was anything but elegant. Can you say "myopic"?

This belief that infection was the providence of the poor and the poor stunk and they lived in putrid squalor and that was that, held sway pretty much up until the end of the nineteenth century. This lofty view of life—where bad things, smelly things, miasmas, happen to poor people but not the well-heeled—has been a repeated, recurring prejudice. It is true that nearly every common medical malady is more frequent in poor segments of society, but not because it is their "providence," but because it is their circumstances. The poor can ill afford to be healthy, to live healthily; they can only afford to be ill.

Ignaz Semmelweis was a Hungarian physician born in 1818 near Budapest. After originally studying law, he switched gears into medicine and then specialized in obstetrics. He received his medical training in Vienna and remained in Austria throughout his medical career. Semmelweis is best known for his study of puerperal fever, also called postpartum infection or childbed infection.

Today, in large part due to Semmelweis, we now know that puerperal fever is a bacterial infection involving the reproductive organs—mainly the uterus—of women who have just given birth, with puerperal fever occurring following the births of both healthy and stillborn children. In Semmelweis's day, puerperal fever was a very fatal and much too prevalent infection, the cause of which was unknown at the time. There was no treatment—antibiotics had yet to be discovered. Either the stricken woman died or did not die according to whatever curve she ended up on, and little could be done in the meantime. It's basically the black hole event horizon of infection, where the person either escapes in the medical equivalent of a burst of Hawking radiation and recovers or spirals toward death. Hippocrates termed it the tipping point, well at least that was one of the definitions of the tipping point.

Semmelweis noticed that women who gave birth at home with a midwife had a much lower incidence of puerperal fever than women who gave birth in a hospital with an obstetrician. That's right—home births in the nineteenth century had a lower incidence of postpartum infection, maybe not a lower complication rate for other childbirth problems, but certainly for puerperal fever. In identifying the key to this puzzle, Semmelweis correctly postulated that hospital doctors and nurses, whose hands were contaminated with bacteria from other patients, including microbe-infested corpses in the morgue, would then bring their non-sterile bacteria-soaked hands into the birthing room and directly into the birth canal. Semmelweis recommended doctors and nurses simply wash their hands prior to delivering a baby (exam gloves did not exist back then), believing that the incidence of puerperal fever would decline as a result. Semmelweis was correct in his thesis, yet it took a disturbingly long, lethal time for the rest of the medical community to catch on. Dogma.

Other sources of bacterial contamination could, can, and still cause puerperal fever, but the light bulb finally went on as to the connection between hands contaminated with bacteria and infection. In this setting it most certainly was not "bad air," it was "bad hands." But unfortunately, as we know, old habits die hard, as do patients—which is perhaps no better illustrated than in nineteenth-century London, where our story now turns.

Cholera is an infection caused by the bacteria *Vibrio cholera*, a bug that mostly attacks the small intestine, which, by the way, begins right after the stomach and is twenty-two feet long, as compared to the large intestine, which is about five feet long. The large intestine ends with your colon, which then ends with your rectum, which then ends with your anus—not Uranus, the butt of many jokes.

Uranus, whose name is inevitably pronounced wrong, I'll get to that, is the seventh planet from the sun. It was given an unfortunate name by its discoverer, William Herschel, in 1781, a curious name that draws out giggles in third-grade science class. All the planets, except Uranus, from Mercury and Venus through Mars, Jupiter, Saturn, Neptune, and even Pluto, are named for an equivalent pagan god as according to the Roman Latin name, not the Greek name for the same god. In case you were wondering, in Greek mythology Mercury is Hermes, Venus is Aphrodite, Mars is Ares, Jupiter is Zeus, Saturn is Cronus, Neptune is Poseidon, and Pluto is Hades. Earth, on the other hand is … Earth. Nothing special here. "Earth" means "dirt," again nothing special with our home planet's name, no gods, no goddess.

We can't know why Herschel, a great and learned astronomer, chose the Greek Uranus over the Latin Caelus, thus condemning that planet for eternity to be the butt of English-language jokes. To be fair to Herschel, the original spelling was Ouranos, pronounced in four syllables with the accent on the second syllable— "O-ur-a-nus"—which isn't a bad name and sounds nothing like "your anus." But when the leading "O" was dropped and its pronunciation morphed into the three-syllable "Ur-a-nus," it was bad luck for that planet and bad luck for Herschel.

Cholera is characterized by profuse watery diarrhea and vomiting—really profuse, like a garden hose—and if one is infirm or malnourished or destitute or, I'll just say it, poor, it often isn't possible to keep up with the fluid losses, and so such people die in droves. As it turns out, death from cholera is not from a fulminant infection as one might imagine, not from end organ failure, like with septic shock; rather, death from cholera is usually due to dehydration shock and electrolyte imbalance, the latter causing fatal heart arrhythmias. When you have profuse diarrhea and lose a lot of fluids via "uranus," you don't just lose volume making one shocky, you also lose much needed, much treasured sodium and potassium electrolytes in your body. These ions are essential to handle all sorts of electrical and transport activity across membranes, and in the particular context of cholera, heart electrical conductivity.

A person who is reasonably healthy and has a robust immune system to counter an attack of cholera will still suffer mightily but likely recover. When it comes to people who are impoverished and as a result malnourished, as well as the elderly and infants, death from profuse watery diarrhea of cholera is most assuredly assured. It is not their "providence"; it is their circumstances. We know today that cholera is transmitted via contaminated water and contaminated seafood—it is not caused by bad air. But that fact was certainly not always appreciated, as our story will soon delve into. Cholera is a waterborne infection, and water can become defiled with cholera from the excrement of an infected person. Cholera can also be transmitted via the fecal-oral route, which basically implies a person with cholera fails to adequately cleanse their hands and then handles the food we eat. Well, you do the math.

In everyday conversation, we toss around the word "shock" as if all its meanings in medicine are the same. They are not. You've got your classic septic shock, volume loss hypotensive shock, allergic shock, and something termed cardiogenic shock—all subjects that will be further defined in *The Second History of Man*, the next volume in this series, so a brief explanation at this juncture is needed.

Septic shock, once called blood poisoning, is shock from an overwhelming, fulminant bacterial infection within the blood, where those bacteria not only defile the blood but hammer the end organs the blood services. The kidneys, liver, heart, lungs, and brain are the main organs that get the brunt of a rogue bacteria septic attack. The drop in blood pressure further compromises the infected end organs.

Hypotensive shock is different: it's caused by a decrease in the volume of blood circulating in a body, the result of such maladies as rapid blood loss due to trauma, internal bleeding from an ulcer, surgical bleeding, and violent fulminant vomiting and diarrhea, such as in our soon-to-be-told cholera story.

Shock can also be allergic in origin, like anaphylactic shock, which is the result of a severe allergic reaction to something like a bee sting, peanut butter, and certain drugs such as penicillin, but hopefully not Godiva chocolate. This form of shock releases a cascade of the immune system's own chemical transmitters that have the unintended effect of overexaggerating their normal duties.

Cardiogenic shock is when the heart can't pump enough to sustain life, most commonly seen in a severe heart attack. Also seen in process around the heart and lungs that compress the heart, like a shrink-wrap.

Returning to cholera, today, intravenous (IV) fluids for the volume loss shock and any number of antibiotics treat cholera. But not back in the John Snow's day—John Snow the nineteenth-century physician, not Jon Snow, born Aegon Targaryen, the dashing and moody heir to the Iron Throne. Our John Snow lived in the 1850s, in the Soho district of London's West End, not SoHo in New York. There are two famous Sohos.

It is thought that the name "Soho" derives from the 1685 battle cry of James Scott, 1st Duke of Monmouth, as he fought and lost his skirmish with troops loyal to King James II. In distinction, New York's SoHo—with a capital "H"—is a contraction of "South of Houston" and refers to a chic shopping district south of Houston Street in the City That Never Sleeps. Some dispute this and say that New York's SoHo name is a tip of the hat to London's Soho, and the whole "south of Houston" thingamajig is just a happy coincidence. And why not? New York is named after old York in England, and New Hampshire is named after old Hampshire in England. It all checks out. Thousands and thousands of American towns have British namesakes. I wouldn't even begin to Birmingham bore you with Manchester examples or Portland towns in the US that have UK equivalents.

But returning to our Soho story in London, John Snow was born in York, England, in 1813. He was the son of a coal yard worker and lived in a poor, unsanitary neighborhood. He climbed himself out of that decay—I dare say that "miasma" or better yet, that "providence"—and by age fourteen was apprenticing medicine in Newcastle upon Tyne under the surgeon-apothecary William Hardcastle. An apothecary back then was a practitioner working with herbs and other medicinal ingredients, perhaps the forerunners of modern-day pharmacists. It was while apprenticing in Newcastle that Snow first encountered a cholera epidemic, and the presentation and gloom of that widespread death so vexed him, so seared his brain, that years later it would embolden him to formulate his singularly unique hypothesis on the infectious transmission of cholera.

After apprenticing with Hardcastle, Snow journeyed to London, where he received his formal medical training at the Hunterian School of Medicine. The Hunterian School of Medicine, like the British Hunterian Society, is named after the internationally famous Scottish surgeon John Hunter, born in 1728. Hunter was an early advocate of empirical medical observation—that is, he was a fan of incorporating Frances Bacon's scientific method, devised a century earlier, into the study and practice of the medical arts—thus bringing medicine kicking and screaming out of the dogmatic Dark Ages, out from under the cloak of Galen, out from under the thumb of religious authority, into the brightness that was the Enlightenment. Hunter created one of those pivot points that helped advance medicine toward the modern practice we know today.

What is interesting about Snow, in addition to stocking up what would become significant fuel for Louis Pasteur's germ theory, is that he was also one of the first doctors to use anesthesia. Snow used both ether and chloroform in calculated dosages to help facilitate surgical and obstetric procedures. He clinically determined correct doses that were neither too small, such that patients would still scream and writhe in agony, nor doses of anesthesia so large that patients would not scream or writhe in agony because they were dead from the anesthesia.

Various chemists around the 1830s, including Justus von Liebig of Germany and Eugène Soubeiran of France, independently synthesized chloroform. The Scottish surgeon and obstetrician James Young Simpson was the first to use chloroform as an anesthetic on a person—or I should say, on people—on November 4, 1847, in what was a not-very-surgical setting: the parlor of his home.

Apparently, Simpson chose to first experiment with chloroform not on an informed patient about to go under the scalpel but rather on several guests at a dinner party he held in his home. We can only imagine the amusement that ensued, the lively times that were had that night, the laughter and bemusement, while the dinner guests were in various stages of drug-induced frivolity, the good doctor no doubt taking notes all the while. We can only assume that what happened in the Simpson residence that night stayed in the Simpson residence.

But to the doctor's credit, several days later, true to form as well as true to chloroform, he then used the anesthetic gas on an actual surgical patient. However, Simpson was not the first doctor in the world to use an inhaled anesthetic, just the first to use chloroform. That distinction belongs to one of two Americans across the pond, but it wasn't chloroform that they used. It was rather ether that was the first anesthetic gas used in the world.

The anesthetic version of ether is actually diethyl ether. It was likely first synthesized in the eighth century by that ancient Persian chemist we've already encountered, Jābir ibn Hayyān, or Geber, although he undoubtedly didn't know at the time of its mind-altering effects.

Before Simpson used chloroform in Britain in 1847, and depending upon who you side with, the first use of ether as a surgical anesthetic was either in 1842 by Georgia physician Crawford Long (as American Southerners claim) or in 1846 by Massachusetts dentist William Morton (according to Northerners). Long did not publish his historical event; Morton did. The debate between North and South over "who was first to use ether" raged for a long time, although the disagreement is not cited as one of the causes for the American Civil War that would unfold less than twenty years later.

The following is a frightening thought for you to consider the next time you go under the knife. Well, maybe not frightening, but a tad perplexing. The exact mechanism of all inhalation anesthetic gases is not known even to this day. From chloroform and ether, both of which have been relegated to the shelf of history, to the more effective inhalation anesthetic gases such as desflurane, sevoflurane, and isoflurane that supplanted them, scientists do not know exactly how they work. That is indeed a sobering thought. Physicians prescribe all sorts of medicines— antibiotics and heart pills and blood pressure pills and intestinal pills, pills after pills after pills— and, for the most part, the mechanism of action is known, sometimes exquisitely so. But this is not the case with general anesthetic gases.

Despite the millions of general anesthetic surgeries that are performed monthly across the globe, this mystery remains. I suspect you might find it a tad troubling to hear—especially in this day and age—that when you go under the knife for your hernia repair, your anesthesiologist doesn't know any better than you how the inhalation gases they're using work. They know the correct dosages, they know how to monitor your heart and breathing, they know that inhaled anesthetic gases induce unconsciousness, block pain sensation, inhibit sight and sound, and generally do a good job at insentience, but your anesthesiologist doesn't know *how* they do all that.

The current prevailing opinion on the most likely mechanism of action for inhalation gases comes from the realm of physical chemistry—not chemical reaction chemistry, but physical chemistry. That is, it explains the "how" of it, not the "where" of it.

When you and I think of chemistry (other than the chemistry between two young people in love), we think of a chemical reaction where things bind and unbind and rebind with each other. You know, sodium (Na) plus chloride (Cl) gives us salt (NaCl). The rust on that rake you left outside all winter is nothing more than iron (Fe) combining with molecular oxygen (O_2) plus water (H_2O) to arrive at iron oxide (Fe_2O_3). Burning fuel, like a log in a fire, is hydrocarbons plus oxygen busted up in the hydrocarbon chain. In classic chemical reactions, atoms are either bonding or unbonding, they're either sharing valence electrons or not.

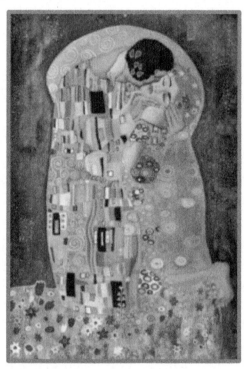

Physical chemistry is slightly different. Especially two molecules do not need to pair bond—do not need to share valence electrons—in order to influence each other, in order to have a reaction or impact upon one another. According to physical chemistry, when it comes to the world of inhalation anesthetic gases, it is believed that the interaction of the gas with its target site in the brain is like a key-lock paradigm. A key-lock might sound like a chemical reaction, but it's not—it is the "how" not the "where" of it. This term describes two compounds accommodating each other without actually binding to each other, meaning no shared electrons and all that sort of rot.

Physical chemistry is sort of like two people in an embrace, like in Gustav Klimt's painting *The Kiss* (1907), in which a man and woman are in a state of, shall we say, physical chemistry. They are physically attached to each other, but there is no chemical reaction between the man and the woman. There is certainly *chemistry* between Klimt's lovers, but that is a different type of chemistry, one of the heart and mind.

Pharmacologists who concern themselves with how anesthetic gases work are pretty sure it must come down to a physical chemical reaction, a key-lock type accommodation, partially based on the obvious and critical fact that gas anesthesia reverses pretty quickly. That is, you turn off the anesthesia machine and patients wake up. But the exact nature of that physical chemistry is not known. They just have no particular idea as to how and exactly where these gases work on the brain. Or, as we have seen is often the case, they actually have several ideas but are unsure which is where the toad of truth sits.

One theory posits that when the gas diffuses into the brain, it physically changes the configuration of fat around neural brain cells without binding to the fat, thereby changing the neural cell itself, which in turn changes its conduction of electrical impulses. This change of neural fat supposedly induces unconsciousness. Another theory suggests that the gas molecule inserts

"I'll be performing the operation, and this is the anesthesiologist."

itself into a neural cubbyhole, like the above-mentioned key-lock, and blocks brain electrical conduction without actually undergoing a binding process. In the end, what we know today is not much, just several untestable hypotheses of which one, all, or none could be true. Pharmacologists thus continue their search for the holy grail mechanism of inhalation gases.

Inhaled anesthetic gases are of course to be distinguished from the local injected, infiltrated anesthetics, like the Novocain a dentist uses, whose mechanism of action is well understood. Any "-caine" compound acts along pain nerve fibers, chemically binding to specific nerve axon sites, thus preventing nerve electrical conduction along that axon. Local anesthetics bind through a classic chemical reaction whereby the local anesthetic molecule forms a chemical bond with a nerve gate, specifically sodium ion channels, thus preventing an influx of sodium ions into the nerve axon. A daisy-chain of these sodium ion gates along a nerve is what propagates the electrical conduction. In order for an impulse to conduct at a gate, sodium ions at that gate must rush inside, termed "depolarization," and after the impulse has passed, the ions are returned to the outside, termed "repolarization." Local anesthetics bind at the gate—gate after gate after gate—preventing the depolarization. This is a reversible chemical bond, which is why a local anesthetics wears off. But before it loses effectiveness, that chemical bond halts the flow of electric currents along the axon. If the axon is a pain fiber and it is blocked, the pain signal can't reach the spinal cord and, in turn, can't reach the brain.

Everyone knows who Sigmund Freud is, or should know who he is: the founder of psychoanalysis—all that id, ego, and superego rigmarole. But Freud actually cut his teeth as a neurologist, starting with the physical aspects of the nervous system before delving into the psychological mysteries of the mind. As a neurologist, Freud is credited with helping to discover local anesthetics through his experiments with the first one ever messed around with, the poster child of all the "-caines"—cocaine.

While today our dentists use procaine, better known as Novocain, and us scalpel-wielding surgeons use lidocaine, the first "-caine" ever used by doctors was indeed cocaine. Long before Freud stepped into the picture, people knew sniffing that particular white powdery substance lit a person up. Cocaine, which comes from the South American coca plant, was known to the ancient Incans, who used it in religious ceremonies. Incans have been chewing the coca leaves since 3000 BC, not to get "high" but to increase heart rate and breathing rate in order to handle life high in the Andes. I guess you might be able to quip that the Incans got high, high in the Andes. But the appreciation of its local anesthetic properties belongs to Freud and to the French chemist Angelo Mariani, whose research actually predates Freud's.

Freud was actually interested in cocaine less as a local anesthetic and more as a cure for opium and morphine addiction. Believing that cocaine was *not* addictive, Freud reasoned that morphine addicts could get that monkey off their back by segueing onto cocaine, and then presumably stopping the cocaine. Freud's most famous test subject was his colleague and friend Dr. Ernst von Fleischl-Marxow, who had become hopelessly addicted to morphine after undergoing a thumb amputation that resulted in chronic, relentless, excruciating pain. The exact history is unclear, but apparently the poor doctor Fleischl-Marxow did not achieve Freud's desired result, becoming not substance free but instead addicted to both morphine and then cocaine monkeys. Fleischl-Marxow died at a young age, presumably from a drug overdose.

One of the most famous surgeons in the world, as the twentieth century dawned, was William Halsted, who worked out of Johns Hopkins Hospital in Baltimore. Johns Hopkins Hospital along with the medical school and nursing school were all three the result of the generous bequeath in Johns Hopkins' estate. Hopkins made his fortune in the railroad industry and the medical school that bears his name was one of the first in the U.S. to admit women. Admitting women into Johns Hopkins School of Medicine was not in the bequeath, but rather, was due to a singularly generous and wealthy woman Mary Garrett who leverage her money wisely. She was the daughter of John Garrett who also made his fortune in railroad, and when the trustees of Hopkins estate came up short building the hospital, she kicked in a huge chunk of change to finish the job, with one caveat, so long as the medical school admitted women. Mary Garrett also founded one of the best private colleges in America, Bryn Mawr.

When Johns Hopkins Hospital opened its doors in 1889, the estate demanded the best doctors be recruited to lead the way. They were known as the "Big Four" like the Four Horsemen: William Halsted surgery, William Osler medicine, William Welch pathology and Atwood Kelly obstetrics-gynecology. Halsted pioneered such operations as the radical mastectomy for breast cancer, perfected the cholecystectomy (gallbladder removal), and is largely credited with establishing the residency training program format that is still used today all over the world. His side kick William Osler founded the traditional medical school format and was a first-class internist. Between Osler and Halsted modern medical school and residency training programs still largely follow their plan.

Halsted became famously, or rather infamously, addicted to cocaine through his experiments to test its use as an anesthetic for surgeries. He had heard about the substance from the European medical world and began to test it out on himself—apparently snorting in addition to injecting it—and, voilà, he was, not surprisingly, addicted.

Halsted's cocaine addiction eventually landed him in the Butler Sanatorium in Providence, Rhode Island, in order to "dry out." At the sanatorium, Halsted was placed on morphine, which cured him of his cocaine addiction—by replacing it with an opiate addiction. Opium, like cocaine from the coca plant of South America, is naturally derived, coming from the poppy plant native to the Middle East and eastern horn of the Mediterranean; Afghanistan comes to mind. Dr. Halsted subsequently and euphemistically "enjoyed" a lifelong dependency on morphine, never able to get that monkey off his back. It was apparently an addiction that did not interfere with his surgical prowess. Imagine a surgeon or any physician at all today being allowed to continue to practice with an active morphine or cocaine addiction. Unlikely.

It goes without saying, but I'll point it out anyway: over in Europe, the remedy for morphine addiction was to switch to cocaine, largely thanks to Freud; and, in America, the remedy for cocaine addiction was to switch to morphine. Go figure.

The German eye surgeon Karl Koller, a contemporary of Freud, was one of the first to employ cocaine as a topical anesthetic in a surgical setting, using it to numb eyeballs before operating on them. Now, I don't know about you, but having a numb eyeball would not be enough to keep me still if a surgeon was about to slice and dice my eyeball. I'd need something else, like a few shots of whiskey or the milk of the poppy. Milk of poppy, known since antiquity, is liquid opium.

Why, you might ask, was cocaine never used as an injectable local anesthetic like Novocain but only as a topical one? Or, perhaps the question that's more relevant today is: Why do drug users snort cocaine but refrain from injecting it? When it's taken topically, like on the surface of the eyeball or snuffed into the nose, cocaine is absorbed slowly and locally, so its intended numbing effect is local; likewise, when snorted, its systemic exhilaration is also a slow-ish, crescendo high. That is, snorting blow results in a paced-out rise of cocaine in the blood, not an immediate all-in surge.

If cocaine is injected into or under the skin, or, worse, into a vein or artery, control over cocaine's far-reaching effects is lost, especially upon the heart and brain. In particular, cocaine's ability to target the heart has dire consequences if it were used as an injectable. Rather than achieve local numbness or brain elation and euphoria, you'll achieve a fairly instant coma and death. Cocaine's cousins—including lidocaine and Novocain as well as many other "-caines"—possess much better safety profiles, meaning that when they're injected under the skin (but crucially not in an artery or vein), they cause less, if any, heart and brain stimulation while providing profound local numbness to the regional pain fibers. No local anesthetic is ever intentionally injected into a blood vessel, except in instances where it's being used as a medical treatment, such as for certain heart arrhythmias, specifically ventricular tachycardia V-tach and ventricular fibrillation V-fib, especially the latter which is incompatible with life.

Now that we've wandered into the realm of Freud, we can't really ignore the id, ego, and superego elephant in the room any longer. That is to say, we should briefly cover all that mumbo jumbo that was the foundation of the Freudian psychic apparatus.

Id, Ego, Superego

The "id," according to Freud, involves the inborn, instinctual drives we possess—you know, sex, anger, pleasure, aggression … did I mention sex?

The "ego" is that part of the psychic apparatus that seeks out, or does not seek out, realistic ways in which to satisfy the id. In simple terms, if you want sex, you take a shower, preen your hair, wear some nice clothes—basically ego yourself up—and go to a bar for a hookup. The ego is also what takes the hit when it is unable to satisfy the id—when that poor schlep of a guy dressed in a leisure suit driving a '73 Cutlass Supreme spends all night at the bar trying and failing miserably to score.

The "superego" describes when the id and the ego are forced to deal with social mores, or family values, or religious values. You know, when the devil is on one shoulder telling you to go out and get some, which would be the id and the ego, while the superego angel is on the other shoulder, telling you that you're behaving irresponsibly. Or, in the words of Austin Powers, "Oh, behave!"

Now that we've run through Freud land, it is high time, not cocaine high, that we return to John Snow and the story of cholera. During the 1840s throughout the 1850s, several disquieting outbreaks of cholera rendered the great unwashed in London very ill—so ill that these epidemics were accompanied by a higher than usual mortality rate. London had grown tremendously during the First Industrial Revolution, a ramping up of industry largely considered to have begun around 1760 and leveling off in the 1840s. From about 1760 to 1830, this phenomenon was mostly a British thing, and subsequently Great Britain became the world leader in manufacturing, culminating in a country soaked in wealth, covered with belching coal chimneys used to smelt iron, plus power looms and spinning jennies for textile production dotting the landscape.

The power loom was a mechanized loom for making cloth, designed by Edmund Cartwright in 1784, the back and forth weaving to make cloth. Our troglodyte ancestors invented string/rope around 40,000 years ago and weaving around 28,000 years back. But it wasn't until a few thousand years ago that a contraption to hold the vertical thread—the warp—while weaving a horizontal thread—the weft—started to dot civilizations. Still requiring the nimble hands of the weaver, making cloth was a slow, tedious process up until Cartwright's power loom invention. The power loom used an engine, usually steam, to move the warp back and forth while the weft was mechanized, guided back and forth without the need for total hands-on. By the 1850s, a quarter of a million power looms were in operation in Britain. "Loom" is from Old English *geloma*, meaning "utensil, tool, machine."

The spinning jenny, invented by James Hargreaves in 1764, a Lancashire man, took up to eight single threads and spun them into a single thread with the necessary overlaps to make thread strong. So even though string or thread was invented 40,000 years earlier, Hargreaves machine made the best thread and made it the fastest. The origin of the name "jenny" as in "spinning jenny" is unclear. Some say "jenny" was derived from "engine," while others say that a woman named Jenny knocked over the protype and the machine still worked sideways, leading Hargreaves to realize that the design could be more compact.

The Second Industrial Revolution, which started in the late 1800s and extended well into the 1900s, was not limited to Britain. Nearly every European country and even those upstart Americans were getting in on the action, sending coal smoke bellowing into the air, making more iron, more steel, more machines, more tools.

Betwixt the two industrial periods, swarms of people migrated into the cities. Everywhere you looked, you saw heaps of tools and machine parts, extensive railroad systems, telegraph systems, attempts at water and sewer lines, and even electric power grids started popping up. This industrial growth led young farm boys and fair farm maidens to leave the countryside in search of work in large cities where the pay was better, resulting in overcrowding, poor sanitation, and hordes of the great unwashed. Across Europe but especially in Britain, the industrial cities could not keep pace with the throngs caused by their own industrial growth.

During the middle the eighteenth century, at the time John Snow was making his medical rounds, the city of London had been popping at the seams for quite some time. Its population had swelled beyond capacity, and its streets and gutters were like its people: unwashed. There was nothing subtle about it—London smelled. It was your basic sanitation nightmare: a network of muddy roads and back alleyways covered in excrement, defilement caused by a bourgeoning multitude outpacing civic abilities to rein in the squalor. It created one of those "perfect storm" situations—a storm in which Londoners were dying in droves from cholera.

The Great Stink, which will be covered a tad later, was an event in the sweltering summer of 1858 where London's River Thames became so saturated with sewer defilement it quit flowing. The situation was so bad that sewage backed up into the city's gutters. Not even the gentlewomen strolling The Strand could hide the stench with Eau de Cologne du Grand Cordon, a popular perfume released in 1857.

Most homes in London had no sewer connection but just foul cesspools located in their cellars, just beneath the main-floor floorboards. I truly can't imagine that. Consider the scene that summer of 1858: Cholera-infected Londoners skulking about their squalid homes, wrapped in flea-infested blankets. The stench of a cesspool emanating from between the creaky floorboards. The muggy summer heat inescapable. Their friends, family members, and neighbors dropping like flies from cholera. You'd like to think that those conditions of cholera and defiled streets would give physicians cause for concern, that such unsanitary surroundings would quickly call into question miasmas as the source of cholera. Sure, the air smelled—but so did the ground. An important point to realize at this juncture is the obvious disproof of the miasma theory that the Great Stink offered: as the inescapable stench wafted throughout London, would not everyone have keeled over from infection? But not everyone did go down in flames. Cholera seemed to confine itself to certain districts while the smell of stench blanketed the entire city of London. There was a huge disconnect in the miasma theory.

In an attempt to mitigate the vicious spread of cholera overtaking the city, Snow developed an alternative theory of infection, which placed him at odds with civic administrators who were trying to deal with the outbreaks in their own way, as well as with his fellow physicians, who held fast to Galen's miasma theory. Snow mulled over other potential avenues by which cholera and like-minded infections could spread. He did not believe in the bad air theory, nor did he accept the "providence of the poor" as a medical premise. Instead, Snow believed infections like cholera and the Plague were from some yet unknown contagion and transmitted by routes other than bad air. While there were still hurdles to overcome, Snow was taking the first steps toward solving the riddle of the London cholera epidemic, and once solved, its answer would, just like an infection, spread across the English Channel and into France, giving fuel to Louis Pasteur for his germ theory.

The summer of 1854, a few years before the Great Stink, is what really propelled Snow into action. That fateful summer brought with it the profoundly infamous 1854 Broad Street cholera outbreak. From summer into the fall, people in London's Soho neighborhood were dying in droves from cholera. Everyone was at a loss as to what to do, except dig more graves. But not John Snow, he was not at a loss. Through meticulous epidemiologic documentation, Snow mapped out the cases of cholera and discovered, to everyone's surprise and disbelief—with an emphasis on "disbelief," especially of the medical establishment—that the cholera outbreak, rather than arising from bad air wafting along the cobbled and muddy backstreets of London, was emanating from one public water pump on Broad Street. Houses in much of London were practically glued together, didn't have fresh water at the turn of a nozzle, and certainly those hovels did not have toilets to flush urine, feces, and vomitus down the sewer to faraway places, places that we don't talk about at cocktail parties.

Today if you need some water to make your coffee, to boil some pasta, or to take a bath, it arrives at the turn of a tap. When you go wee-wee or boom-boom in the toilet, it flushes away to places we rarely think about. Not true in Soho in 1854. If mother needed fresh water, she'd dispatch one of her grubby children to trudge to the nearest public water pump for that bucket of water.

Snow was zeroing in on the bedrock of the epidemic—that Broad Street water pump he believed was quite mysteriously providing contaminated water chock-full of cholera. All he knew for sure was that the water source had become corrupted somehow. Snow was particularly confident in his theory because neighboring districts that obviously shared the same air that blew across London—the same miasmas—did not share the same water pump nor subsequently shared the same epidemic of cholera. If the source of infection was miasmas, the demarcation around the area of cholera infection in Soho would not have been so sharp, so abrupt. Rather, it would have blurred into neighboring communities and involved other wards. But it didn't. The outbreak was situated around the Broad Street water pump.

In standard bureaucratic fashion, Snow's recommendation to shut down the Broad Street pump was met with skepticism and resistance. Typical public officials. After all, those officials knew about the providence of the wretched poor and their smelly fate, and everyone in the medical community knew that the 1,600-year-old proclamations of an ancient physician couldn't be disputed, so why would they listen to this Snow guy? What must have been especially frustrating for Snow was that he had quantitative proof that neighborhoods adjoining Soho were not experiencing the same level of cholera deaths. I guess everyone else figured those corrupted miasma vapors knew the invisible boundaries between neighborhoods and dare not cross such imaginary lines. Public officials rejected Snow, the medical community rejected Snow, and he seemingly had no one left to turn to. Or did he?

The Broad Street pump remained open for business. John Snow, having failed to convince public officials and fellow physicians, did something surprising for a man of science: he turned to religion, specifically to the good Reverend Henry Whitehead. Fortunately for Snow, and for those yet to die from the cholera outbreak, Rev. Whitehead of St. Luke's Church in Soho, a believer in the miasma theory, granted an audience with the young physician, subsequently appreciating the merits of Snow's epidemiological evidence.

An important lesson is to be learned here—a theme that pops up now and again within the pages of this narrative: men of the cloth can be both piously religious and versed in science. Especially in Snow's era, the nineteenth century, many spiritual leaders were part-time science dabblers. Rev. Whitehead was such a person, a servant of God and a learned man. Various spiritual leaders today—especially in America, with our wholly uninformed, uneducated, unimaginative evangelical Christian leaders—could take a page from their historical brethren.

Together John Snow and Rev. Whitehead convinced local authorities to shut down the Broad Street water pump and investigate. Almost from that very day, the incidence and subsequent deaths from cholera declined. Precipitously. Apparently, no new cases of cholera occurred—sort of like a switch had been turned off. Of course, a switch literally was turned off: the Broad Street water pump

City engineers eventually located the source of the problem, which, as it turned out, was under a nearby hovel where cholera had been wildly prevalent within a single family. The engineers discovered that the cesspool under the floorboards of that single home was leaking into the earth the foulest of defiled stool and vomitus—excrement no doubt teeming with cholera bacteria, though neither Dr. Snow nor Rev. Whitehead and definitely not those engineers would've known it was bacteria. The association between bacteria and infection had not yet been conclusively made by Pasteur. As they dug in the stench of that defiled earth, what was supposed to be a clean water valve leading to the Broad Street pump was discovered to be faulty. The valve was soaked in defilement. It wouldn't take a rocket scientist to figure out at that point that fresh water contaminated with poop might not be a good thing to ingest, and specifically to realize that what was making people ill with cholera symptoms was something other than "bad air."

This discovery connecting cholera to a probable waterborne pathogen—not airborne, not miasma—gave credence to the soon-to-be-established germ theory, which would push the doctrine that tiny microscopic pathogens are the source of infection, not vapors. The world was about to change.

The discovery vindicated Snow, whose reputation in the meantime had been debased, much like the Broad Street pump had been debased. The number of lives Snow saved is incalculable. The doctor himself would die just four years later, in 1858, at the young age of forty-five from a stroke. He is buried in Brompton Cemetery, in the well-to-do borough of Kensington and Chelsea, not interred in poor Soho. The *Lancet*, Britain's premier medical journal, barely noted Snow's passing, as his contributions to medical science were at the time still largely unnoticed and unappreciated. Old habits of crusty old men die hard. In 2013, 155 years later, the *Lancet* corrected its original obituary of Dr. John Snow, giving him his just dues. It's never too late to make amends.

If you were to visit Broad Street in London today, you would find a replica of that defiled pump, a memorial to the outbreak's victims as well as a tip of the hat to Snow and his determination to help push through a paradigm shift in our understanding of how infections are spread.

Interestingly, and important to understanding infection, neither Rev. Whitehead nor any of his staff got cholera, even though their church was within a stone's throw of the Broad Street pump. Apparently, the good reverend and his ilk did not drink water, at least not water drawn from the Broad Street pump; they smartly drank beer, likely from a nearby brewery. But wouldn't that nearby brewery also have drawn its water from that same Broad Street pump? Wouldn't their beer also be defiled with cholera?

We don't know where that brewery obtained its water, and we'll never know. It's possible it used another water pump or had a different source. But whatever the case, another lesson arises here that supports our soon to be discussed Louis Pasteur's germ theory his pasteurization concept.

Using fermentation to make alcohol has been known to humans since at least 6000 BC. It quite likely was the first deliberate chemistry our Neolithic ancestors learned. The basic concept involves taking a starch—grapes, barley, rice, any fermentable cereal—adding water, sprinkling in either some yeast (which is a fungus) or some lactic acid bacteria, and letting those little devils digest the fruit or cereal into alcohol, which is just smaller organics broken down from the larger starch molecules. That is the basic process. Not too complicated. The yeast or bacteria want the starch for food and leave behind the alcohol by-product. It is the same process mentioned earlier, when 3.8 billion years ago the first Stage 1 bacteria on earth—before switching to photosynthesis—used fermentation as their energy source.

However, what is important to note here is that, when it comes to fermentation, nasty pathogens like cholera, if present in the water supply, cannot typically survive the acidic environment created by the fermentation. Fermenting yeast and fermenting bacteria love the acidity, they produce the acidity, but pretty much any of the pathogenic microbe find an acid environment inhospitable. If the acidic environment doesn't kill off the pathogens, the alcohol content most likely will finish them off. And if that doesn't the heat of some fermentation processes might do the trick. The point is, fermentation climate might be suitable for some bacteria but not others.

Whereas fermentation generally does not involve much heat (except the heat produced by the chemical reaction itself), still the fermentation heat to make beer is likely enough to wipe out some bacterial load, and failing that, pasteurization does the rest. Fermentation is an exothermic reaction, which as you recall from biology is a chemical reaction that releases heat. Pasteurization is basically using brief much, much high temperatures to kill things, to kill microscopic things. At a more detailed level, pasteurization uses high temperatures for brief moments to denature any bad bacteria without significantly denaturing the product, that is, without altering the flavor of the liquid being pasteurized.

With all this pasteurization talk, now we have finally arrived, as you might have guessed, at the story of the great Louis Pasteur and his germ theory. Accordingly, we'll need to leave 1850s London, cross the English Channel by boat, and travel on to Paris.

16 THE GERM THEORY

Louis Pasteur was born in 1822 into an impoverished family near the eastern edge of France, an area known as the Jura department, near the border with Switzerland. In fact, you might say Pasteur was one of the world's providential poor. At an early age, Pasteur showed academic promise, so much so that despite his meager upbringing and his financially strapped family (his father was a leather tanner), he was admitted to the prestigious École normale supérieure in Paris when he reached university age. It was one of the postsecondary institutions of higher education founded in the aftermath of the French Revolution of 1789–99 and in the wake of the Enlightenment as it swept across Europe, espousing an educational philosophy that higher education should be for the masses, not just for the well-heeled.

I do not wish to delve too far into this, but with the abandonment of the old ways of monarchs and religious doctrine, as the Enlightenment gave way to, more or less, governments "of the people, by the people, for the people," the right to an education was determined to be fundamental. So it would not be surprising to learn that the once rarified concept of receiving an education and especially attending university, previously reserved for the upper crust and for those entering divinity, were becoming more and more available to the commoner. Germany especially was a pioneer in in public education, but France, Britain and other European nations were in lockstep.

Trained in math and chemistry, Pasteur initially became a professor at the University of Strasbourg, France, before landing a professorship back in Paris at the École normale supérieure, which is where he produced much of his Nobel-worthy work on germ theory in the 1860s. I say "Nobel-worthy" because Pasteur never received the actual prize, since, quite simply, he died in 1895, six years before the Nobel was established.

Pasteur's research on microscopic organisms first demonstrated that the then prevailing opinion that life occurred from "spontaneous generation" was indeed false. Spontaneous generation is one of those theories that had been around since ancient Greece, a concept that posited life arose fully formed from nonliving matter. The classic example was that of maggots, which is the larval stage of the fly. Leave a piece of meat lying around and maggots will appear, and for thousands of years humans believed those maggots arose from the dead meat. Since the microscope had not yet been invented all those centuries earlier, observers failed to appreciate that those annoying flies buzzing overhead decaying meat would lay eggs on the meat, eggs that could not be seen, microscopic eggs, from which sprang maggots, and from those maggots emerged the annoying flies. Being that fly eggs were microscopic and scientists myopic, the dead meat springing forth the maggot was touted as a perfect example of "spontaneous generation."

Other classic (false) illustrations of spontaneous generation from antiquity that persisted well into the 1800s were that fleas arose from dust, that from the sun's heat upon the Nile river sprang mice and rats, and that eels came into existence when another eel brushed up against river rocks, dislodging certain rock particles, from which emerged the new eels.

A similar incorrect theory to spontaneous generation is "equivocal generation," which states that some species could arise from unrelated living organisms. This theory was based on the observation that some animals would "spontaneously" produce another organism, such as a tapeworm within the gut of a pig. Observers clumsily deduced that the tapeworm grew from the swine's intestinal lining, rather than from the pig having swallowed a microscopic worm larva that then grew into the mighty macroscopic tapeworm.

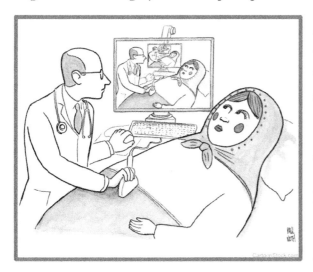

Yet another incorrect but cherished theory of organism development is "preformationism," the concept that tiny versions of the adult form, termed animalcules, exist inside the adult. This theory states that animals, including humans, are formed from teeny tiny "preformed" versions of the adult, neatly tucked inside the already full-grown specimens, sort of like those Russian nesting dolls.

According to the preformation theory, sex between a man and a woman triggered the tiny preformed animalcule to grow inside the woman's womb and, when it was large enough, it was expelled through childbirth. Some argued intently that it was the semen that carried this tiny version, and others argued, equally excitedly, that the mini-form was already housed inside the eggs of the female womb. Arguments became heated, fists were likely thrown, and imagine that—people arguing and being passionate about a theory that was totally wrong on both sides of the argument. Logicians might call this a "balance fallacy," where two sides argue their case with neither having proved the merits of its own theory.

From the time of Hippocrates, it was assumed that the human male had a "seed" and the human female had an "egg," although how they combined or if they combined at all and which did what with the other was hotly debated. But either way, it was believed a tiny human was triggered to grow from sex, like a balloon expanding. It was believed that other animals, and even plants, had "his" and "her" parts that worked together in some way to trigger the already existing tiny versions of that organism, so why not humans, too. A veritable mini-me Russian nesting doll.

Regnier de Graaf, born in 1641, was a French physician and anatomist and the first scientist to conclusively prove in 1672 that the female "testicles," later to be called "ovaries," contained the female egg, the oocyte. His discovery was made with rabbits—and, what the heck, there is definitely a joke in there. Many believed that the human female egg contained the preformation animalcule—the mini-me—and the male seed, the sperm, if it did anything at all, inspired the preformed fetus to materialize. The concept of fertilization—egg plus sperm co-mingling—was still off the table. There's another joke there about co-mingling on or off the table.

As we already know, the non-scientist Van Leeuwenhoek also at this time identified the male seed, the sperm, swimming around in what we can only assume was his own spilt seminal fluid. (Others claim the seminal fluid he observed was not his but provided by, and I'm not sure how it was procured, a "friend who had lain with an unclean woman," whatever the significance of that would be.) Unfortunately, Leeuwenhoek, through the power of his discovery coupled with the power of his microscope, added to the power of his ignorance in science—since he was never ever formally trained—was convinced that the fetus arose not from the egg but from the semen, those tiny sperms he saw swimming about, and the female "vault" was merely the depository for the "preformationism" of the sperm to unfold. Whether or not the human female had eggs was of little concern to Van Leeuwenhoek.

It took 150 years to finally put all the pieces of the puzzle together, or, in this case, put the male sperm gamete and the female egg gamete together to form the zygote, the solitary fertilized egg on the road to the fetus. One of the most important finishing touches came courtesy of the German scientist and explorer Karl Ernst von Baer, who did not prove the sperm fertilized the egg but did identify, in 1827, the next best thing: a fertilized egg, the zygote. That zygote under the microscope did not have a mini-me or a mini-you or any mini preformed animalcule inside, but it definitely looked different than an unfertilized egg. It was a single fertilized cell that would eventually divide along the germ layers.

The final piece of the puzzle was slotted into place in 1875, when Richard Hertwig, a German zoologist, observed under the microscope a male sperm gamete actually penetrating a female egg gamete to produce the fertilized egg zygote. Hertwig was working with sea urchins or some such lowly creature, but, hey, what's good for the goose is good for the gander. It has even been shown in the human model, that the female egg, ripe for penetration, releases some chemical, apparently progesterone, that is a clarion call for the sperm to "come hither." That signal chemical causes sort of languid sperm to firm up—there has to be another joke in there—and their tails whip faster. In other words, observed sperm not in the environment of a ripe egg, are floppier, lack a rigid shape, swim slower.

The human version of the animalcule has a special name: the homunculus. Seriously, since the days of Aristotle and well into the eighteenth century, many science folks, let alone everyone else, believed that humans grew inside their mothers from a little *homunculus*, from a tiny preformed version of the adult form. In an episode of *The Big Bang Theory* called "The Gothowitz Deviation" (season three, episode three), upon Penny revealing to Sheldon why she and Leonard are having noisy sex in Leonard's room, in Sheldon and Leonard's shared apartment, rather than across the hall in Penny's apartment, where she lives alone, it is revealed that Penny's bed broke after an earlier interlude of amorous sex:

> **Sheldon**: That doesn't seem likely. Her bed is of sturdy construction. Even the addition of a second normal size human being wouldn't cause a structural failure, much less a *homunculus* such as yourself [referring to poor Leonard].
> **Penny**: A homunculus?
> **Leonard**: Perfectly formed miniature human being.

Of course, men always complain nothing is ever big enough, and women are understandably quick to concur.

I feel for the diminutive Leonard. I was one of the more vertically challenged boys in elementary school, which I was annually reminded of during class pictures, where I always ended up in the front row with the other boy homunculi. You'd think the photographer, or the teacher would've taken some pity on us not-very-tall preadolescent boys. They did not. Fortunately, by the end of elementary school I had grown out of the front row—not very *far* out of the front row, mind you. But I was happy enough to make it into the second row for class pictures. Incidentally, the second row was where most of the pretty girls were, including my fifth-grade crush, Julie, who never knew she was my fifth-grade crush, and since she never knew, it made our inevitable breakup easier to bear. Shyness, like gravity, can be a heartless biatch. Makes you feel small.

"Kind of makes you feel large and significant, doesn't it?"

With that, we've covered the pathway to Pasteur's overturning of the "spontaneous generation" theory, and a few other similar pretender theories, but we still need to cover a bit more ground before we reach the point where Pasteur was able to put forth his germ theory.

In the 1820s, the German biologist Christian Heinrich Pander identified the (three) germ layers of the chick embryo using a microscope. In other words, he found that the chick embryo from the fertilized ovum seemed to have divided into three distinct layers, each of which respectively evolved into predetermined inner, middle, and outer parts of the full-grown chicken—

same is true with all animals. Today we call those three embryo layers "endoderm," for the inside, like the intestines and lungs; "mesoderm," for the middle, which includes heart, muscle, kidneys, reproductive stuff, and the skeleton; and "ectoderm," for the outside, like the skin and oddly the brain and neural network.

If you find it odd that the outside ectoderm layer would give rise to both skin and also something so exquisitely internal and complex as the brain, I can assure you I once thought so too, but the reason why this is so is really quite simple. When you think about it, what the brain essentially does—besides hopefully engage in thought—is absorb information from the "outside" world, such as processing the five basic senses of touch, sound, smell, taste, sight. This makes the brain almost an extension of everything on the outside of the organism, which is why the central nervous system is an ectoderm derivative despite being so internalized. And that is about all I have to say about that. Or is it?

Pander's work on the chick embryo was soon followed up on by his student, who we previously met: the Russian biologist Karl Ernst von Baer. In addition to identifying the fertilized egg (the zygote), Baer also proposed the mammalian egg cell concept of embryology. That is, Baer concluded that the female egg gamete and male sperm gamete *unite* to form a zygote, and through cell division over and over and over again, with the folding of layers here and there, voilà, you end up with an organism. In slightly more complex terms, what happens is that the zygote first splits to form what's called a "blastula," a grape-like bunch of cells that still lack differentiation—that is, all cells in the blastula have yet to declare their embryologic fealty. These undifferentiated cells then cross over into a point of no return, where they've declared their allegiance, dividing and folding into the three germ layers, partially assisted by those epigenetic switches previously discussed. The blastula also has a layer that forms the placenta for us placental mammals. Eventually those predetermined three layers with their predetermined cells go down certain cell-division paths, and at each step declare more and more specificity, eventually arriving at the end product. Baer's was the first theory to challenge the spontaneous generation and preformation theories.

Baer not only challenged the existence of homunculi, but he thoroughly described how a dividing fertilized egg would form general characteristics (spine, head, trunk, heart, gut, extremities) before special characteristics (eyes, ears, mouth, tongue, fingers, toes, and so on). He further went on to propose that a dividing embryo can only go down a certain path—later to be known as its DNA—and not share or converge with any other species. In fact, that is the classic definition of a species: it is the endpoint of an evolutionary branch that can successfully breed only with its own kind. A single species, over the pressure of time can eventually give rise to new species owing to evolutionary influences (Darwinian evolution), but those new distinct species that arise will now no longer be able to mate with each other nor with the "mother" species from which they speciated, only able to mate with their own species, or …

Returning to *The Big Bang Theory*, specifically this time its pilot episode, when physicist roomies Leonard and Sheldon first meet Penny, that beautiful waitress and aspiring actress. Penny is just moving into the apartment across the hall from the two nerds, with Leonard already displaying amorous thoughts toward Penny:

> **Sheldon**: That woman in there is not going to have sex with you.
> **Leonard**: Well I'm not trying to have sex with her.
> **Sheldon**: Oh, good. Then you won't be disappointed.
> **Leonard**: What makes you think she wouldn't have sex with me, I'm a male and she's a female?
> **Sheldon**: Yes, but not of the same species.

As for humans and the possibility of new species splintering off from our own evolutionary branch: many people who concern themselves with the possibility of human speciation consider current human evolution to be at a dead end. There is nowhere for us to go. We will continue to maybe get smarter, something like that, but not speciate into a different species. There are not enough environmental influences on Earth to force humans to speciate further. We control our world, our fate. We'd have to go to an off-world colony where perhaps, over thousands and thousands of years, that new world with different environmental influences and challenges might force speciation. Which means if the ancestors of those space-going humans ever came back to our Earth a million years from now, their DNA mismatch with our DNA would preclude successful mating.

Consider this, modern day woman could mate with a Cro-Magnon Man of 50,000 years ago since both are *Homo sapiens*—same species matched DNA (or enough matched)—but she couldn't successfully mate with *Homo erectus*, the bloke who predates *Homo sapiens* as their DNA would be too much of a mismatch for successful fertilization. And yes, there is a joke in there, too, modern-day women nowadays complaining some of their blind dates have been little more than Neanderthal erectus.

As for the forced breeding of humans, like Hitler envisioned, we don' need to go there again. Suffice to say, any program like that would be a disaster, biologically and in many other ways. So, for now, on this world, we might be close to "the best a man can get," as the Gillette razor advertising slogan goes. Humans are done speciating, we are the final product. Man is not going to evolve much more. As for woman, well, she already is perfect.

You might be wondering why I keep referring to the sitcom *The Big Bang Theory*, even though I'm aware that many readers might not have seen a single episode of the fantastic Chuck Lorre and Bill Prady sitcom (though you should). This is a book partially about science and scientists, who we could classify as your basic nerds—although some are nerdier than others. But even the nerdy scientists among us have a widely relatable side right under that layer of dorkiness. Sort of like Spock from *Star Trek*—he has his logical Vulcan side and his dryly humorous human side. Nerds are not, in fact, androids, not synthetics, and Lorre and Prady convey this with humor, and they convey it quite convincingly, in *The Big Bang Theory*. Not that I was a nerd in high school, I was not. Wasn't smart enough. But in more recent times I did an online quiz—the veracity of such quizzes in doubt—to determine which *The Big Bang Theory* character I was most alike. The verdict: Bernadette. Which makes sense, other than she is a female, but because she was a non-nerd scientists which I suppose best typifies me. Its final season was sadly in 2019. Like when the comic strip *Calvin and Hobbes* by the masterful Bill Watterson ended its decade run in 1995, I went into a melancholy when *The Big Bang Theory* ended its run.

Having explained that the textbook definition of "species" is an organism that represents a branch endpoint that can breed only with itself, it is necessary for me to also explain that there are a few exceptions. There are always exceptions in life, and some species can in fact interbreed. A tiger and a lion can produce a "liger" or "tigron," depending upon your lexical preference. Sheep can breed with goats, and their offspring is called either a "geep" or a "shoat." Horses with donkeys gave us "mules," which explains why they're so stubborn—they aren't sure who or what they are. Polar bears are known to crossbreed with grizzlies, apparently not picky when it comes to mating. The bear wants what the bear wants.

Recall our earlier discussion when I was in grade school there were two kingdoms, plant and animal (no longer the case), followed by the memorized subgroupings of phylum, class, order, family, genus, and species. Reciting this sounded like a poem, the taxonomy poem: "*kingdom phylum class order family genus species.*" And that poem was also about all I learned of taxonomy when I was a knucklehead grade school kid, partly because teachers in grade school, probably in high school, too, and the book they chose made it all so unnecessarily complicated, and boring. It need not be that way.

And as also mentioned earlier in this narrative, today we have two super kingdoms, or domains: Prokaryota (no nucleus) and Eukaryota (nucleus), with the eukaryotes divided into at least four kingdoms: Protista, Plantae, Fungi, and Animalia. Viruses are not cellular, so they are not a part of this deal. They have their own deal.

But what the heck does the next parts of the rhyme—"*phylum class order family genus species*"—mean? First off, I should say that taxonomist don't always agree—they also probably throw punches at each other at their international meetings. But I'll try to explain the most agreed-upon meaning of each line of the taxonomy poem with a single sentence, mostly using the dog lineage as our example, plus at little of us humans thrown in for good measure.

"Phylum" means all members have a similar body plan.

It's that simple. Members of a phylum share major structural similarities. Take the phylum Chordata in the Animalia kingdom: it is defined by a front end or head, a rear end or tushy, a tail or tail remnant, a middle body with similar internal organs and symmetrical side appendages. All Chordata possess a dorsal spine along the back, where the spinal nerves connect to a brain, and which is the *sine qua non* for the phylum's name, chordate equals dorsal nerve cord. This phylum includes dogs and bears and lions oh my, but also includes wascally wabbits and dolphins and fishes and amphibians and reptiles and birds and sharks and crocs oh no, and especially includes primates and man.

I know what you're thinking: humans look nothing like fish. But we do, in that our overall body plans are similar: head where food and air go in, tushy where waste goes out, a middle section with internal organs, and four side appendages, whether they're legs or fins, and most importantly, a dorsal spine. In other words, man looks more like chordate member crocodile, than it does look like a jelly fish (phylum Cnidaria) or a snail (phylum Mollusca) or a starfish (phylum Echinodermata) or an insect (phylum Arthropoda). When it comes to phylum just think overall "body plan."

"Class" is where we see a more similar grade of organization, level of complexity, layout of internal organs, and type of external features.

Our pet dog and us humans share the same class, called Mammalia, where things like fishes and birds and lizards and amphibians are excluded, breaking off into their own classes. Us mammals are left to our own devices, from rodents and bats and shrews to horses and pigs and elephants, and not forgetting all the carnivores: seals, dolphins, lions, tigers, and bears oh my, primates, man, woman, our pet dog, and "I tawt I taw a puddy cat." All of us are mammals, and not only are we all similar as far as our front-to-back, side appendages, dorsal nerve cord and internal organ layout, we are also all warm-blooded, which is a defining characteristic of mammals. All mammals also have a four-chambered heart, to pump that warm blood through the lung circuit, and all us mammals have fur or hair. Most mammals have a placenta except a few outliers, the marsupials and monotremes. Although birds are warm-blooded, they lack hair among a few other things, so they have their own class.

"Order," the next level of differentiation in taxonomy, is where members start to look more and more like each other.

Pretty much all birds look like birds and all fishes look like fishes, all primates look like primates, men look like men, and women look prettier. Our pet dog now splits off from us, falling into the order Carnivora, which also includes the cats, wolves, bears, hyena, racoons, seals, mongooses and skunks, and things like that. As for us humans, we enter our own order, called Primates.

I know what you're thinking now: given the manner in which humans consume meat, you'd think we'd be allowed to stay within the class Carnivora, but, interestingly, our ancestors, the primates, were not much into eating meat. Primates are for the most part vegetarians. Humans decided to eat meat later on in our evolution, long after our primate ancestors parted ways with the carnivores.

The next line in the taxonomy poem is "family," and not only do members of a family look more and more like each other, they also behave similarly.

Behavior is one of the key characteristics of family rather than anatomic appearance. Continuing to trace our pet dog, after the order Carnivora comes its family, Canidae. Canidae drops off things like rats and mice and shrews and bears and dogs' archenemy, cats. But it maintains your basic wolf, fox, jackal, and the wild dog. As for us humans, after Primates come about sixteen families, and we landed in the family Hominidae, which includes the great apes.

Next up is "genus," and whether plant or animal or fungus, there are no hard and fast rules at this level of taxonomy—just grouping based on organisms that look even more alike and behave even more alike.

There's often not a huge difference between family and genus, at least to the casual observer like you and me. I'm sure taxonomists can see a huge difference, and likely huge disagreements about these differences lead to those fist fights. But, hey, they're taxonomists, they probably slap more than punch, so it can't be all that bloody dangerous. Our pet pooch's genus, *Canis*, provides a good example of the blurred lines between family and genus. In the family *Canidae*, we have a bunch of different-looking wolves, foxes, jackals, and wild dogs, and in the genus *Canis*, we still have a bunch of wolves, foxes, jackals, and wild dogs, but now they look more like each other: similar size, similar appearance, and other subtle things you and I would probably not pick up on.

As for us humans, our genus is *Homo*, the likely first *Homo* was the species *habilis*, who arrived on the world stage about two million years ago, give or take. The bridge species between the family *Hominidae*, the great apes, and the narrowed-down *Homo* genus was likely the clever chap *Australopithecus africanus* and a few very close friends, who are occasionally referred to as "the missing link."

"Species" is the narrowest branch in the taxonomy tree, and for lack of a better single-sentence description, we'll go with: species mate with each other and no other.

Tracking our pet dog, we arrive at the species *lupus*, the ancestral gray wolf of which I am told there are thirty-eight subspecies. So where precisely is the household pooch? Largely due to earnest human domestication of the wolf beginning perhaps 14,000 years ago, our pet dog is in a subspecies known as *lupus familiaris*. Wolves started hanging out near human campsites perhaps 100,000 years ago, and for sure 50,000 years ago, but the intentional domestication and breeding of the wolf commenced around 14,000 years ago.

"Darn, it's another bald-crowned big-bellied male."

As for us humans, our species is *sapiens*, as in *Homo sapiens*, arriving on the world stage to take charge perhaps 300,000 years ago, and the extant species, likely Cro-Magnon man or *Homo sapiens sapiens*, arrived around 50,000 years ago. That's not a typo, nor an echo, the "*sapiens*" is repeated for the extant species.

And so our dog's full taxonomy poem after Eukarya goes something like this: Animalia Chordata Mammalia Carnivora Canidae *Canis lupus familiaris*. And our human refrain is Animalia Chordata Mammalia Primates Hominidae *Homo sapiens sapiens*.

Returning to and building upon the chick embryo work of Christian Heinrich Pander and Karl Ernst von Baer, Louis Pasteur was able to propose that "life came from life," and, at that, from the same species. A great deal of Pasteur's original scientific work was with molds, a type of fungus, which hardly offers a romantic prelude to germ theory. Pasteur and others, such as the German scientists Robert Koch and Ferdinand Cohn, demonstrated that molds—which they observed growing on a growth medium in a Petri dish—do not arise spontaneously from that growth medium. Rather, the Petri dish has been contaminated with invisible fungal spores that waft through the air. A half-century after Pasteur did his work, in 1929, a similar serendipitous wafting of mold spores contaminated a Petri dish sitting in a lab in London, which led to the discovery of penicillin by Alexander Fleming.

Pasteur's proof that microscopic life was indeed at the core of the generation of life is strongly associated with the image of his famous swan-neck glass beakers, a really simple yet elegant exercise in scientific design. Pasteur used two types of bulb-shaped glass beakers that sort of looked like upside-down standard light bulbs, but whose tops finished in a curving taper reminiscent of a swan's long neck. One of Pasteur's swan-neck beakers had a tapering neck that opened up toward the sky. The other's neck curved down toward the ground. Just think of one swan looking up at the sky (why a swan would do that I have no idea) and another swan looking down at the water (more believably looking for food).

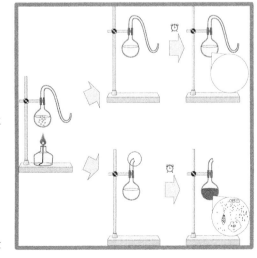

Pasteur set both beakers, each filled with a nutrient-rich broth, in spots where spores of mold were most assuredly floating about and added a little heat just

217

to prime the medium. And now we arrive at the key event, the observation that shows life did not come from nonlife:

- Mold did not grow in the downward swan-neck beaker, simply because mold spores were not able to float into that beaker.
- Mold grew in the upward swan-neck beaker, simply because mold spores were able to float into that beaker.

Simple. Elegant. Proof.

Voilà! Louis Pasteur had disproved the theory that life arises from nonlife and rather proved that life comes from life—thus burying for good dogmatically held beliefs such as spontaneous generation and preformationism.

The final stamp of approval in support of germ theory was provided by, of all things, an Irish physicist, John Tyndall, who was an admirer of Pasteur as well as of Charles Darwin and his theories of evolution. Tyndall cut his teeth working on physics problems, such as his pioneering work describing diamagnetism, which is the ability of a magnetic field to repel metal materials. The magnet you use on your metal refrigerator door to remind you Sally's soccer game is on Wednesday is a ferromagnetic material, usually iron, and it loves metal because it has at least one unpaired electron in its outer orbit seeking a magnetic metal to balance its frustratingly unhappy electron. An iron magnet is one type of paramagnet. A diamagnetic material, on the other hand, the type that Tyndall described, has all its electrons paired and therefore has no interest in, or even is repelled by, metals like your refrigerator door. Examples of diamagnetic materials include gold, copper, and mercury.

One of the criticisms leveled against Pasteur's "life from life" theory was that, as a few skeptics had shown, even if a Petri medium rich in bacteria was heated to kill off any bacteria already living in the medium, some bacteria would still grow. Since humans cannot tolerate much in the way of hot water—generally, we'll die if our bathwater goes much above 115°F, which is 17° above body temperature—it was incorrectly assumed by these skeptics that similar heat would kill bacteria (and molds, too). So, when the heated medium cooled again, some bacteria grew, suggesting to the skeptics that the bacteria arose from the nonliving broth rather than any bacteria that had survived the jacuzzi. This, they claimed in their myopic view, was in fact evidence of spontaneous generation.

Tyndall so strongly believed in Pasteur's work that he took time off from his physics pursuits to demonstrate that heating killed some bacteria but did not kill *all* types of bacteria. It turns out that some unicellular organism, like mold, like bacteria, can withstand higher temperatures than us fleshy humans. Which is why grandma tells us to make sure we take that canned soup to a simmer before we eat it, you know, to annihilate botulism. The heat not only kills the bacteria, but likely denatures any botulinum toxin.

Archaea are a division of bacteria that went down a separate pathway than those bacteria that we commonly associate with today. Some archaea can live in extreme conditions not suitable for any other form of life. They have been discovered living happily on the edge of deep-sea hydrothermal vents that can reach 248°F, above the temperature needed to boil water, and they seem equally pleased when things dip below 0°F just a few feet away. They have been observed to purr along in an acid pH of 1 or an alkaline pH of 11. Some archaea deep in the Mariana Trench sit pretty at pressures of over 1,000 atmospheres. These organisms that can handle extremes of temperature, handle both acid and alkaline worlds, handle crushing pressures, handle Chuck Norris, are known as "extremophiles."

John Tyndall was born in 1820 in County Carlow, Ireland. He began his career as a land surveyor but eventually headed to Germany to study chemistry, physics, and mathematics at the University of Marburg. Tyndall then moved on to London, quickly becoming a leading scientist of the era. In addition to advancing diamagnetism and helping to prove Pasteur's "life from life" theory, Tyndall also described how Earth's atmosphere has the capacity to retain heat. He determined the greenhouse properties of our atmosphere by working out the physics of the absorptive heat qualities of atmospheric nitrogen gas and carbon dioxide gas. Recall from that earlier discussion about the high heat capacitance of water, or that of antifreeze for our cars, the low heat capacitance of metals (they get hot quick) making them ideal for saucepans. Tyndall demonstrated the heat capacitance of atmospheric gases. Today, the heat capacitance properties of atmospheric gases discovered by Tyndall are at the heart of the climate change conversation. Although on the one hand it means the atmosphere does indeed have a high heat capacitance, it doesn't mean that as the capacitance is driven up adverse events won't happen. Which brings us to a short detour regarding the capacitance (also known as tolerance) for some pious scientists, juggling their religious beliefs with their scientific discoveries.

When it comes to that intersection of religion and science, the 1800s was a lively time indeed. Some of the great scientific minds of the nineteenth century attempted to reconcile their scientific discoveries with their strong religious beliefs, such as James Prescott Joule and his mathematical depiction of a unit of energy (termed the joule); James Clerk Maxwell, who we visited with earlier and his elucidation of electromagnetism; and Balfour Stewart and his work with radiant heat and weather prediction.

You wouldn't think that someone such as Maxwell, a man who defined one of the four fundamental forces of the universe, would be so orthodox on matters of religion that he would abridge his scientific discoveries with a flexible capacitance so as to align his scientific observations with his religious beliefs. But he didn't. He remained true to his religion and true to scientist and I guess compartmentalized each. Even when evidence contrary to religious doctrine smacked such scientists right across their face, they navigated carefully those stormy waters. Religious beliefs are strong, profound, and not easily discarded.

But still many nineteenth-century scientists found it difficult to accept concepts that ran contrary to the Bible, accounting for some of their resistance to Pasteur's life from life theory and the even-harder-to-swallow theory contained in Darwin's *On the Origin of Species*. What temerity these religious scientists would exclaim towards Darwin, to allege humans descended from the humble ape! Pasteur and Darwin put forth theories that certainly ran contrary to the book of Genesis. Fortunately, other liberal-minded and brilliant nineteenth-century scientists, such as of course Tyndall, the great comparative anatomist Thomas Henry Huxley, and the earlier mentioned rock star of geology Charles Lyell, came to the defense of the Pasteurs and Darwins of that era. Eventually, the scientist naysayers—the ones who often also held strong religious beliefs—came around, tucking their religious beliefs just far enough into their coat pockets so as to be unseen, allowing these religious men of science to accept empirical evidence over scripture.

Charles Lyell in his geological narrative demonstrated that Earth was millions and millions of years old, not the mere 6,000 years ascribed by the scriptures. As you'll recall, Lyell's demonstration that rock sedimentation needed millions of years to layer in the way it did simultaneously gave Darwin the timeline he needed for his theory that life likewise evolved slowly over millions of years. Darwin now had the ammo needed to explain how a favored trait—natural selection—would keep a species going in the struggle for life.

And where did Charles Lyell learn most of his early geology? From William Buckland, an ordained priest, who is perhaps most famous for his study of the Kirkdale Cave in Britain, a hyena den that contained the bones of hippopotamus and elephants, two mammals thought never to have roamed Britain. Biblical scholars used that hyena den as proof of the Biblical flood narrative, that those bones had floated into the cave all the way from Africa. Buckland demonstrated that those bones were dragged into the cave at a much earlier point in history by the hyenas (the fossilized bones had gnawed marks), when ocean levels were lower during the last ice age (ocean waters locked up in the polar ice caps), and Britain was connected to mainland Eurasia. Reverend Buckland is also credited with giving the first full description of a dinosaur in 1824, the Megalosaurus.

The first dinosaur bone ever discovered although he didn't know that was what it was, was the poor, hapless Gideon Mantell in 1822, a tooth resembling a modern iguana, so the extinct owner of that tooth became known as Iguanodon—that is, when it was realized a few decades later it was a dinosaur. Between the fossil hunters Buckland, Mantell and another bloke, the ill-tempered Richard Owen, that entire group of ancient extinct reptiles became known as "dinosaurs," a name coined by Owen.

In today's world, the same sort of balanced thinking cannot be bestowed upon "science deniers," many of whom are religious fundamentalists who take their cues from the Bible, even in the face of overwhelming evidence to the contrary. It may be true that everyone is welcome to their opinion, but opinion and fact are not equivalent. People are not welcome to their own set of facts when those facts are really nothing more than opinion masquerading as fact.

The geologic record is chock-full of evidence that chance and mutation gave rise to biodiversity, all the way up to and including man, and the even more advanced species—woman. That same geologic record also preserves examples of species that once existed, species that once roamed the savannah, species that once swam to the far reaches of the seas. Species that failed to move on to the next stage of evolution, subsequently were sent mercilessly into extinction by lack of a favored trait, global changes, such as climate changes, continental shifts, and even the asteroid strike that doomed the dinosaurs into extinction.

In short, the same geologic record that Lyell demonstrated had taken millions of years to evolve is the same geologic record that shows the evolution of both plants and animals, as well as reveals forms of life that once existed and were then vanquished. The obvious question that arises when put in the context of religion is: Why would a benevolent God make all that biodiversity, which we can see in the geologic strata a million years back, only to decide later on at various stages of Earth's history to doom certain creatures and plants to the fossil record while allowing others to pass on through the next door of evolution? Such a God wouldn't do that. There would be no reason, no purpose for God to have done that. Consider this: about 99 percent of all creatures that arose during before and during the Cambrian Explosion 540 million years ago, that swam in the oceans and crawled along the land, are extinct. Did they all sin and God as a result punished their entire species by sending them into the fossil record? Did Noah forget to take them two by two on to his ark?

Along the geological timeline, early humans appear in the very top layer of the fossil record, arriving on the scene rather late in the game. The trilobite, on the other hand, an early multicellular marine arthropod life form, an extinct life form, is found in layers of sediment that are 520 million years old. It and most of the rest of the immensely diverse array of species produced during the Cambrian Explosion are now gone, most of it lost during the subsequent Permian Extinction event 250 million years ago. The Permian sent 96 percent of marine life and 70 percent of terrestrial life that came out of the Cambrian Explosion into the fossil record. Even insects, which one would think could avoid such annihilation, were not spared from Permian elimination. We can see evidence of the Permian Extinction in the dirt, and we can scientifically date that dirt with radiometric dating. The Permian Extinction was caused by a change in the environment: Oppressive heat raised atmospheric and oceans temperatures. This sauna-like world was deadly for many organisms., especially marine species. Then the oppressive heat eventually gave rise to an ice age, which extinguished from Earth countless more species that, having survived the brutal heat of the Permian couldn't deal with the gloomy cold of the Ice Age.

Normal global warming, in and of itself, can trigger an ice age—and when I say normal, I mean not caused by human industry. How? As the earth warms, equatorial waters evaporate, head up into the atmosphere where they turn north or south flowing toward the poles. There the colder air turns that arriving moisture into snow that then blankets the poles. The snow then turns to ice, and this ice not only does not melt, it also *reflects* sunlight, a double whammy, making things around the north and south poles darn right frigid. Like a dog chasing its tail, the cycle of warm equatorial waters eventually blanketing the polar regions with ice sheets. The sound bite is this: equatorial warming makes the poles colder and as the ice sheets advance, reflecting more radiant sunlight back, global temperatures drop. Ergo, you have an ice age.

Add to that a combination of runaway volcanos spewing overwhelming amounts of ash into the air plus a few large meteor strikes here and there, and pretty soon the sun is blighted. This lack of radiant sunlight from atmospheric ash, euphemistically dubbed a nuclear winter, further advances global cooling. Between global warming triggering icing of the poles and the volcanos and meteors mucking up the sky with ash and soot, Earth enters an ice age.

The phrase "nuclear winter" came out during, and this is too rich, the "Cold War" between the US and the USSR with the buildup of nuclear arsenals. It was based on a theoretical nuclear war between the two nuclear powers, where, if enough nuclear bombs were launched at each other, and enough smoke and debris was hurled into the atmosphere, the sun would be blighted, global temperatures would drop, and Earth would descend into an ice age, a cold nuclear winter. Not to mention all those dead, vaporized bodies. As for the phrase "Cold War," it was precisely not derived from the prospects of a cold nuclear winter from an all-out nuclear war. Instead, it was coined by American financier Bernard Baruch, an adviser to President Harry Truman as an explanation of the "chilly" relations between the US and the USSR after World War II.

Those species that did not enter the fossil record from the hothouse earth lead-up to the Permian Extinction were jettisoned into the fossil record from the subsequent icehouse glaciation that followed. Nearly all life was wiped off the face of the planet between cooking and freezing. Earth almost became a dead planet. Almost. But life on this planet—some life—fortunately had the determination to push on.

Today many scientists are working hard to educate the rest of us to appreciate science, and to encourage science deniers to take a moment of pause and consider fact over opinion—which is where our story now turns, a great American scientist.

17 THE SAD STORY OF PLUTO

Neil deGrasse Tyson was born in 1958 in New York City. His mother was a gerontologist (a medical doctor specialized in aging) for the federal government and his father was a social welfare advisor for New York Mayor John Lindsay's administration. Tyson is a product of the New York City public school system—a perfect example of the importance of public education—graduating from the Bronx High School of Science. Tyson then received his undergraduate degree in physics from Harvard College in Cambridge, Massachusetts, his master's degree in astronomy from the University of Texas in Austin, and a doctorate in astrophysics from Columbia University, back in New York. That is quite a résumé—and he was just getting started.

Tyson is famously, if not infamously, known for being on the international committee that in 2006 demoted Pluto from planet to dwarf planet status, or, worse yet, to a trans-Neptunian object TNO. Why? Simply, Tyson and the other astronomers on the committee thought it made no sense for Pluto to be considered the ninth planet.

Pluto was discovered in 1930 by Clyde Tombaugh at the Lowell Observatory in Flagstaff, Arizona. That there was a ninth planet beyond Neptune was suspected based on some perturbations of Neptune's orbit, a planet dubbed Planet X, a planet that was supposed to be a real whopper. The Lowell Observatory was built and funded in 1894 by the wealthy Percival Lowell of the Boston Lowells, who made their fortune in shipping and railroad industries. Their Brahmin eliteness is evidenced by the John Collins Bossidy poem "Boston Toast":

> And this is good old Boston,
> The home of the bean and the cod,
> Where the Lowells talk only to Cabots,
> And the Cabots talk only to God

The Boston Brahmins or Boston elite are an old upper class, terribly wealthy Protestant group. Oliver Wendell Homes, Sr. coined the phrase "Brahmin Caste of New England" in an 1860 story for *The Atlantic Monthly*. "Brahmin" stems from the highest caste system in Hindu India, likely originating from Brahma, the Hindu god of creation. The only thing higher than a Brahmin would be God. And which is why only the Cabots talked to God.

The Hindu caste system was likely borne out of a need to control the masses, more so than out of any religious dispensations from their gods. Brahmins were priests, the highest, then followed in order by Kshatriyas, your basic warriors and rulers, Vaishyas were farmers and merchants, Shudras or laborers, and the Dalits or outcastes who swept the roads and cleaned latrines. Although the Indian constitution banned the caste system in 1948, it is still widely practiced to greater or lesser degree even today, depending upon where you venture in that country of 1.35 billion.

Other than offering faith and hope, there are many who believe all religions formed out of a necessity to control people, to keep them inline. If you strip away the religious and god façade, strip away the customs and relics, beneath it is an owner's manual of how people should live and live together, chock full of dos and don'ts.

When not making money, Percival Lowell was an avid amateur astronomer who not only built the Lowell Observatory but ironically, was on the wrong side of science when he declared Mars had canals made by Martians so as to direct water. Those canals were optical illusions due to the upper resolution limits of telescopes at that time.

But returning to the down-and-out story of Pluto. In our solar system, first come the four terrestrial planets: Mercury, Venus, Earth, and Mars. They're called "terrestrial" planets because each has an identifiable surface, or "terra," upon which you could theoretically stand. After Mars, and after you've waded through the asteroid belt beyond the frost line, come the four gaseous planets: Jupiter, Saturn, Neptune, and Uranus. "Gaseous" because there is nothing solid, no definable surface upon which to land a spacecraft, nowhere where you could stand and plant a flag of discovery.

While you theoretically could stand on Mercury or Venus, if you actually attempted it, you'd melt in a New York minute—they're too close to the sun. We will never see any manned mission to Mercury or Venus. Can't happen. Mercury's surface temperature during the day can reach 800°F, and since Mercury doesn't have an atmosphere, at night that same spot plummets to -290°F. Venus isn't any more inviting, daytime temps reach 864°F, and because Venus *does* have an atmosphere—sulfuric acid clouds that retain heat—it is just as hot at night. A manned flyby around Venus is possible, but landing is out of the question. You'd need to crank up the AC way too high, and besides, your nifty NASA spacecraft would melt on approach.

Mars is the only planet in our solar system where a manned mission to land is possible. You could stand on Mars so long as you were wearing a spiffy space suit, since Mars has no atmosphere, and in particular no O_2 for us oxygen-breathers. And you would need to be standing somewhere particularly hospitable on Mars, the equator, like along the edge of Schiaparelli, an impact crater where the temperature during high noon would be manageable, reaching a balmy 70°F. Not bad. As for spending a night along Mars's equator—that would be a nightmare, unless you're in a toasty warm habitation module, as the temperature outside can easily plummet to -

100°F due to the lack of an atmosphere. The average length of a Martian day, its daily spin, is actually similar to an Earth day: twenty-four hours and thirty-seven minutes. But its orbit around the sun, that is entirely different: a longer, lonelier journey of 668 days for one Martian year.

Mercury in Roman mythology (Hermes to the Greeks) was the son of Jupiter (Zeus). Mercury is the patron god of commerce, financial gain, poetry, eloquence, luck, trickery, and messages and communication. That latter "messenger" duty is the reason why Florists' Transworld Delivery or FTD uses the god Mercury, with his winged boots, as its logo: to send messages of love with their bouquets of roses.

Mercury's most celebrated moment in mythology was when Saturn, the father of Jupiter, tasked him with killing the 100-eyed monster Argos in order to free Saturn's daughter Io, which is also the name of one of the many moons of Saturn. Mercury also had the daily task of escorting those condemned to Hades, to their final destination. Mercury would deliver their poor wretched souls to the boatman Charon on the River Styx, who would then escort the dearly departed into the underworld. At that time in ancient history the "underworld" was not the "hell" as we know it; the concept of "hell" arose with the rise of Christianity, as will be momentarily discussed. When you watch a Hollywood movie and two coins are placed on the eyes of a dead person, it's to pay Charon for their crossing of the River Styx. As a messenger, Mercury delivered both good news (flowers) and bad news (death and an assignment to the underworld). That's quite a spread in duties.

Which of the planets is the ruler of the underworld according to Roman mythology? That would be Pluto, of course. Its two main moons are Charon the boatman and Styx the river to cross. Pluto has five moons, and because Charon is almost as large as Pluto, they are sometimes considered a binary planetary system, since the center of their orbits around each other do not lie within either body. The other three moons of Pluto are Nix, Hydra, and Kerberos.

You might be wondering why, in ancient times, did everyone who died all cross the River Styx into the underworld? Why didn't any of them get to go to heaven? Was everyone bad then, and so Hades was the destination for all? In antiquity, the concept of "heaven" existed (it is mentioned in the Hebrew scriptures), and of course both Greek and Roman gods lived in heaven. But that is just it—in antiquity, heaven was the domain of the gods or of God, Earth was the domain of the living, and the underworld was just that, the land of the dead. In Judaism, heaven includes all three spheres: God above, Earth, and the underworld. In other words, before Christianity, the underworld was not hell, the concept of hell as a place of punishment did not exist. The underworld was just where all dead souls went.

With the rise of Christianity, probably beginning in the first century AD, the "Kingdom of God" concept emerged, and the idea of going to heaven came into being. That is where Saint Peter oversaw entrance into Heaven, into the Kingdom of God, through the pearly gates, or, where he decided you deserved the alternative: eternal damnation in hell. Pretty clever. Great way to keep the masses in check. And so, the underworld became the exclusive domain of the devil and human wrongdoers, whilst heaven was to enter the glory of God.

Venus—or to the Greek, Aphrodite—is the goddess of love, beauty, sex, fertility, seduction, and victory. It is said that Venus formed from the sea when Saturn—the god of the new gods, before being overthrown by his son Jupiter—castrated Venus's father Uranus, Caelus in Greek. The blood from his severed testicles dropped into the ocean, forming foam, from which sprang Venus. Ta-dah! The planet Venus was so named because it is just so gosh darn beautiful when viewed from Earth.

Venus had two husbands, Mars and Vulcan. It is said that Vulcan tricked Venus by trapping her in his bed with a net, but since Vulcan's rape of Venus was loveless, it bore no children. In today's world, trapping a woman in a bed with a net would not only be loveless, it would be a felony. Unless you're the fictional BDSM enthusiast Christian Grey and the woman consented to the net-trapping, then I'm not sure what shade of gray it is. Venus had children with Mars, the husband she actually chose, as well as with a few other gods—she was, after all, the goddess of sex and made the rounds. Those other paramours of Venus included Phobos and Deimos, both moons of Mars.

As for Venus's rapist-betrothed Vulcan, no planet or moon is named for him. At least not a real one—just a planet in *Star Trek* where Spock hails from. However, at one time astronomers thought there was a real live Vulcan planet lurking in our solar system. In the mid-1800s, when it was noticed that Mercury had a slight wobble in its orbit around the sun, stargazers hypothesized that another planet must be out there, a planet even closer to the sun than Mercury, that was causing that gravitational wobble. Named Vulcan even before attempts were made to find it, dubbed as such because Vulcan was the Greek god of fire, reasoning that any planet nearer to the Sun than Mercury would surely be on fire. The search was on, but no such Vulcan planet was ever found. Since then, Mercury's rocking perturbation motion has been attributed to Einstein's general theory of relativity, with the sun's massive warping of space causing Mercury's wobble.

The next terrestrial planet out from the sun after Mercury and Venus is of course Earth. Earth is not a special name. It has no mythological precedent, no cool Greek or Roman goddess to brag about. Whether you're using the Germanic, Old English name of Earth or the Greek Gaia, our planet's name simply means "dirt, soil, dry land." How boring. Especially slightly erroneous since two-thirds of Earth is covered by oceans, not dirt. Our moon doesn't have a special name either; derived through various languages, it means simply "month." And our Sun? Nothing mythological there either. "Sun" is a word from an old Germanic root *sonne* that means what it means: sun.

It would be nice, I believe, to give our Sun, our Moon, and our dear planet Earth more colorful names, such as Sol for sun, Luna for moon, and Pale Blue Dot for Earth.

Sol is the Latin for "sun," but it just sounds way more eloquent. *Luna* is likewise the Latin for "moon," and Luna was also the Roman goddess of the moon, the consort to the sun. Luna is furthermore associated with the Italian word *amore*, "love and affection." So, I'd much prefer to call our sun Sol and our moon Luna. Wouldn't you? And no I'm not a "lunatic" which is also a word derived from the Latin *lunaticus* meaning "moon-struck," as in werewolf and full moons.

As for Earth, if you were riding in a spaceship near Jupiter and looked back at Earth, it would appear as a "pale blue dot," as the astronomer Carl Sagan famously described a photograph of Earth taken by Voyager 1 from six billion miles away. Ours is the only planet in the solar system with a blue hue. Why? Because oceans are covering two-thirds of its surface: ocean water scatters the blue wavelength making it visible while absorbing all the other wavelengths, imparting Earth with a pale blue hue visible from outer space. Renaming Earth as Pale Blue Dot would be a lot better than calling it Dirty Dot, which is essentially what "Earth" means: dirt.

After Earth comes Mars, who was the god of war and agriculture. Known as Ares in Greek, Mars was the son of Jupiter and Juno and had four siblings: Vulcan, the god of fire; Minerva, the god of wisdom; Bellona, the goddess of war; and the mighty Hercules, the god of strength and heroes. Mars's interesting family doesn't stop there: he was also father of the twins Romulus and Remus. According to myth, Mars raped the vestal virgin Rhea and she begot Romulus and Remus, who founded Rome and eventually the Roman Empire.

Mars also fell in love with his half-sister, the beautiful Minerva, only to have his advances spurned. Mars, not to be deterred in his incestuous passion for his half-sister, sought the help of an old goddess, Anna Perenna, who herself secretly fancied Mars. So, Anna Perenna subsequently disguised herself as Minerva and not only bedded Mars but married him. I guess Mars never got to enjoy that incestuous interlude with his half-sister Minerva. Along with being dubbed the "red planet," because it can appear fiery red in the night sky, Mars lends its name to the month of March.

We've now arrived at the four gaseous planets—Jupiter, Saturn, Uranus, and Neptune—which consist of ever increasing densities of various gases as one travels toward the core of each planet, mostly hydrogen gas. The average temperature of these four planets at the point where a "surface" could be considered to exist—that is, where the atmospheric gases give way to the planet gases—ranges from a nippy -234°F to a nippier -350°F. That's cold even for Oymyakon, Russia, the coldest permanently inhabited place on Earth, where the average winter temperature—that's right, *average*—is -50°F and the average annual temperature a balmy 3.2°F.

Jupiter, Zeus in Greek, was the king of the new gods, as well as the god of the sky and of thunder, which is why most renditions of Jupiter show him holding a lightning bolt with an eagle. His father, Saturn (Cronus in Greek), gives his name to the next planet out, who was the first ruler of the new gods before his son Jupiter usurped him.

How and why did Jupiter take down his father Saturn? Before Saturn had children, he was warned that someday one of his sons would become king of all the gods, and the power-hungry Saturn was not about to give up his throne, even to a son. Gods are immortal, so their reign is forever. Whenever Saturn's wife, Ops (Rhea to the Greeks), gave birth, Saturn would—I am sorry to say—swallow the babies whole as they popped out, so as to prevent the prophecy from unfolding.

By the time Jupiter was born, Ops had had enough of her husband devouring their children, so she tricked Saturn, moving the newborn Jupiter to the island of Crete and instead handing Saturn a rock wrapped in a swaddling blanket. Saturn, without verifying what was in the blanket, swallowed the package whole, never the wiser.

When Jupiter came of age, he successfully challenged his father to become king of the gods, not only taking the throne but also forcing Saturn to expel from his stomach all the other babies he had previously swallowed. Out came Neptune and Pluto, as well as Ceres, Vesta, and Juno, who lend their names to large asteroids. It is unclear if Saturn also expelled that swaddled rock and or realized how he had been tricked. Mythology is rather unclear on the subject.

Ceres is the largest known object in the asteroid belt, which is that circumstellar swath of leftover rocky schmutz between Mars and Jupiter, debris left over from when our solar system first formed. Because Ceres is such a large asteroid, it is sometimes referred to as a dwarf planet. Vesta is the goddess of the hearth and has a large object in the asteroid belt named for her; perhaps it's even another dwarf planet. Vestal virgins were ancient Roman priestesses of Vesta, sworn to chastity, their sole duty to keep the sacred flame of Rome from going out. Which is odd because Romulus and Remus, founders of Rome were the result of Mars raping the vestal virgin Rhea; yet Romans decided Rome needed the other vestal virgin Vesta, sworn to chastity to keep the flame of Rome burning. There is some sort of irony in that, or hypocrisy. As for Juno, she is special counsel to the gods and the protector of women, and also spends her endless days and nights circling the sun with Ceres and Vesta in the asteroid belt.

Saturn was both an old god and a new god, bridging the two worlds, and in each world he was the god of plenty, wealth, agriculture, birth, dissolution, and renewal. Before his wife, Ops, with whom he had (and ate) children, his consort was Lua, the goddess of destruction and of bloodied weapons—that is, weapons obtained from destroyed enemies. Oddly, Saturn was a good god to mankind, not demanding much from us puny humans. It was a sort of golden age of the gods and, as time would reveal, the father, Saturn, was better liked by humans than the son, the usurper Jupiter, would be. Jupiter was a task master. Talk about a game of thrones.

Jupiter was not as easygoing on us hapless humans as his dad. So, out of allegiance to the old king of the old and new gods, the ancient Greeks and Romans continued to celebrate the festival of Saturnalia, which fell in December, likely to coincide with harvesttime and the winter solstice, the shortest day of the year. This long-celebrated festival from antiquity also ended up, as discussed previously, being hijacked by Christendom in the fourth century AD, when the clever Church moved the birth of Jesus from summertime to December 25, to pull Romans away from the likes of Saturn and Jupiter and toward the teachings of Jesus. It, of course, worked.

Saturn's name is used in yet another way. Saturnism or saturnalia poisoning is a historical name for lead (Pb) poisoning. Lead poisoning, which continues to be a modern-day issue through environmental exposure to leaded gasoline, lead-based paint, and lead pipes—the element lead represented by the letter Pb is from the Latin *plumba* from which is derived plumber who would use lead pipes— was also a problem in antiquity, although they did not know what was going on. In preparations for Saturnalia, revelers would collect huge amounts of wine and beer, storing their alcohol in wood caskets made waterproof with a paste of lead, which is a soft earth metal. If you think the Irish drink a lot, it's child's play compared to the habits of the ancients, a well-pickled group.

The lead paste in the barrels would leach into the alcohol. Little did the merrymakers know they were on the road not only to becoming well-fortified with alcohol but to becoming well-fortified with lead poisoning. Before the connection between lead-paste barrels and lead poisoning was established, the drunken stupor was called "Saturnalia poisoning," the near-comatose revelers likely assuming Saturn had a hand in making the alcohol more "potent" than usual. Fast-forward to more modern times, when physicians finally figured out that lead causes poisoning: in a moment of sentimentality, they named the condition "saturnism." Old names die hard, as did those revelers—unless they were one of the lucky ones "saved by the bell."

The phrase "saved by the bell" indeed finds its origins in saturnism, or at least some historians believe so. Because lead poisoning can render a person comatose—quite distinct from your garden-variety blackout drunkenness—those who observed the apparently lifeless saturnism victims, your undertakers, morticians, and the like, often assumed these (unknowingly) poisoned people were dead. Their lead poisoning comatose was so deep as to be confused with being dearly departed. So, they buried them. What a horrible fate, to be first poisoned only to awaken hours later in a pine box under six feet of Dirty Dot earth. Because of this unique aspect of saturnism—making one appear deceased when not, as well as a few other comatose states—bells were tied to the ankles or big toe of presumed dead revelers while preparations were made for their interment. The slightest wiggle of the bell, even when made in an unconscious state, would alert the undertaker that perhaps the person wasn't dead after all. They were—lucky for them—saved by the bell.

Next comes Uranus, who was the god of the heavens and the sky. Uranus, as we know, is the only planet in our solar system that uses the Greek mythological name, rather than the Roman, Caelus. Uranus would eventually procreate with his consort Gaia, but for reasons that are not clear, Uranus hated his children—all eighteen of them. Maybe it's understandable, really. With eighteen children running amok, you'd probably grow to hate every last one of them. Besides the twelve Titans, Uranus's children included the three one-eyed Cyclopes and the three Hecatoncheires, which were 100-handed, fifty-headed giants. In other words, three of the children of Uranus, each with fifty heads makes for 150 heads, and each with 100 hands makes 300 hands. And that was just three of his children; no wonder he hated them. Grubby hands and prying eyes. The need for alone time becomes ever more understandable.

Gaia was not happy Uranus hated their children. Mothers are after all very protective, instinctively vigilant. To seek revenge, she fashioned a sickle, a semicircular blade, and, along with her youngest Titan child, Cronus (Saturn in Greek), ambushed Uranus, castrated him, and threw his testicles into the sea—which caused, as we've already seen, the goddess Venus to spring forth. This was different from the 1993 Lorena Bobbitt affair: Lorena cut off her supposed abusive husband John's dong, not his testicles, while he lay asleep in bed and threw it out the car window into a field. Nothing sprang from that Bobbitt appendage except for arrests and trials, he for repeated rape and her for mutilation. Both were acquitted. The dong was successfully reattached.

The last of the four gaseous planets is Neptune, or Poseidon to the Greeks, who along with his brothers Jupiter and Pluto presided over the realm of man after the three of them overthrew their father, Saturn. Neptune is the god of both the salty sea and of freshwater, as well as the god of earthquakes. He possesses the mighty weapon of the trident, a three-pronged spear. After the three brothers Jupiter, Neptune, and Pluto took command of the human world, they divided it into a triumvirate—and apparently fared better than any of the Roman triumvirates. Jupiter took command of the sky, Neptune the sea and land, and Pluto the underworld. Poor Pluto—what hell has been continuously cast upon you?

Which brings us back to why Neil deGrasse Tyson and his colleagues on that international committee demoted Pluto to dwarf planet status.

To understand Pluto being exiled from the planet gang, we must leave mythology and turn to cosmology. When our sun was born, condensing and flinging debris into the void, that debris nearest to the sun, where it was warmer and where gravity was keener, formed the four terrestrial planets. The higher gravity forced gases into their liquid states at the core of these four planets— molten nickel and iron—and into solids like rock in the regions closer to the crust.

Farther out from the sun, right before the frost line—between Mars and Jupiter—gravity only provided for the oddly shaped asteroids to form, including those very large asteroids sometimes referred to as "planetesimals," planets that just didn't quite make it. Everything before the frost line is pretty much solid. And beyond the frost line near Jupiter, it was cold enough for volatile gases like water, methane ammonia, carbon dioxide and carbon monoxide to condense into their liquid states. Every planet beyond that snow line, beyond the frost line are gaseous, although not their moons, which, like asteroids, are solid satellites that were captured.

The massive size of Jupiter caused gravitational perturbations that prevented those planetesimals in the asteroid belt from coalescing into a real big-boy planet, dooming them to baby status planets forever. Although those asteroids sometimes misbehave, leaving their assigned station within the solar system jettisoning centripetally away from the sun or careening centrifugally towards the sun. One of those big bad boy asteroids is what doomed the dinosaurs into extinction, the Chicxulub asteroid that hit off Mexico's Yucatán Peninsula 66 million years ago. That geologic signature—a layer of dust with high levels of iridium ejected into the atmosphere, became part of the earth's crust—is variously termed the Cretaceous-Paleogene (K-Pg boundary) extinction event or the Cretaceous-Tertiary (K-T boundary) extinction. Before that signature geologic boundary *T. rex* and friends were free to roam, after it they practically vanished in a flash.

"WELL, THERE'S SOMETHING YOU DON'T SEE EVERY DAY."

Beyond the frost line, which is an imaginary line drawn in the sand of the cosmos, just before Jupiter, is where gaseous debris coalesced into the four gas planets. Although these four behemoths might have some type of solid metallic core, by and large they are either gas towards their surface and compressed gas into liquid (like your propane tank used to fire up the grill). These gravitational condensates were too cold to grab much in the way of larger elements like iron and nickel, much less gravitationally compress those gases and liquids into much resembling solids. Subsequently, we have the four gas giants: Jupiter, Saturn, Uranus, and Neptune, collectively known as the Jovian planets.

Which brings us full circle to Pluto, where you can see, stands out of sequence. It is a *solid, terrestrial* body, not gaseous, and it is actually more ice than rock. Simply, it doesn't follow the planetary pattern beyond the frost line, not to mention that its orbit around the sun sometimes takes it closer to the sun than Neptune's ecliptic, such an errant solar orbit would be more typical of an asteroid or comet than a planet. In other words, Pluto is less like a planet and more like Raj Koothrappali's (from *The Big Bang Theory* sitcom) trans-Neptunian objects (TNO) floating out there with other icy schmutz balls in the Kuiper Belt

The other thing about Pluto's orbit is that it's not on the same ecliptic plane as the eight planets—a plane determined by the gravity of the sun, which has its own equator. The gravitational influence of the sun's equator forces all eight planets to orbit in the plane of that equator. Pluto does not do this. If you were to look at an artist's image of our solar system, you'll see or you should see that all eight planets are in the same orbital plane, whereas Pluto should not be.

All this is why poor Pluto was redesignated as a dwarf planet, or a planetoid, and as such falls into the ranks of trans-Neptunian objects. Now you know Pluto was demoted for reasons far beyond the control of Neil deGrasse Tyson, so don't go blaming him—and especially don't blame Mike Tyson, a modern-day Hercules who had absolutely nothing to do with Pluto's demotion. The only connection Mike Tyson has with Pluto is that, if you faced Tyson in the boxing ring, there would be Hades to pay.

As mentioned, Pluto was discovered by the brilliant Clyde Tombaugh, a good ol' Illinois boy. Five of the planets other than Earth—Mercury, Venus, Mars, Jupiter and Saturn—were known from antiquity, so no one individual gets credit for their discovery. Uranus was discovered by the British astronomer William Herschel in 1781, and Neptune in 1846 by John Couch Adams, also a Brit. Pluto was the only American-discovered planet, and when it got demoted, America even lost that. Tombaugh died age ninety in 1997, fortunately before his beloved Pluto was demoted in 2006.

Let's now delve a little deeper into the differences between a planet and an asteroid. A planet is of course large, but, more importantly, planets orbit a star in a pretty precise ecliptic orbit. The plane of that orbit is predictable and the time it takes for one solar orbit is consistent, rarely varying over millions and millions of years. For Earth one solar orbit is 365.25 days, never varying, and which is why we have a leap year, introducing February 29 into the calendar every four years. An asteroid by comparison is smaller, and although it also orbits the sun, the orbital plane can vary widely and wildly, as can the period of its journey around the sun. In other words, a "year" on an asteroid is not always fixed, nor is its journey, nor is its ecliptic. Planets are predictably predictable; asteroids and comets are often predictably unpredictable.

Mercury and Venus do not have moons. The nearby sun simply had too much gravity to have allowed any cosmic leftovers to form into moons that could then have been captured by Mercury or Venus when our solar system was taking shape.

From our previous discussions, one might think Earth's moon formed from the coalescing of schmutz into an earth orbit, just as how Earth too formed from cosmic schmutz. And if one thinks that, who could blame you. But such was not the case. It is believed that Earth's moon formed from an asteroid impact around 4.5 billion years ago, jettisoning debris into orbit around Earth that eventually coalesced into our solid moon. Scientists believe this to be the case because our moon is made of material remarkably similar to Earth's crust, something we learned from the Apollo moon missions, material that would not have existed otherwise on the moon unless the moon came from Earth. So romantic.

There are other theories as to the origin of our moon, such as the "capture" theory, which proposes the moon formed elsewhere in the solar system and was captured by Earth into a gravitational orbit. The "condensation" theory states Earth and the moon formed independently but in proximity to each other from condensing gases, with the smaller moon eventually seized into orbit by the larger Earth. Of all the options, the "asteroid impact" theory remains the most plausible explanation for Earth's moon—maybe not other moons in the solar system but our moon—due to those similarities of crustal composition.

Earth's moon sort of orbits about the Earth's equator—it is off by 6.7 degrees—and that orbit does vary, based on the 23.5 degree tilt of Earth. That tilt of Earth to 23.5 degrees is believed to have happened when that same asteroid that struck creating our Moon also lurched earth into an incline. Every 18.6 years, the moon goes from one extreme angle off the equator to the opposite extreme angle, but always within 6.7 degrees. The moon orbits Earth counterclockwise, and both Earth and the moon spin on their axes counterclockwise. Why is the Earth tilted 23.5 degrees on its axis but nearly all the other planets apparently not? Well, that asteroid strike just mentioned is believed to have caused earth's tilt, but the other planets escaped such a calamity. Neptune is a strange bird, too, it's not tilted but its spin is 90 degrees off the plane of its solar ecliptic, almost appearing as if it fell over on its side. Maybe it did. If so, I'm sure the other planets got a huge chuckle out of that! That's like having your shoelaces tied together when you're not looking and fall over on the first step.

This next part is cool. The Earth and moon are in synchronous orbit, which means from Earth we can only see, and forever will only be able to see, the near side of the moon. The dark side of the moon is called precisely that because it has never and will never be seen from any point on Earth. Ever. In fact, the dark side of the moon was first seen by man during the Apollo missions.

After Earth, the rest of the planets, including Pluto, have moons. Mars has two moons, Phobos and Deimos, likely the result of Mars's gravitational pull forcing debris into two solid orbiting satellites. Jupiter has fifty-three known moons as of this writing, including Ganymede, the largest moon in our solar system, and is bigger than the planet Mercury. Other Jupiter moons include Europa, Io, Callisto, and Thebe. As of this writing, Saturn has sixty-two known moons, Uranus has twenty-seven, Neptune fourteen, and the non-planet planet Pluto has, as mentioned, five moons.

If you were to think that since Pluto has two large moons orbiting it—Charon and Styx—let alone three more, that that should give it back its planet status, think again. Those three smaller moons in orbit around Pluto—Hydra, Nix, and Kerberos—are likely not moons but captured asteroids. In fact, in the asteroid belt, if an asteroid is large enough, it might have smaller asteroids in orbit around it. This has not only been appreciated in the asteroid belt, it is suspected in the Kuiper belt as well. Gravity does that.

What is perplexing is that Jupiter, Saturn, Neptune, and Uranus are gas giants, but all their moons coalesced into solid spheres. Sort of odd, really—gaseous planets with solid terrestrial moons. Go figure. But it's likely due to the gravitational forces of the respective planets forcing their moons to become solid while the mothership planets remained gaseous.

The equatorial rings around Jupiter, like the rings around Saturn, Uranus, and Neptune, are precisely that: rings. And they are equatorial. They formed from chunks of dust, ice, and rock that were unable to either form into moons or be sucked into the planet or be jettisoned into the great void. Since the four gaseous planets' gravitational fields are not uniform, and the greatest gravitational influence is along the equator, all that debris orbiting those four planets eventually tracked to the middle equatorial region, forming those magnificent rings. The reason none of the planets, and the sun for that matter, are not perfect spheres, is because they spin, and in so doing, they are fatter at the equator.

Because bodies that are spinning tend to have a wider equator than circumference top to bottom—Earth is fatter at its equator than pole-to-pole, as if Atlas got tired of holding the Earth up and sat on it for a spell—the gravitational influence along the equator can be remarkable. The sun forces the eight planets into an ecliptic around its equator, our moon nearly orbits our equator, and the moons of Mars orbit about its equator. But when you get to the gaseous giants, all bets are off. Some of their regular satellite moons orbit near the equator, some irregular satellites do not. Probably because there wouldn't be enough room for them, and probably because they influence each other in orbit.

Not only a celebrated astrophysicist, Neil deGrasse Tyson is keenly skilled at teaching, able to converse in scientific jargon with colleagues and then switch gears to talk up astronomy to those among us who slept through third-period science. Tyson, like Carl Sagan before him, is able to reduce complex concepts to a comprehensible level, recognizing his audience yet not talking down, not being condescending. That is quite a skillset. For this reason, Tyson was the host of the highly successful science show *Nova ScienceNow* (2006–11) as well as the host of *Cosmos: A Spacetime Odyssey* (2014), the successor to Sagan's fabulously successful *Cosmos: A Personal Voyage*, a PBS special that aired thirteen episodes in 1980.

Tyson also coined the slogan "science is not a liberal conspiracy" as part of one of his more recent efforts to defend scientific knowledge against those who wish to marginalize it, who would use their fundamentalist biblical beliefs to lord over scientific fact. These folks are what you'd call your proverbial book-burning crowd.

Actual book burning is fortunately not a common ritual these days, although it was carried out in very, very public displays in the recent past, by those who wished to censor material that ran contrary to their ideological, cultural, religious, or political beliefs. Nazi Germany practiced book burning. I sometimes fear I live in an America where the reigniting of the book-burning bonfire is just a gas can and lit match away.

Short of book burning, but in a dubious move nevertheless, several schools in America have removed fabulous books from their library. A religious school in Nashville removed J. K. Rowling's *Harry Potter* series, and the classic *To Kill a Mockingbird* by Harper Lee has been removed from schools in Tennessee and Minnesota. A few school districts have also nixed these absolutely wonderful novels from their library: *The Adventures of Huckleberry Finn* by Mark Twain, *Of Mice and Men* by John Steinbeck, and *The Catcher in the Rye* by J. D. Salinger.

And with that, we've brought ourselves full circle from the dominance of religious dogma, to exploring Earth's original biodiversity as evidence of evolution, to a little sprinkling of astronomy, to Pasteur's "life from life" germ theory. This last topic is where our story now jumps off from: the microscopic world of human infection discovered after Pasteur.

18 A LEVEL PLAYING FIELD

If you ever took a biology class, you might have grown something in a round, shallow, clear glass dish with a growth medium inside, which we call a Petri dish. The dish is named after the German microbiologist Julius Richard Petri, who was a student of Robert Koch, another great German microbiologist. Petri received his medical training at the Kaiser Wilhelm Academy for Military Physicians in Berlin, and then worked in the same city at the Charité Hospital as well as in Koch's microbiology laboratory.

Koch continued on where Pasteur left off, going on to define the postulates that would bear his name—Koch's postulates—which are designed to help establish whether a specific microscopic organism is indeed an *infection-causing* microscopic organism. An important difference exists between the three classes of bacteria: those bacteria that don't cause infection and don't even get near humans, those bacteria that enjoy a mutualistic relationship with humans, helping us purr along, and those bacteria that do cause human infection. Koch provided a way to make that distinction, which we'll explore momentarily.

But before that, we need to introduce a few more players in this story. Walther Hesse was a German microbiologist who also worked in Koch's lab and who had developed a bacteria growth medium he called agar. The name "agar" is derived from the Greek and describes a range of red algae that grow on certain plants, such as seaweed. More specifically, agar is the jellylike substance produced when algae are heated, releasing the gelatinous material that had formed the cell wall structure. Algae are an odd group of life, mostly characterized as eukaryotes placing them firmly in the plant kingdom, yet blue-green algae are prokaryote and include the world famous cyanobacteria that gave us our oxygen atmosphere. Why that is so remains a mystery, but bacteria love growing on agar nevertheless, and if its cannibalism of a cousin its cannibalism.

For reasons that escape history, Hesse's wife, the American-born Angelina Fanny Hesse née Elishemius, who was also a microbiologist and worked alongside her husband, suggested to Petri and Koch that they should use her husband's agar as a growth medium. Petri looked into Hesse's algae agar concept and then ran with the idea, thus developing the standard glass dish filled with agar used for culturing bacteria, which has the essential nutrients any growing bacteria could possibly want. And this is why the little shallow Petri dish bears the Petri name.

Why was the Petri dish such an important invention? Standardization in microbiology was becoming ever more critical, as the ability to grow bacteria consistently and reliably in labs around the world was a pressing concern. After all, being able to identify bacteria such as *Streptococcus* as the pathogen that caused strep throat in Berlin and strep throat in Paris and strep throat in London and strep throat in New York was decidedly advantageous. Having a standard growth medium allowed microbiologist to compare notes by leveling the playing field, sort of like how football fields, soccer fields, and basketball courts are standardized. Baseball, as it were, is a different story altogether, as its playing field is not quite standard.

Sure, the infields of a baseball field are all the same, but the standardization ends there. If you can believe it, that elusive home run wall varies. The Cincinnati Reds' Great American Ball Park, the Baltimore Orioles' Oriole Park at Camden Yards, and Yankee Stadium in the Bronx have shorter distances to the home run wall, thus favoring the batter. Ballparks like Dodger Stadium in LA, the St. Louis Cardinals' Busch Stadium in St. Louis, and Angel Stadium, home of the Los Angeles Angels, have deeper home run walls, where the pitcher is favored.

Let's spare a thought, though, for how a Los Angeles baseball team decided to name itself "The Angels." *Los ángeles* is Spanish for "the angels," so the Los Angeles Angels baseball team is basically named the The Angels Angels. Unless angels have their own angels, it is a tad redundant.

When Robert Koch proved in 1882 that tuberculosis (TB) is caused by a specific bacterium now named *Mycobacterium tuberculosis*, it marked an important milestone in medicine. It was the first time a specific human infection was directly connected with a specific bacterium that had been grown in a lab on agar medium. Bacteria had been grown before and identified with names before, but Koch connecting TB on the Petri dish to the TB bacteria was a first of its kind. TB was the first infection where the dots were connected. It not only paved the way for the better diagnosing of tuberculosis, by culturing the coughed-up sputum of the afflicted, it also inspired a slew of microbiologists to connect the dots between other genera and species of bacteria with known infections. Koch is also credited with identifying and connecting the dots for cholera and anthrax.

Streptococcus was identified as the culprit for strep throat and *Staphylococcus* was pinned as the common critter that gives rise to skin infections like God's sixth plague—boils—and so on. The ability to pair a particular bacterium with a particular infection with a particular constellation of symptoms also eventually allowed physicians to pair a specific antibiotic with that specific bacterium to treat that specific infection … that is, once antibiotics were discovered.

When Koch published his paper in 1882 connecting the lung infection tuberculosis to the bacteria *Mycobacterium*, he gave absolutely no credit to either Walther or Fanny Hesse for recommending the algae agar used to make his discovery. Nor did the Hesse family receive any royalties from Julius Petri for agar Petri dish sales. I don't think Koch gave Petri much credit either, for creating the Petri dish in the first place. Koch apparently liked to be centerstage with no other actors hanging out in the wings. No matter for the Hesses—turns out Fanny came from money. Her father, Gottfried Elishemius, was a wealthy New York import merchant.

It is unclear how and where Fanny, a New Yorker, met her husband, Walther, a German. Some accounts say the two met in 1872 when Walther, who was working as a German merchant ship's doctor, was docked in New York Harbor. Others claim the two met in Germany while Fanny was traveling abroad. Either way, they were married in 1874 in Geneva, Switzerland.

As we have previously discussed, while many microorganisms are killed in the process of fermentation, like that used to make beer and wine, due to the acidic environment it creates, plus some exothermic heat produced by the fermentation, not all of them critters are killed. It was also known in the late nineteenth century that many highly desirable foods, such as cow's milk—which does not undergo fermentation unless it is spoiled—in which case you throw it away—often came with deadly bacteria. Some of the infections connected to consuming raw animal milk that were more or less verified in the late 1800s through microbiology include brucellosis, bovine tuberculosis, listeriosis, salmonella infection, E. coli infection, campylobacteriosis—it's a long list. These infections also can be acquired through other avenues besides unpasteurized milk, but unpasteurized milk is an infection pathway they share. Louis Pasteur in 1862 was able to discover that higher temperatures kill even more pathogens than fermentation does, so much so that the bacterial load after direct heat is applied is either nil or not substantial enough to cause infection.

It's possible for cow's milk to harbor agents infectious to humans even if the cow it comes from is not ill. These culprits include the just mentioned *Mycobacterium*, *Listeria*, *Salmonella*, and *Campylobacter*. The key to killing bacteria in milk is using heat to render the bacteria dead without rendering the milk tasteless. Otherwise, what would be the point. Pasteur demonstrated that heating milk to temperatures around 140°F (60°C) for twenty minutes will scorch pretty much all the bacterial vermin yet not ruin the milk sugars and milk fats—the importance of tastiness not lost on Pasteur.

In honor of its creator, the process of heating milk to kill microbes was named "pasteurization." Today, several industrial processes exist for the pasteurization of milk, such as 161°F (72°C) for fifteen seconds and 284°F (140°C) for four seconds. Besides purging milk of unwanted infectious guests, pasteurization also prevents spoilage, that is, it prolongs the shelf life of milk, beer, canned foods, dairy products, juices, low-alcohol beverages, syrups, wines, vinegar, nuts, and even water itself. It kills bacteria that might not want to harm us but most certainly if sour will give us a sour gulp. Pasteurization is what allows us to have that expiration date on cartons of milk and whatnot, which we all carefully check

LOUIS PASTEUR LOOKS BACK

Of course I'm very pleased with pasteurized milk, but naming a wine or beer after me would have been a lot more charismatic.

lest we swallow a mouthful of curdled sour milk and spit it out all over the floor.

The milk spoilage that occurs after milk has been pasteurized is due to some beneficial bacterial load remaining after pasteurization. These bacteria—lactose bacteria, or *Lactobacillus*—are generally considered friendly, but given enough days or insufficient refrigeration, our *Lactobacillus* pals will turn against us, converting what was once a nice bottle of milk into a foul-smelling sour glop of gloop. The *Lactobacillus* in sour milk won't cause infection if swallowed, but the sour taste—one difficult to eradicate with toothpaste and mouthwash—will turn your day sour. Which is why the smart person not only looks at the expiration date on a carton of milk but does the sniff test, too. The sour of sour milk is due to those *Lactobacillus* breaking down the milk sugar to lactic acid, which like many acids is sour to the taste and to the scent. The chemistry definition of an acid is a chemical that donates a proton (H+), but not all acids donate protons. However, one thing is for certain: pretty much all acids taste sour. Don't try it—just take my word for it.

Which brings us full circle to the original postulate of Girolamo Fracastoro in the 1500s that a nonliving "spore" was the cause of an infection—a radical departure from Galen's miasmas—was partially validated when nineteenth-century microbiologists, like Pasteur and Koch, demonstrated that not nonliving spores but similarly sized *living bacteria*, which bore a resemblance to the definition of "spore" (a tiny seed), caused infection. Essentially, the only major item Fracastoro missed in his theory was that the nonliving "spore" concept of his was indeed living. Interestingly, Fracastoro received his medical education in the ancient city of Padua, a city founded by Prince Antenor in 1183 BC after the Greeks destroyed Troy during the Trojan War, a war made famous by Homer in *The Iliad*.

Prince Antenor was adviser to King Priam of Troy, in Anatolia (now Turkey). Antenor, along with Priam's eldest son, Hector, advised that the Trojans should return Helen—coveted and kidnapped by Hector's younger brother, Paris—to her husband, the Greek Menelaus. We all know how the subsequent ten-year Greek siege upon the impenetrable walls of Troy ended, after Priam failed to release Helen: Odysseus's Trojan Horse scheme and the fall of Troy. It also ended with the death of the greatest warrior to have ever lived, Achilles, killed by an arrow to his heel, the only vulnerable spot on Achilles's body. The arrow was shot by Paris but guided by the god Apollo. Achilles's misfortune is, of course, how we arrived at the phrase "Achilles' heel" to describe a weak point.

When Achilles was born to his mother, the immortal sea nymph Thetis, and his father, the mortal Peleus, king of the Myrmidon Greeks, a prophecy was foretold that Achilles would die young. This prophecy so troubled Thetis that she subsequently took her baby Achilles to the River Styx which not only crossed over into Hades but had special waters that offered powers of invulnerability. Thetis dipped Achilles into the waters, holding him by his heels, which subsequently were the only part of the infant's body not washed over by the magical Styx waters. Achilles's heels remained vulnerable—yet no one knew this, including Achilles. Well, almost no one. Apollo was aware, since gods know things, I guess.

Paris of Troy, not an especially skilled warrior or archer, shot an arrow at Achilles as his city was falling to the Greeks (Paris was quite upset that Achilles had earlier killed his brother Hector). That arrow wouldn't have killed Achilles save for the intervention of Apollo, who guided the otherwise errant bolt straight to the heel of Achilles. Apollo, you see, had a bone to pick with Achilles for an earlier desecration of one of his statues.

Although the Trojan War is largely myth, most scholars believe there is some validity to such a conflict actually having occurred. Having said that, though, it is unlikely the real Trojan War was started over Menelaus's kidnapped wife, Helen—notably the most beautiful woman in the world at the time—as Homer suggests in his epic. The Greeks, and especially their overall king, Agamemnon, wanted to conquer Troy, so used the kidnapped Helen as an excuse to rally the Greek troops and go to war against the Trojans. I suppose a kidnapped wife is as good a reason as any to go to war—and perhaps a better, or even the best, reason. As for a kidnapped husband … I mean, who would really care to go to war over that? Perhaps there would be a celebration with lead-laced beer, a saturnalia festival. But *conflict* over a kidnapped husband? Unlikely.

Fracastoro is known not only for his spore theory of infection but also for being curiously at the center of a novel infection that just so happened to be raging and ravaging sixteenth-century Europe, an infection that was unintentionally and circuitously named by Fracastoro himself: syphilis. It is truly bizarre how that disease came to be called "syphilis."

The venereal infection, imported into Europe from the Americas by European sailors beginning in the 1600s, originally went by other names, the most common of which was "the great pox." Other names included "lues," which is short for the Latin *lues venerea* or "venereal plague," and it was also called the "French pox," simply because one of the early descriptions of syphilis was during the French invasion of Naples around 1495. Other names for syphilis were the "Spanish disease," with a similar story to its French counterpart, and "Cupid's disease," which really needs no explanation, as well as still other descriptives. But what we're interested in here is why the medical world settled on the name "syphilis."

In 1530, Girolamo Fracastoro wrote an epic pastoral poem—I suppose I should mention he was both a scientist and a poet—entitled *Syphilis Sive Morbus Gallicus*, which loosely translates to *Syphilis or the French Disease*. There was no love lost between the Italians and the French at the time. The poem tells the tale of a presumably French shepherd boy named Syphilus who insults the Greek god Apollo and is then punished with the horrible disease then ravaging Europe. Although syphilis was not just a "French disease," seeing how it was ravaging all of Europe, Fracastoro either didn't know or didn't care about that fine little detail. The French had recently seized Naples, and Italians like Fracastoro were none too happy.

Fracastoro's poem clearly laid blame for the great pox on French lovers. Of course, this is historically incorrect, as it was in fact mostly Spanish and Portuguese Age of Discovery sailors, plus French, Italian, British, and all manner of explorers who brought syphilis back into European ports from the New World during the Age of Discovery. Christopher Columbus and later sailors might have struggled to find that ocean crossing to the Far East's Spice Islands, might have failed to find gold to bring back to Europe, but they did find syphilis. And when sailors disembarked at their home ports, syphilis disembarked within their genitals.

Syphilis arrived in Europe from the New World as part of what is known as the Columbian exchange—an exchange of commodities, mostly plants and animals, between the Americas and Europe during the Age of Discovery. Some of the commodities brought from Europe to the Americas, commodities that did not exist in the New World, included olives, sugarcane, pears, peaches, coffee, oranges, lemons, bananas, cows, sheep, pigs, and horses, and horribly, tragically, enslaved African people. Commodities that flowed from the Americas that did not exist in Europe included beans, chili peppers, tobacco, corn, tomatoes, potatoes, cotton, and turkeys. But, as we know, it didn't end there.

Other exchanges during the Age of Discovery occurred on the microbial level. An exchange of infectious diseases. Spanish and Portuguese explorers, and later on Italian, English, and French colonialists, arrived on the shores of the New World and disembarked along with cholera, mumps, measles, smallpox, typhus, whooping cough and the flu. These pathogens might make a European back home merely ill, but, without a whit of immunity, Indigenous Peoples in the Americas were killed in droves, as they lacked any immune history with those pathogens. If a population has never been exposed to a specific bacterial or viral scourge, such as smallpox, it will run through entire villages like wildfire. The world is currently experiencing how a novel infection can rip a new one when there is no natural immunity, the currently raging COVID-19.

As for European infections laying waste to Indigenous People of the New World, the Good Book says, an eye for an eye: Indigenous Peoples exchanged like for like with the European colonizers. Sailors returning home to European ports after plundering the New World, including raping the women who lived there, brought syphilis back to every seaside town in Europe. Syphilis was a mild disease to Indigenous people but decidedly not mild to Europeans, who lacked any immunity to what was then to them a novel pathogen. Payback is a biatch.

Whether syphilis had become more prevalent in some European countries than others hardly mattered to Fracastoro. As far as he was concerned, it was a French disease. No one knew then it was imported from the Americas. For reasons that are unclear, he named the boy in his poem Syphilus, a name that stuck in the minds of Europeans, just like the syphilis pathogen stuck in their genitals. From that day until this day, that notorious sexually transmitted disease has shed its many other names—the great pox, lues, venereal plague—and been called "syphilis." All because of a disgruntled Italian physician-poet who despised the French.

(As an aside, some etymologists think Fracastoro chose (and may have misspelled) the Latinized Greek word for pig-lover, ending up with Syphilus).

Along with Fracastoro's spore theory, Pasteur's work also validated Agostino Bassi's earlier mentioned silkworm experiment, which in 1835 demonstrated that a certain vegetable parasite causes the infectious disease muscardine in silkworms. Muscardine was at the time devastating the French silk industry. (Muscardine by the way is a fungal disease to be distinguished from muscadine without the "r" which is a type of grape.) The French, who obviously liked their silk, had long ago stolen the secrets of the silkworm caterpillar from the Chinese, much in the same way the British stole the secrets of tea cultivation from China and brought them to British India. For the previous few thousand years, silk had made its way into Europe only along the ancient Silk Road, which is why it was called precisely that.

To cut out the middleman and that long, treacherous journey of silk out of China along the Silk Road, Europeans and apparently specifically the French, smuggled the silkworm out of China and began processing their own silk. The first people to take silkworms out of China were two Byzantine monks, who in 550 AD hid silkworm eggs in their bamboo walking sticks. I mean, if you can't trust a monk not to steal your silkworms, who *can* you trust?

It's a similar story for tea. The British East India Company stole Chinese tea plants and brought them to India, which had appropriate soil for growing tea. This is why tea was not cultivated directly in Europe—the earth, the dirt, the climate wasn't right in Europe, but it was in India. The thief in this instance was the botanist Robert Fortune, who in 1848 disguised himself as a Chinese peasant—how local Chinese officials could not ascertain a Brit was not Chinese is curiously absent in accounts of the theft—wandered into China, studied the secrets of tea cultivation, and quietly exited China with some tea plants, all on behalf of the British East India Company.

You might be wondering why Europeans didn't steal spice plants, too, which they were equally fond of. The highly treasured cinnamon, cassia, mace, nutmeg, cardamom, ginger, turmeric, and saffron come from the Moluccas Islands in Indonesia, otherwise known as the Spice Islands, and they could only grow on those few islands in the Banda Sea and nowhere else—not even India and certainly not Europe would those spices survive. I'm sure attempts were made to cultivate those treasured spices outside the Spice Islands but were met with repeated and disillusioned failure.

As for the silkworm, I guess it doesn't care if it is in China or some silkworm farm outside Paris, so long as the temperature is within range and it is fed the only food it loves, the mulberry leaf. That was the catch for stealing the silkworm secrets from China; it wasn't the little critter; it was the little critter's specific dietary needs. Sounds like going out on a first date with a woman who's a vegetarian, lactose intolerant, likes green bell peppers but allergic to red bell peppers, sulfites give her headaches, nitrites even worse, and of course, gluten intolerant, who while ordering dinner asks so many questions of the waiter that he begins rolling his eyes. In the midst of her litany of questions about the menu, the guy downs his scotch & soda and with pleading eyes, and a slight tip of his highball glass, nods to the waiter "may I have another."

The silk we wear is manufactured from the protein fibers, mostly fibroin, that the silkworm produces during its larval stage. The silkworm is actually not a worm; it is the caterpillar stage of the domestic silk moth that eventually morphs, obviously, into the silk moth. It spins fibroin silk to form its cocoon. Although silk can be produced from the cocoons of several species of agreeable caterpillars, the most coveted comes from *Bombyx mori*, or mulberry silkworm larvae, which these days are grown in silkworm farms around the world. The word "silk" is from the Old English *sioloc*, derived from the Greek *serikos*, apparently derived from the Chinese *sirghe* or *sirkek*, all of which mean about the same thing: "soft, shimmering."

As we previously learned the British East India Company successfully swiped tea cultivation from China and setup shop on agriculture plantations in British India, where the soil and climate were similar to China. For the silkworm, the soil and cooler climate of Europe were not an insurmountable problem, just grow the little guys indoors, but the mulberry tree, that was another matter altogether. It was not native to Europe and had no intention of growing in European soil and climate. At least not without some coaxing. So, it would take time for the silk industry in Europe to cultivate the mulberry tree in enough places in order to feed the insatiable and picky appetite of the silkworm. The specific mulberry tree, the one that was needed for the finnicky silkworm is native to warm, dry soil regions in Asia, Africa and parts of the Americas. Fortunately for the French silk industry, the mulberry tree that was needed for the mulberry silkworm could be found outside of China.

Once your silkworm makes you protein fibers for its cocoon, how then is the silk fabric made? Well, we've already touched upon how our caveman ancestors invented string some 40,000 years back, and crude cloth weaving 28,000 years ago, all that horizontal weft thread weaving through the vertical warp thread magic. The key was to first make thread, and when humans discovered that overlapping and staggering hair or fiber not only created thread, but that thread were exponentially stronger due to the overlap, well, that was a pretty nifty invention, right up there with the wheel and the iPhone.

Sometime around 5,000 to 6,000 years ago, probably in Egypt but also in India, weavers began taking all sorts of material to make fabric. Sheep gave wool, flax plant gave linen, hemp plant gives hemp (and also something to smoke for those who were so inclined), and cotton plant gave cotton. And silkworm cocoon fibers gave silk. No longer did fashions, especially for women, require the latest fad in deer or camel skin. Once you had linen, and then cotton, you could dye the fabric into all sorts of trendy colors. The discovery of such dyes figures later in this story, alongside the discovery of some antibiotics.

As for the French and the devastation of their silk industry, correctly deduced by Bassi to have been caused by a parasite that was infecting the silk larvae: it turned out that the parasite was a vegetable pathogen fungus—a fungus now named after the man himself: *Beauveria bassiana*.

In the end, Pasteur extinguished Galen's miasma theory, which had been hanging around like, well, a miasma since the time of the ancient Greeks. Others besides Pasteur contributed greatly to the advancement of germ theory—John Snow and the Broad Streep pump cholera epidemic—and to the fledgling field of microbiology, although none more than the above-mentioned Robert Koch and fellow German scientist Ferdinand Cohn.

Koch was born in 1843 in the mountainous Harz region of Prussia (modern-day Germany) and received his medical education at the University of Göttingen under Professor Jakob Henle, whose own theories on microbiology helped mold Koch—no pun intended. The Harz mountains, incidentally, would become the last stronghold of Nazi Germany's Western Front during the waning days of World War II, before the Allied forces crushed the Wehrmacht.

In addition to Koch isolating and demonstrating that *Mycobacterium tuberculosis* causes tuberculosis, he also proved *Bacillus anthracis* caused anthrax and that *Vibrio cholera* gives rise to cholera—the bacteria at the center of the story of that 1854 Broad Street pump corruption.

As we briefly touched on earlier, Koch also developed what are now known as Koch's postulates, four criteria that determine whether a certain type of bacteria is the cause of a specific human infection, as opposed to one that is not infection causing. These postulates might seem intuitive in hindsight but given the climate at the time, breaking from Galen's miasma theory and considering bacteria had just been discovered as a cause of infection, being able to distinguish between good, friendly bacteria and bad, evil bacteria was a new and pressing problem.

Koch's four postulates are quite to the point. First, any bacteria that cause human infection must not be present in healthy people (actually that's now known not to be entirely true, as some colonized bacteria can turn rogue on us hapless souls). Second, that bacterium must be isolated from an infected person and be cultivable on a Petri dish. Third, that cultivated bacteria, when introduced into another person, must cause the same exact infection (in today's world that would constitute illegal experimentation). And, finally, that same bacteria must be reisolated and recultivated from that second person and be identical to the bacteria in the first person.

Koch's postulates seem so simple, but they laid the foundation for scientists to be able to separate bad bacteria from good bacteria. Which is where our story now turns: a scene where a wet-behind-the-ears medical student is navigating the world of infection in a Denver emergency room, or so he thought.

There I was that day in third-year medical school, stationed in the Dog House ER, having been assigned Gurney #3, who was complaining of flu-like symptoms. However, while Gurney #3 had symptoms like those caused by the flu, which as we know is a viral infection, she should not have had intestinal symptoms, which she did, as vague as they were. Children with the cold or the flu can get an upset tummy, but adults generally do not. Respiratory colds and flus are different than intestinal flus—an entirely different family of viruses. The nausea Gurney #3 had fell outside the constellation of typical adult respiratory flu symptoms. In retrospect, it was a subtle red flag, and I totally missed it. Initially.

For us medical students, the way an ER rotation worked—and I assume still works—was quite simple. We were assigned a safe patient, meaning one not in the throes of a true emergency, a patient not standing on the edge of a cliff. It would be bad form indeed to assign a newbie medical student a patient on the edge, lest that patient fall over the brink and start circling a whirlpool right before the student's well-intentioned but uncomprehending eyes. That would not be good. To be fair, medical students know more textbook medicine than anyone on the planet, including residents and including full-fledged doctors, but *practical* medicine—that is a different set of skills altogether and only comes with experience. So, there I was, a medical student on ER rotation at the Dog House, assigned the safe patient of Gurney #3—mere flu-like symptoms. She didn't have a primary care physician (PCP) to visit, and she, like many Americans without a PCP, used the emergency room as a walk-in clinic. And who could blame them.

As you'll recall, the name of the game for medical students at the Dog House was to do an initial interview and a complete history and physical exam, the true goal of which was impressing the attending ER physician, to whom we reported our findings and who graded our performances on that rotation. The polite way of describing that would be acting in a very overt obsequious manner; the vernacular would be brown-nosing. I had to make a good impression on the attending with Gurney #3, and so I dutifully shimmied through all the hoops.

Gurney #3 was in her late fifties, was born in Mexico, and had immigrated to the US as a child, upon which she became an American citizen. Her children and grandchildren were born in the States, and although her English was somewhat broken, I could understand her just fine. Her family was bilingual, fluent in English and Spanish, a skill I lament not possessing. But I'm clumsy enough with English as it is, so you know, there is that. Gurney #3 reported she had not been feeling well that morning, maybe the night before, and the symptoms included a general "ill" feeling, shortness of breath, chest fullness, an occasional cough, and some nausea. She was not running a fever. The flu often presents with little in the way of a fever, truer still with a cold.

I obsequiously reported my findings to the attending physician, who then asked me a series of questions, in a manner like how Socrates might have interrogated his students as they gathered at his feet on the steps of some Greek temple, prodding them along. The attending ER doc was poking at me to see if I had performed a complete history and physical exam—all those seemingly inane questions I was required to ask about Gurney #3's work experience, what her husband did for a living, the ages of her children, where she had recently traveled … her favorite color … what she thought of John Denver … I think you get the gist. She wasn't feeling well, and some dumb medical student was asking her all sorts of useless questions. The attending's other objective as he was prodding the history and physical exam I had performed was to pull me along to reach a "differential diagnosis." That was the teaching part, buried in the minutiae.

In medicine, a differential diagnosis is a set of possibilities based on presentation—signs & symptoms, but as you drill down a differential can include lab tests & imaging—where each possibility has its own constellation of characteristic findings, but the differential also has extreme overlaps. The key, or one of the keys, is to filter out the nonoverlapping clues that, as unique findings unequivocally, undeniably point toward the correct diagnosis. Sounds great in theory, but it seldom pans out that way in the clinical setting. Even computers would be little better at it.

Computer programs have been written—they even existed back when I was in medical school—to see if a computer algorithm could stand in for a doctor, take a history—patient enters the data—and arrive at a diagnosis. The concept then, as it is now, was fraught with failure. Due to the symptom palette of many illnesses overlapping, the patient's subjective bias of their own symptoms and the total absence of a physical exam, computers can do little more than offer a differential diagnosis, something any self-respecting physician is quite capable of performing just as faster if not faster. I think it is safe to say that cyborgs are not soon to replace physicians on the frontline since a computer cannot take a physical exam, not an exam that removes the patient taking their own physical exam. Sure there are some things that can be monitored remotely, like blood pressure, heart rate, heart arrhythmias and blood oxygenation. But a complete physical, ain't going to happen. Even modern-day robotic surgery is not a robot conducting the surgery, the surgeon controls the robotic arms, which, it is true, provides access and scalability to nooks and crannies within the body. But still the robotic arm is just an extension of the surgeon's hands.

Entire books have been written about differential diagnoses, and, as preparedness would have it, I indubitably possessed one those tiny booklets stuffed into that not-very-clean, waist-length white medical lab coat I wore. Another pamphlet I had on me was a cheat sheet on ordering tests that would help prove or disprove the diagnosis of a multitude of diseases; a third piece of literature was a general "how to treat a gazillion and one diseases" type thing; and the fourth pocket-sized booklet was a nifty tiny pamphlet on drugs: their indications, doses, contraindications, drug-drug interactions, and precautions. I had in total four palm-sized guides stuffed into my grimy but spacious lab coat, the type of medical bibles any self-respecting, aspiring dolt of a medical student possessed for navigating the turbulent waters of hospital rotations. These little publications contained the things they really don't teach in the first two years of medical school. Today's dolt medical students no doubt use their iPhone and perform a Google search.

When you present your findings to an attending physician, and are being graded on that presentation, you can't very well plunge your hand into one of the pockets of your grimy waist-length jacket and ask the attending to "wait a sec" while you thumb your way through cheat pamphlets looking for the answer to what diseases cause blue urine—nor can you launch a browser on your iPhone for the same purpose. That would be bad form. We were expected to know on the tip of our tongues the differential diagnosis, the test to order, how to treat the problem at hand, and the drug of choice, including dosage and precautions—for every disease under the sun. That's the medical school student's cross to bear.

By the way, blue urine is a real affliction. It is caused by an X-linked (males only) recessive genetic disease where the body fails to absorb across the gut lining the much needed tryptophan, an essential amino acid, from the diet. The unabsorbed tryptophan in the gut is then partially digested by gut bacteria into indole, which is blue, and which then which is ironically absorbed further down the intestines. The absorbed into the blood indole is then dumped into the urine, giving newborn babies blue urine: blue diaper syndrome. Long term, the indole can damage the kidneys, so a special diet low in tryptophan is the order of the day, forever, only taking in enough of that essential amino acid as absolutely required. Tryptophan, an amino acid, is used to make various protein neurotransmitters like serotonin.

When an attending doctor takes you down lines of inquiry, when they're pimping you to see what you know or don't know, the one thing worse than saying "I don't know" would be to make something up. Not knowing something might lose you a few points but lying in medical school is an unforgiveable transgression. You can always look something up. But being dishonest is a big deal—it needs to be rooted out. Deceit is a far worse sin than not knowing. Of course, repeatedly not knowing would not bode well either.

When I was in medical school, officially published cheat pamphlets as I described above were actually few and far between. They really didn't exist in great numbers, and, worse yet, you needed *multiple* pamphlets stuffed into those grimy pockets, as no one bothered to merge them. So, I decided to take that task upon myself, writing my own "all-inclusive" little black book, about the size of my hand—but fatter—that contained nearly an entire textbook of medicine written in small block print only someone with good vision could decipher. If I had any brains my little black book would have had phone numbers of cute girls. All these years later, I still have that little book, the medical one, not a cute girl phone number one. It is sitting there, on a shelf, not more than a few feet from me now, a preserved ink-and-paper version of my young medical student brain. Being a specialist, I have no need for that tiny book in my practice, nor could I now read that tiny print—although at my advancing age I still don't require reading glasses. It is a fond memory of days gone by, now collecting dust.

After relaying to the attending physician my initial encounter with Gurney #3, and after the exhausting dog and pony show of a Q&A, that rapid repartee between the attending and me, to see what I *didn't* know rather than what I *did* know, we settled on a preliminary diagnosis of "infection-like." I'm a surgeon, I like to fix things, so "infection-like" was sort of a huge let-down, something hanging out there on a limb. One of the reasons I didn't go into a medicine specialty, like internal medicine as did my father, is because you can point to many moments like these where there is a lot of talk and not enough decisive action. In medical school, medical rounds to me were an endless roll call of disease possibilities and laboratory tests to order, an academic orgy. I always found it exhausting. I'm a get-it-done type: I like to fix things, usually with the help of a scalpel, and then I like to move on. Medical school surgery rotations were more suited to my personality, a surgical problem to fix. It is true that some presentations had a differential and needed a work-up, but most were self-evident and needed a fixer-upper.

I gave the lab test orders to the ER nurse who was assigned to Gurney #3, and she, knowing I was a neophyte medical school student, smiled and nodded, all the while likely thinking to herself: "What a putz, another medical school student who doesn't know shit from Shinola." Actually, medical students, who know a lot of stuff, but mostly useless stuff, not the art of medicine, need to start somewhere.

"Shit from Shinola" is quite the phrase, one that dates to 1940s America, where Shinola was a popular brand of shoe polish. The figurative meaning is a person who is stupid, ignorant, or lacks common sense, apparently based on the literal concept that such a person could have their shoes shined with shit rather than Shinola and they wouldn't know the difference.

After giving the lab orders to the nurse, they miraculously appeared on the Big Board. I was so proud. I agree with the nurse, what a putz I was. It wasn't a particular impressive spread of orders: a blood count, a few blood chemistries, a urinalysis and a chest X-ray. But they were my orders, so I was chuffed. It takes a few hours to get results back, or at least back in those days it did. Things are faster now. In the meantime, I jumped in on other things happening in the Dog House ER, you know, the gunshot down in Trauma 1, the forehead laceration on Gurney #7, the broken arm in Ortho 2, and the guy who feigned an illness so he could get a free ambulance ride to the hospital to fill a fake prescription for narcotics. If I recall, the prescription was written for "mofeen."

While I was distracted with all of the comings and goings of the ER that day, waiting on the labs I had ordered, Gurney #3 was evolving. Evolving right before my eyes, my uncomprehending, clueless eyes. I checked on her from time to time, noticing that she was more restless. Uncomfortable lying down, she much preferred to sit up on the edge of that gurney. Now, it is common practice in hospitals and especially emergency rooms for patients to remain on their gurneys with the siderails up, lest they fall off, break their neck, and get admitted to the neurosurgery service instead. Prime pickings for an ambulance-chasing lawyer to sue the hospital.

Despite those hospital rules of keeping the siderails up, every time I checked on Gurney #3, she had lowered the siderails and was sitting up—I would remind her to lay back down and keep the siderails up—when I checked yet again, the siderails were down and she was upright. This seesaw was not a communication problem due to her broken English. Something was screaming at me right in my face and I was deaf to it. It was a "shit from Shinola" moment. Not my first, not my last.

Had I been a bit smarter, had I been more experienced, her insistence on sitting up would have told me something. It was an observable clinical sign—the sort of clue why a cyborg will never replace a physician—a hint that I was initially missing. I might have been a slow learner that day, the wheels sometimes do turn slowly, but as she was becoming more ill, I was becoming more learned.

In Koch's era, in the late 1800s, it was maybe okay to use humans as test subjects to prove a medical experiment. You know, try something out on your basic prisoner, transient or homeless person, psychiatric patient, or someone of diminished mental abilities. Even children remanded to state homes were experimented upon. The horror of it all. Today, human experimentation must follow a strict ethical rule book with an overseeing board and the full informed consent of the test subjects. "Full informed consent" by definition necessarily excludes all those above categories I just mentioned, at least in the United States. People of diminished capacity or suffering from mental demise, children, and prisoners cannot give full consent, as any consent obtained is considered to have been given under duress, tainted, making it invalid.

It is sad to realize yet mixed in with that sadness is a debt of gratitude, that with the rise of medicine, certainly since the dawn of the Scientific Revolution in the eighteenth century, subsequent to and coincident with the Renaissance that arose in the fourteenth century, medicine advanced through experimentation on people. Whether it was Koch deliberately giving a bacterial infection to a willing, or not so willing subject, or even Edward Jenner's discovery of the smallpox vaccine, knowledge was obtained by enterprises that would not meet modern ethical standards.

What was even worse, much worse, than the experiments conducted on humans, was the experiments conducted on animals. It was no doubt horrifying. If you think a slaughterhouse was inhuman, at least until regulations were enacted, I dare say they pale in comparison to what was likely performed on living animals to advance science and medicine. I don't want to go into detail, but the word "torture" comes to mind.

In medical school, to learn about how the heart medicines worked—increase a heart rate, decrease a heart, you get the picture—it was a big deal for the pharmacology department to have the medical students inject dogs with various cardiac drugs in a true dog and pony show. Fortunately, medical students were allowed to opt out—I did opt out as well as half the class, especially the women medical students. I did not need to inject a cardiac med into a dog's vein and watch an EKG monitor hooked-up to pooch to understand the physiology of the drug. You don't always need to observe something in order to appreciate its mechanism of action. And it's not simply taking their word for it. Once you reach a certain point in your education, high school maybe, certainly college, one should have the wherewithal to extrapolate from context.

Britain took the lead in protecting animals subjected to research, when in 1822 they passed the first animal protection law, followed by the 1876 Cruelty to Animals Act, rules and regulations for the sole purpose of enacting humane treatment of animals, especially when it came to medical and science testing. In the US, the American Society for the Prevention of Cruelty to Animals (ASPCA) was established in 1860s in order to force similar laws, and the ASPCA continues to do their good work. In the US the Animal Welfare Act was enacted in 1966 and has continued to be amended and further defined.

Vivisection is surgery and dissection preformed on a living organism that has a central nervous system (and presumably can perceive pain), in order to study alive internal structure and function. Nowadays, such investigations need to be performed on a fully anesthetized animal; I shudder to think what it was like a few hundred years ago when the animal was fully awake and strapped down. Interestingly, the laws governing research on animals actually doesn't apply to all animals; pretty much only to mammals, birds and a few others deemed worthy. Research on insects is not covered when it comes to cruel and unusual experimentation.

From Koch's hopefully benign experiments on humans, giving them bacterial infections, in order to establish his four postulates, we now turn to one of the first-ever recorded bacterial infections. When God delivered boils upon the Egyptians, he was not, according to Moses, experimenting, although he was deliberately infecting people. That is, when Moses went before the pharaoh requesting, "Let my people go," and the pharaoh refused, God sent down his sixth plague of boils. We know today that boils are caused by bacteria, and not just any bacteria, but usually *Staphylococcus aureus*. Staph is the most common causative agent brewing within skin boils—also called furuncles, abscesses, carbuncles—as well as staph species involved in other gross skin infections. Staphylococcus is also one of the more common causes of such unfortunate infections as septic infected joints, toxic shock syndrome, and blood poisoning or sepsis. Some species of staph release an exotoxin in defiled foods, giving rise to one type of food poisoning.

An *exo*toxin is a chemical manufactured inside bacteria and then released so it can travel about an organism, unleashing all kinds of havoc. An *endo*toxin is similar, a chemical that wreaks havoc, but instead of being manufactured inside the bacteria cell, it is a component of the bacteria cell wall, released when the bacteria ruptures. Why do some infectious bacteria release exotoxins or endotoxins? Simple: to extend the local infection and to extend the bacteria lineage. Exotoxins and endotoxins, in addition to facilitating the local spread of infection, also cause symptoms that help the bacteria leave that host and infect the next host. Hacking and sneezing and vomiting and diarrhea are not only a person responding to the invader, but those symptoms are the invader's *modus operandi* to find its way into another host.

With the current COVID-19, and the CDC recommendation of facial masks, hand washing, hand sanitizers, and social distancing, what those are intended to do is to take away from the coronavirus avenues for extending its lineage. Viruses are actually not very good at jumping from person-to-person as one might think; the only reason viruses entered the human theater 5,000 to 10,000 years ago, was there was enough of us living in close enough proximity to provide avenues for spread. And the symptoms the virus cause are precisely that: avenues of spread. In the absence of natural immunity or an anti-infective drug, what you have left is prudent precautions: face mask, handwashing, hand sanitizer, social distancing. People, presidents, can poo-poo it all they want; but they are on the wrong side of history.

The word "staph" comes from the Greek and describes a "cluster." The "cocci" part of the bacterium's name also comes from the Greek and means "round" or "spherical" structure. Taken together, *Staphylococcus* bacteria appear under the microscope as "spherical clusters," like a bunch of grapes. And that is exactly what they look like under the microscope: a harmless-appearing cluster of grape bacteria. They are, however, usually anything but harmless.

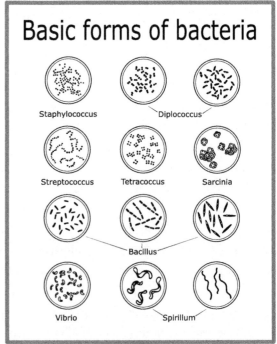

Streptococcus, which causes strep throat, is derived from the Greek *strepto*, for "string," plus the above "cocci." Through a microscope, strep looks like a "string of spheres," sort of like a string of pearls—deadly pearls, ones you wouldn't want to wear around your neck.

Pneumococcus it a common bacterium behind pneumonia, "pneumo" from the Greek meaning "lung," and again we have "cocci": "lung spheres." *Pneumococcus* is also called *Diplococcus*, from the Greek *diplo*, or two, meaning "two spheres" or "paired spheres," which, again, is just how they look under a microscope: two bacteria cells hanging out together. Other types of bacteria include *Tetracoccus*, or four bacteria hanging out together, and generally these do not cause infection in humans.

Then there is the *Bacillus* group, however, which is not so benevolent toward us vulnerable humans. This genus includes rod-shaped bacteria, the more well-known scoundrels of which are: *Mycobacterium*, causing tuberculosis and leprosy; *Clostridium*, behind tetanus, botulism and the gut-infection-causing *Clostridium difficile*; the mostly intestinal troublemakers *Listeria*, *Salmonella*, *Klebsiella*, *Shigella*, and *Pseudomonas*; and, if ever there was a nasty fellow, there is the *Yersinia* bacteria that unleashed the previously discussed fourteenth-century Black Death, wiping out 25% of the world population. All these bacteria are not round cocci but rod-shaped bacillus. Sarcina bacteria is both a shape and classification for the Clostridia characters. At least half a dozen other rod-shaped bacteria cause infections in humans, including one bacillus shaped tribe termed a bit on the nose: *Bacillus*, its most famous tribe member is *Vibrio*, that cholera culprit, although technically *Vibrio*, albeit rod-shaped is a curved rod shape.

Moving on, the *Spirillum* bacteria are spiral shaped and include *Campylobacter*, a common intestinal pathogen; *Helicobacter*, associated with stomach ulcers; *Leptospira*, which also attacks our pet dogs; *Borrelia*, as in Lyme disease; and the fan-favorite *Treponema*, that syphilis scoundrel.

By the way, unicellular bacteria starting to hang out in clusters and strings was the beginning of multicellular life on Earth. When those groupings of bacteria got big enough, some in the group decided to organize into a division of labor, where maybe the members closest to the outside took in nutrients, while the inside members processed those nutrients for the entire group. The first Stage 1 bacteria appeared on Earth 3.8 billion years ago, LUA 3.5 billion years back and cyanobacteria at least 3 billion years ago. The first multicellular organism, where the division of labor among cells clearly emerged, arose perhaps 600 million years ago. Therefore, it took 2.9 billion years to go from lone ranger bacteria to sea sponge like creatures. Some researchers claim multicellular life appeared much earlier in the fossil record, perhaps 1.5 to 2 billion years ago, when a grouping of single-celled algae started hanging out together, assigning division of labor amongst their array. And if these little creatures actually existed, no doubt they were too small to appreciate without the aid of a microscope.

My earliest recollection of using a microscope was in eighth-grade biology. Our teacher assigned us a lab exercise using some water from a nearby murky pond. We were to look at one drop of full-strength pond water under the microscope and count the bacteria we saw in a single view. We were then instructed to dilute one drop of pond water with twenty drops of tap water in a small specimen cup, mix it up with a toothpick, then place one drop of diluted pond water on another microscope slide, and count the bacteria in that 1/20 dilution. Not sure what the point of the exercise was. Even in middle school it didn't take a rocket scientist to guess how dilution affects concentration, an example of extrapolating from context. But, you know, in a grade-school science class, such dog and pony shows were designed to keep the students entertained. Maybe the point of the exercise was to learn how to use a microscope. All these years later I have no idea.

My classmate Gary did not follow the instructions. Instead of diluting in a small mixing cup, he placed one drop of pond water directly on the microscope slide, to which he then added twenty drops of water onto that same microscope slide, mixing with the toothpick as the pond water dripped over the edge onto his fingers and his lab bench. Gary then placed that farkakteh slide on the microscope stage, adjusting the optics right into the bubble of dilute pond water, contaminating the objective lens. Oy' vey! Pretty dumb, right? The thing was, Gary had perhaps the highest IQ in our class. Four years later, he graduated first in our high school and went on to MIT. Go figure. Gary was a genius at theoretical stuff and math, possessing one of those eidetic memories (photographic memory for us lay folk), like Sheldon Cooper from *The Big Bang Theory*, but Gary was a complete disaster at the practical stuff.

Gary like the fictional Sheldon Cooper, was apparently not much into biology. He was a numbers guy, which might explain that ridiculous display of pond water dilution. Perhaps Gary's lack of interest in biology also parallels this exchange in the same sitcom between Sheldon and his girlfriend, the neuroscientist Amy in the episode "The Comic Book Store Regeneration" from season eight:

> **Sheldon**: When I was doing string theory and hit a dead end, why didn't you try to help me?
> **Amy**: I did. You said the only math biologists know is if you have three frogs and one hops away, that leaves two frogs.

A common *Staphylococcus* infection besides skin boils is the spreading skin infection cellulitis, which means infection of skin cells. Cellulitis can be caused both by staph species and by strep species—and less commonly other bacteria. Staph is also the most common microbe behind wound infections, such as those suffered after trauma, a laceration or after surgery, as well as a common source of bacterial pneumonia, heart valve infection (termed endocarditis), bone infection (or osteomyelitis), the impetigo skin infection that often occurs in children, lung abscesses, toxic shock syndrome, and food poisoning.

In food poisoning, whether it is from staph or other critters like *E. coli*, it is actually not the bacteria that makes a person sick and hammers the intestines as much as it is that toxin, exo- or endo-, that the bacteria releases. Heating food contaminated with toxin-producing bacteria might kill the bacteria, but that heat might not be sufficient to denature the toxin, that is, to break the toxin's chemical bonds into nontoxic substrates. Those exotoxins and endotoxins at the root of food poisoning are more heat resistant than the little rapscallions that produced them in the first place. Even boiling an endotoxin laced can of soup might kill the bacteria but not destabilize the toxin.

Staph bacteria are also often the source of toxic shock syndrome. Like food poisoning, the milieu of this deadly affliction is more the result of the released toxin wreaking havoc than the bacteria itself. Toxic shock is an infection we will visit with in more detail later.

A boil is an abscess, and like almost all abscesses (the few exceptions include liver and brain abscesses), the primary treatment is not antibiotics but rather draining the infected pus through a usually simple procedure termed "incision and drainage," or I&D. The reason for this is quite simple. An abscess by design does not let blood flow into its interior, into its core, which is precisely where the burden of bacteria is hiding. Without blood flow, antibiotics, along with the usual tactics of the immune system, are practically useless. They might very well help to wall off the infection and contain it to the abscess, but they struggle to destroy the nest, or nidus, of the infection. It is a standoff, a siege, the immune system cells and antibiotics outside the bacteria castle trying to enter the fortress, and the bacteria defenders resisting. The only way to kill the bacteria is to tear down the portcullis gate, climb over the castle wall, use catapults, things like that.

For a liver abscess or a brain abscess, the procedure is different. These abscesses are not usually drained straightaway, but first treated with antibiotics to ensure the infection is contained to within the castle, allowing the immune system to build its own siege wall around the infection. Delaying the drainage of a developing brain or liver abscess until a formidable siege wall has been constructed prevents spillage which in turn allows protection from collateral damage when the abscess is eventually drained. This process is not always used for brain and liver abscesses but, for the most part, it is often the go-to solution.

There is a rule among surgery residents—Rule #53, no wait, Rule #23: Don't let the sun set on an abscess. It means exactly what it says: any abscess of significance—liver and brain excluded—should be drained that day. You should not wait until the sun sets; you should not wait until *mañana*. Or, to quote Shakespeare: "In delay there lies no plenty." Draining that fortress of invincibility that is the abscess using a scalpel is the surgeon's Trojan Horse, as not only does it expel the burden of infection, sweeping away the bulk of bacteria culprits, but drainage also reestablishes necessary blood flow to the area, allowing reinforcements in the form of immune cells and antibiotics to aid in the battle.

There isn't actually a Rule #23 in surgery residency, or any set of "cardinal rules" for that matter—at least not like the codified ones that govern crashing a wedding in order to pick up girls, as illustrated in the classic comedy *Wedding Crashers* (2005), starring Owen Wilson, Vince Vaughn, and Will Ferrell. The movie also stars the beautiful and talented Rachel McAdams. When I first watched *Wedding Crashers*, I assumed the rules the movie's characters live by were made up just for that movie, but apparently real "wedding crasher rules" are actually written down out there. Google it. They exist. Although it's unclear, at least unclear to me, who compiled the list. I wonder why there isn't a wedding crasher's rule book for women, or do single women attend weddings looking for a potential husband, only to find a delusory date that inevitably will crash and burn, like an airplane. Or as the saying goes, a good man is hard to find.

I know you're curious, so I'll tell you that there are 110 rules in the wedding crashers' bible. From Rule #1: Never leave a fellow wedding crasher behind, to Rule #110: Never walk away from a crasher in a funny jacket, plus the 108 rules in between, guys who crash weddings to pick up girls have a set playbook to follow. Rule #14: You're a distant relative of a dead cousin. Rule #18: You love animals and children. Rule #27: Don't overdrink; the machinery must work in order to close the deal. Rule #34: Be gone by sunrise. Rule #40: Dance with old folks and kids, the girls will think you're "sweet." Rule #63: Bring an extra umbrella when it rains; courtesy opens more legs. Rule #106: Eat plentifully, digest your food; you'll need the energy later

"Tough break, Marcy."

Which brings us to a guy who built an entire company based on catching rule breakers. Allan Pinkerton was born in 1819 in Glasgow, Scotland, and when his father died, the young ten-year-old lad was forced to become a worker in the cooper business—not copper but *cooper*: people who make barrels and utensils out of wood. In 1842, now twenty-three years of age, Pinkerton immigrated to the United States and settled in the Chicago area. In addition to continuing his cooperage trade, he became active in the abolitionist movement in America. His Chicago home was a destination point for the Underground Railroad, one of a series of safe houses and secret pathways that Black Americans used to escape slavery in the South to the Northern free states prior to the Civil War.

When I was at Antioch College in Yellow Springs, Ohio, there was this old Victorian-style home perched on the southernmost edge of town, accessible along Route 68, a road that meanders up from Cincinnati toward Columbus. Ohio was initially a non-slave border state when the Civil War broke out, but eventually sided with the Union over the Confederates. Its immediate neighbor to the south, Kentucky, was a slave state. You can imagine the strategic importance of that quiet, unassuming Victorian home in Yellow Springs, as once-time enslaved people worked their way along the Underground Railroad from Mississippi, Alabama, and Georgia, through Tennessee, across Kentucky—all slave states—before crossing the Ohio River to reach Cincinnati, and their freedom.

The Underground Railroad was a network of safe roads and trails in the woods, coupled with safe homes, which often featured a widow's walks, the network formed in the late 1700s. The "workers" on the Underground Railroad included white Southerners who kept the fact that they were against slavery a secret so that they could more effectively move slaves out of the South, and also of course Northerners, who accepted and redirected those freedmen and freedwomen. After the Civil War, the Underground Railroad was no longer needed. It is unclear to history how many enslaved people reached their freedom through this network but estimates easily exceed 100,000.

That Victorian design Yellow Springs house—I don't know if it still stands, it was pretty rickety back when I was in college—had what is known as a "widow's walk" or a "widow's nest," a balustraded platform on the rooftop used to look down that southern valley—now called Route 68—for Underground Railroad passengers winding their way up the dirt road (Route 68 is of course paved now). Such rooftop widow's walks were commonly installed on New England seaport homes so that the wives of mariners could watch for their husbands' safe return, or not safe return (thus the "widow" part), from the sea. Being a fisherman then, and even now, was the most dangerous profession around, and many trawlers never made it home—and so that simple rooftop perch or cupola gained its tragic name.

But in Yellow Springs, there is no sea, no lake, not even a respectable pond, but there was a small stream running alongside that dirt road towards Cincinnati, as that dirt road winded its way through the narrow valley to Yellow Springs and eventually over to Columbus, Ohio. From there, freed people could head north to Detroit, Chicago, or Canada, or travel east to Philadelphia, New York, or Boston. In other words, that widow's walk atop that Yellow Springs house served one purpose, and one purpose only: it was not a lookout for mariners at sea, it was a lookout to help bring fleeing souls to safe harbors.

Allan Pinkerton's home was once such safe house in Chicago although his home did not boast nor need to boast a "widow's nest.". He simply couldn't abide the practice of slavery, but it certainly wasn't the only type of wrongdoing he wouldn't stand for.

One day when Pinkerton was wandering the woods near his home looking for coopering wood, he came across some bandits involved in what turned out to be a counterfeiting operation. He observed their movements for some time, and then reported his findings to the local sheriff with impressive clarity. Pinkerton, after having played a key role that day in halting what turned out to be a major counterfeit ring, had one of those galvanizing moments. Abandoning the cooperage business, which he apparently had no love for anyways, he became that professional detective we still know him so well as today, opening up shop as the Pinkerton National Detective Agency. It's motto? "We Never Sleep." Sadly, this is my motto, too. A suffering, and perhaps sometimes insufferable, insomniac. Pinkerton's crew became especially adept at thwarting and capturing train robbers.

In 1864, when Pinkerton was sixty-four years old, he tripped on something of no consequence and bit his tongue. His tongue became infected, formed an abscess, and he subsequently died of bacterial sepsis. Perhaps had that abscess been drained in a timely fashion, Pinkerton might have "robbed" death.

Cellulitis is a skin infection most commonly caused by the *Streptococcus* genus of bacteria, but other culprits can also lead to such a spreading skin septicity, especially staph bacteria. With cellulitis, the skin turns bright red and a distinct visible border usually develops between the involved infected skin and the as yet uninvolved, noninfected skin. A well-demarcated advancing infection, sort of shaped like an inkblot, except red. Cellulitis is usually not a surgical infection, like an abscess, because there is nothing to drain; instead, the infection sort of percolates within the upper skin layers. When you dip a paper towel in water, the water seeps onward and upward through the interstices of the paper, which is sort of how cellulitis crawls and seeps through the interstices of the skin epidermis, maybe a tad into the dermis. Luckily, unlike with an abscess, antibiotics can follow bacteria into the many lairs of cellulitis. In treating this skin affliction, we use the waxing and waning of that well-demarcated border to assess the success or failure of the selected antibiotic. That is, either the red ink blot gets smaller, which is good, or larger, which is not so good.

Necrotizing fasciitis or flesh-eating bacterial infection involves the next layer below the skin, the fascia, which is under the dermis and on top of the subcutaneous fat, which is itself on top of the underlying muscle. But as flesh-eating infection advances, it ascends upwards to infect the skin dermis and then epidermis and descends downwards to infect fat and then muscle.

And which is where our story now turns: What exactly are the signs and symptoms associated with an infection? And, on the flip side, what are the signs and symptoms associated with the immune system attempting to thwart an infection? Pathogens cause symptoms most often as a way to extend their lineage into the next host, and the immune system causes further symptoms in its attempt to thwart those pathogens. The symptoms of the former and of the latter, as it were, often oddly overlap.

19 ONOMATOPOEIA IS NOT AN ONOMATOPOEIA

That phone rang.

We went to an early New Year's Eve dinner in Tampa, but my thoughts were troubled. I was distracted, elsewhere, 2,500 miles away in Los Angeles, wondering how Deborah's emergency room visit was unfolding. I knew the seriousness of her condition, and without frightening my wife and mother-in-law half to death, I attempted to impress upon them the gravity of the situation without too much gravity. I measured my words.

There's that popular idea that food can be a comfort to those who are depressed. You know—when you are in the doldrums, suffering the sorrows of life, so you eat or attempt to eat your way out of the melancholy with potato chips or Häagen-Dazs ice cream. Then there are those souls who, when depressed, basically starve themselves. At the other end of the emotional spectrum, there are those who gorge themselves when they're happy, sort of like a treat. I'm not sure if there is such a person who starves themselves when they're happy, though. Why would they?

I'm one of those knuckleheads who gets it both ways: when I'm depressed, I eat; when I'm happy, I eat. I should be like one of the Macy's Thanksgiving Day Parade balloons, all Michelin Manned out. But I'm not. I eat like a pig sometimes and I'm barely overweight. Perhaps "barely" is too strong of a word; but I'm not fantastically overweight. Must be my genetics; neither of my parents were ever overweight. My eating habits when I'm depressed and when I'm happy are entirely different than when I'm stressed. Stress is a different beast altogether than gloomy or cheery. When I'm stressed, when I'm contemplative, when I'm lost in thought, when I'm worried, when I'm troubled, not only do I *not* eat, but I end up playing with my food like a ten-year-old not wanting to eat his broccoli, pushing it around the plate as if tilling soil.

I was not eating that night in Tampa. I was tilling.

Probably from my days as a surgery resident, I adopted an unhealthy eating trait: I didn't then and don't much now care for food in my stomach when I'm operating. Sort of like my sports days, when I also greatly preferred an empty stomach. As a result, I tend not to have much in the way of breakfast, a habit I began in residency. Just coffee and I'm good to go. Even on days when I don't have surgery, like weekends, like while on holiday, I still skip breakfast, still skip lunch, and gorge at dinner. It's not a healthy habit but, well, there you have it. So usually by the time dinner rolls around, I eat and eat with much gusto. Like I said, I can be a pig.

Interestingly, it is highly doubtful our species was designed to have three squares a day. Having breakfast, lunch and dinner is probably a construct we grabbed, like grabbing a bag of potato chips, as our ancestors domesticated plants and animals which in turn provided a relative overabundance of food. I suspect our specie is best designed for one meal a day, maybe two, but not three. It is probably why the ketosis diet of skipping breakfast—maybe some cream in your coffee—a light lunch and then a reasonable dinner with little to no carbs during the day, works for so many who are attempting to lose weight. In that scenario, by the time morning rolls around, whatever carbs are in the system have been consumed, your body then presumably goes into a mild ketosis and the liver begins breaking down fat stores for energy. I guess it is also termed a fasting diet. I am told fasting is good for our bodies, purges things, but what do I know.

But as I sat in that lovely Tampa restaurant that New Year's Eve, on December 31, 1995, tilling my food, having neither an appetite nor the desire to force food down my throat, my thoughts were focused. I probably did have a beer though—there's always room for one drink. I could not wait to get home and have my mother-in-law ring Alan to see how Deborah's ER visit went. Mobile phones were not terribly common in 1995. Although I've had a cellular device since 1990, pretty much no one else in my family had gone mobile at that point. Landlines still ruled the day, so calling someone meant calling them from home, from what is retrospectively called a "landline," or from a telephone booth if you could find one. It seems odd that in today's wireless world we refer to wired telephones as "landlines." They weren't called that then—just "phones"— but they are called landlines now to distinguish that aging stationary technology from the new mobile world.

I still remember our rotary dial phones growing up, one in the kitchen, one in my parent's bedroom, and one in the hallway on the wall between us children's bedrooms. We actually had our own phone number, just for the kids, since my father was a physician and in constant need of

being accessed by his answering service. Heck even pagers didn't exist then. Back in the 1960s, if my father's answering service needed him on a Sunday while he was playing golf, the starter would have to grab a golf cart, careen out onto some fairway and retrieve him to the golf shop.

For you older readers out there, remember our seven digit home phone number started with two letters? The phone numbers in those bygone days having two letters followed by five numbers, the five numbers were the subscribers actual phone number and the two preceding letters—which were really just numbers disguised as letters—were the exchange at the

ATT and Ma Bell switchboard to which the phone call was routed. Our home number when I was a punk kid was DE3-1006, the DE was the exchange and the remaining 3-1006 our actual number. I still remember it. This is why old-style landline phones beginning around 1955—rotary then push—and even early cellular phones maintained on the dial the three-letter block over numbers 2-9, so as to give the phone company all those two letter exchanges like our "DE.". Figuring us doltish customers could not remember the two letter prefix/five letter number format, AT&T/Ma Bell recommended that the two letter exchange be given full names to better jog our collective dimwittedness. Our 333-1006 was pawned off as DE3-1006 but best remembered as Dexter3-1006 because it rhymed, MA Bell figuring the rhyme made it all so easier to remember.

At my urging, we skipped dessert and hurried home so Barbie could landline call Alan … multiple times, because, well, there was no answer. Alan was not answering the landline. Had Alan had a mobile phone, I wonder if it would it have made any difference. In this story, probably not. That phone just rang, or to quote Glen Campbell's cover of the Jimmy Webb 1965 song *By The Time I Get to Phoenix*: "The phone keeps right on ringing and ringing and ringing and ringing and ringing and ringing / Oh, and ringing off the wall," which I thought was a good thing—they were in the emergency room. I was relieved, comforted in the belief that Alan had taken his very ill wife to the ER and she was now receiving emergency care. Relieved that Deborah was getting plugged into the system, being dialed back into good health by the very best that modern medicine could provide. We called several times over several hours, and the phone just rang "off the wall," reassuring me all was right throughout the land.

I went to bed, wrapping my mind in a blanket of comfort that Deborah was being cared for. That whatever acute or chronic or delayed or recurrent or secondary infection that had stricken her was being beat back into oblivion. Deborah's "confusion"—that had been the main symptom that drove my anxiety—was not just a red flag, it was a red flag on fire. Nothing about confusion is good. I fell asleep, feeling safe with the notion that Deborah was safe.

Human infection doesn't limit itself to bacteria, which is the underlying frame of this narrative. We also have to deal with other pathogens, some microscopic, some not so microscopic. These pathogens mainly make their home in the nucleus-sporting eukaryote domain—bacteria are in the no-nucleus prokaryote domain—eukaryotes include parasites, fungi, and worms, but not the viruses. Viruses were never a member of the prokaryotes, never a member of the eukaryotes. They have their own kingdom of sorts, really more of a classification system since they're not cellular.

Those are the five big partitions of human pathogens: bacteria, parasite, fungus, worm, virus. Pretty much it is only the plant kingdom that doesn't attack us humans with infection, poison ivy yes, and poisonous mushrooms most definitely, but those aren't infections. At least not directly. One type of bacteria called *Pseudomonas aeruginosa*—a nasty fellow if there ever was one—can cause infection in plants, a sort of rot, that can transfer to humans through contact or possibly consumption. But it's really not the plant causing the infection, the plant is only the carrier, much like mosquitos carry malaria to humans.

Other than classifying human infection by the buggers themselves, there are other ways to look at infection, beginning with classic symptoms, then mode of transmission. We'll then venture on to parsing infection with such descriptives as clinical, subclinical or latent infection; primary, secondary or opportunistic infection; acute, chronic, relapse or recurrent infection; contagious or noncontagious, and last but not least, outbreaks, epidemics and pandemics.

When we think of communicable contagious diseases—the cold, the flu, sexually transmitted diseases—we generally think of person-to-person contact, and when we think that we are often correct. But not all communicable diseases are spread person-to-person without a circuitous route of some sort. For instance, in that 1854 Broad Street cholera epidemic we discussed earlier, the cholera bacteria left one infected person, soiled the earth underneath a Soho hovel, and in a circuitous journey contaminated the freshwater supply, entering the gut of the next and numerous unsuspecting victims. This still qualifies as a communicable infection, albeit in a rather distant manner. A less circuitous transmission of cholera would be one ill person preparing food for their household and passing the pathogen along through the food, a rectal-to-hands-to mouth sort of nauseating thing.

Cruise ships are especially susceptible to this mode of transmission, and subsequently go to great lengths to avoid dysentery outbreaks, but occasionally cruises experience a spread of norovirus or buggers of similar ilk, an intestinal flu virus, which can turn what was going to be a fun-filled Caribbean vacation into a veritable floating Petri dish of vomit and diarrhea. Which is the topic we're heading toward now: the classic signs and symptoms of infection and understanding of which is desirable to appreciate modes of transmission of infections.

Coughing certainly is annoying, both for you and for those who are forced to listen to you, but coughing can be your friend. Coughing is the way the body protects the lungs from unwanted guests such as irritating fumes, fluids, particles, other irritants, and of course pathogenic microbes. Since coughing actually serves as a protective function, it has rightfully earned the title of "watchdog of the lungs" as it expels uninvited guests. It's only when it gets carried away that we curse it and take measures to banish a cough, or at least rein it in.

I'd be remiss if I did not mention that two of the most common symptoms of a cold or the flu—cough and sneeze—are also intended consequences of the pathogen, a construct we'll visit a few times as a reminder. Cold and flu viruses don't make you cough and sneeze just for the heck of it, it's not their pastime to make you miserable. Coughing and sneezing is how many respiratory viruses and bacteria learned to extend their lineage. They intentionally—unknowingly but intentionally—evolved to cause their host, which would be you and me, to cough and sneeze their pathogenic progeny all over tarnation, aerosolizing infectious droplets that are then breathed in or land on touchable surfaces so that the next host, again you and me, become infected.

The act of coughing has three distinct phases. The first is the inhalation of air into the lungs, followed by phase two, a forced buildup of air against closed vocal cords, a closed glottis, and ending with the third phase, the sudden opening of the vocal cords, an open glottis, with the violent expulsion of air along with whatever evil was lurking in the lower reaches of the respiratory tract. The violent opening of the vocal cords with the rush of air across them is what produces the cough sound: *cough ... cough.*

An onomatopoeia is a word that imitates the sound of the thing or action it describes, such as how the waterfall "plops" into the pond, the breeze "swooshes" all around us, or the chains go "clank" as she handcuffs him to the bedpost. "Ahem" is an onomatopoeic word for clearing the throat. "Ahem" is also a word used to draw attention to oneself, as in, "Ahem, where's that kiss you promised me?"

"Cough" is an onomatopoeia spelled to sound like the sound a cough makes and appears to be from Old English. As for the word "onomatopoeia," it is itself decidedly not an onomatopoeic word but is Latin derived from the Greek *onoma,* meaning "name," plus *poiein,* meaning "to make," to arrive at "word making."

Even though coughing potentially spreads the virus, suppressing a cough is not always a good idea, because it's guarding our lower respiratory latitudes from foreign invaders. So it also protects our own lungs. Coughing (and sneezing) is inevitable; spreading germs is not. Cough into your elbow or cough into a hanky, just don't cough into your hands or cough into the wide blue yonder. Don't you just love it when people are hacking away without any effort to cover up? Covering up is called a "best practice": it's one way we prevent a flu from becoming an epidemic.

But when a cough goes rogue, when it gets carried away, when some poor hacking sufferer is losing sleep from excessive coughing—when his wife is also losing sleep from that interminable hacking—when we have an irritating tickle in the back of the throat that drives us nuts, the benefit of suppressing the cough in order to get some shut-eye can outweigh the cough's protective function. It is your classic risk-benefit ratio. Especially true if the infection has long since been vanished and the cough is merely the aftermath inflammation on the battlefield. The key dialing, however, is to not suppress the cough *completely.* The management of a respiratory illness is a balancing act. When cough suppressants are used, they should be used sparingly, judiciously, based on the argument that much needed rest gained by suppressing a cough helps fight infection, too. Which is where Deborah steps back into the story.

In the weeks leading up to that phone ringing, when Deborah had a mere cold, she might have consumed a great deal of cough suppressants in an attempt to navigate the workday and to avoid sleepless nights. Over this distance of time, we'll never know whether she liberally consumed cough suppressants and how that might have played a role in the evolution of her infection, or the emergence of a secondary infection, but the possibility does exist. If the cough is not able to protect the lungs, if upper respiratory secretions soaked in bacteria are allowed to trickle down below the vocal cords and creep into the lungs, it is a setup for pneumonia.

Cough medicine comes in several varieties. The classic true suppressants are prescription narcotic opioid derivatives such as hydrocodone, which is also the active ingredient in the common pain reliever Vicodin. Narcotics block the cough reflex at the neural level in the brain. Nonprescription over-the-counter cough suppressant medicines contain dextromethorphan, a synthetic compound that is chemically related to opiates, and it too works at the level of the brain to suppress the cough reflex, albeit not nearly as well as narcotic cough suppressants do.

Expectorants are different than suppressants. They work by thinning lung secretions to make it easier to cough things up, easier to clear the airway, based on the assumption that thicker secretions are much more difficult to expel from nooks and crannies, too tenacious to be easily dislodged and cast asunder. The classic expectorant contains guaifenesin, a mucolytic agent, meaning it is lytic, that is, able to lyse or cleave or thin out thick phlegm. It does this by chemically breaking the protein component of mucus.

The good ol' fashion way to help loosen secretions in both the upper and lower airways is steam. A full steam shower can work wonders, as the heat mobilizes secretions, absorbing the water vapor which thins the mucus, making it easier to expectorate. The steam heat especially is harmful to viruses, like the cold and flu, because these tribes of vermin are heat labile, but steam heat also can annihilate heat sensitive bacteria. Our parents didn't know it when they shoved our faces over a simmering saucepan of water on the stove with a towel draped over our heads, the steam clearing the sinuses of thick secretions was not the only thing being accomplished. The heat was likely killing the pathogen.

Viruses possess an outer capsid, like a cape, that is mostly made of fat; the steam heat liquefies that fat, rendering the virus noninfective. Basically, the capsid is what allows a cold or flu virus to recognize and attach to a respiratory cell in order to inject the virus's genetic material into the cell for replication. Steam makes the capsid cape liquefied, no viral cell entry, no viral replication. It is also one reason why winter time is flu season time, cooler weather firms up the viral capsid making it virulent; warmer summer weather softens the capsid making it less virulent; making winter-summer the waxing-waning of flu season.

Antihistamines and decongestants mostly work up in the higher reaches of the respiratory passages, in the nose and sinuses, thus decreasing secretions from the get-go, secretions that would otherwise block an already engorged nasal passage from the milieu of the infection, secretions that flow down the back of the throat toward the airway triggering the cough reflex. Postnasal drip is precisely that—your garden variety, highly annoying nasal secretions dribbling down from the back of the nasal passage and into the throat.

Expectorants, antihistamines, and decongestants still allow coughing to continue if needed, but aid in moderating coughing associated with postnasal drip and excessive secretions. The narcotic-type medicines, on the other hand, those truly stop coughing dead in its tracks and should be used with caution and restraint, should be used sparingly. Coughing during a respiratory infection is ultimately your friend. Just cover up and don't infect the world.

Sometimes we develop a lingering cough well beyond the initial respiratory infection, a cough that is neither from postnasal drip nor from a more deep-seated infection that has been vanquished. That lingering, persistent, annoying cough that drives you and everyone around you nuts is most often the aftermath of the battle fought to kill the cold or flu pathogen. A cough caused by an inflamed voice box, trachea, or main bronchi, which are right below the upper airway, can linger for days, even for weeks. An important take-home lesson, a chronic *noninfectious* bronchitis after a respiratory infection is common, and results when the smooth muscle and mucosa lining the trachea and bronchi become irritated, become inflamed. It is at this juncture that a lingering cough is often confused with either a persistent infection or a new infection, neither of which is usually the case. It is this period after the infection has departed but the inflamed bronchi—a veritable war zone—sparks spasmodic coughing spells that people, understandably, assume they're still infected.

A quick anatomy lesson is needed at this juncture. Below the vocal cords is the trachea, which then branches into the right and left main bronchi, respectively servicing the right and left lungs. Most bronchitis infections and inflammations involve the trachea and the bronchi. As the bronchi travel further into the lungs, they continue to branch, to arborize, like a beautiful oak tree, eventually narrowing down into bronchioles. Further still into the chasm of the lungs, these bronchioles branch into air sacs, called "alveoli." It is at the level of the alveoli that oxygen gas exchange takes place: where oxygen is captured and carbon dioxide liberated. Anatomically, pneumonia, whether bacterial or viral, is usually at the level of the alveoli. That's not to say bronchi and bronchioles are not also infected during a pneumonia—they can be, too. It's also possible to have an infection of the bronchi (bronchitis) and bronchioles (bronchiolitis) that has not yet involved the alveoli (plain old pneumonia sometimes called pneumonitis).

Often, all too often, patients who have this lingering, stubborn non-infection coughing seek antibiotics under the belief that they must still be suffering from an infection, and physicians oblige by prescribing the requested meds. The infection would have been the classic Nyquil "nighttime, sniffling, sneezing, aching, coughing, stuffy-head, fever, so you can rest medicine" that has departed, leaving only the "coughing so you can't get rest" inflammation. And as the doctor prescribes a regimen of antibiotics, that will do nothing to resolve the cough and everything to promote drug-resistant bacteria. The other kicker is this: the cough is likely in its waning days anyway, with or without treatment, on its way to complete resolution but prescribing antibiotics at that juncture only reinforces such antibiotic-seeking behavior, the cougher believing the antibiotic did the trick when the trick was tincture of time. In some cases, the first course of antibiotics doesn't relieve the cough—precisely because it was never going to anyway, as the cough is not bacterial—and then a second, different course of antibiotics is often prescribed, which then so happens to be coincident with the cough resolving on its own. You see the problem here? Or rather, *problems*? It is the incorrect correlation of cause (taking antibiotics) and effect (cough goes away), reinforcing a false correlation, since in reality the cough went away on its own.

Taking narcotics for weeks on end in an attempt to abate the cough of a noninfection chronic bronchitis will also accomplish nothing, at least nothing good: it could potentially create an addiction to opioids and possibly make matters worse, losing the protective "watch dog of the lungs.". If that chronic cough needs to be suppressed here and there, like for sleep, there is a better—much better—alternative: asthma inhalers. That's correct, asthma inhalers work for things other than asthma.

That chronic bronchitis cough is due to inflamed bronchial muscles spasming into a coughing spell, often termed a "spasmodic cough," sort of as if your bronchi are having a seizure. Asthma inhalers, such as albuterol, termed beta-blockers, relax those bronchial muscles, thus preventing the spasmodic cough. Just because you have a lingering cough taking its sweet patooties to resolve doesn't necessarily mean you still have a lingering infection, doesn't mean you need antibiotics, doesn't mean you need powerful narcotic cough suppressants. Most likely your irritating cough is just that—irritated, inflamed bronchi, with muscle spasms convulsing into a cough. Asthma inhalers might just be the ticket, not steroid inhalers but beta-blocker inhalers.

Sneezing, like coughing, is a protective reflex mechanism designed to expel unwanted visitors attempting to pass through the nose, such as smoke and small particles. Sometimes merely bright sunlight hitting your face can trigger a sneeze, which is actually another built-in response along with the cough reflex to protect our airway, known as the "photic sneeze reflex." This reflex is likely an adaptive mechanism from when our primate ancestors started walking upright. Imagine *Homo erectus* strolling around two million years ago: their nasal opening more upright, more exposed to things floating about into their nose. As the early *Homo* species evolved from lumbering on all fours to walking erect, the nose necessarily had to evolve too, our *Homo* ancestors began sporting noses with the openings facing downward in a protected position. The follies of walking upright.

"Look! Now they're walking upright! — I'm *tired* of trying to keep up with the Joneses!"

The sneeze mechanism results from a stimulant irritating specific nerve endings along the nasal lining, which then release histamine, a neurotransmitter, in the local nasal tissues. The histamine in turn triggers a different set of local nerves that travel to the brain stem, where yet another set of signals are sent cascading back to the smooth muscles of the nose and throat—setting off a sudden, violent, rhythmic contraction from inside out. That contraction begins in the lower reaches of the upper airway, right at and above the vocal cords, and like the wave during a football game, that contraction expels air through the mouth and nose. The sneeze and the cough are both great spewers of droplets, and if those droplets are laced with viruses or bacteria, well, everyone around gets sick.

Unlike for coughing, we have no known "sneeze suppressants." Which is probably good, because, like coughing, sneezing is our friend. While having said no sneeze suppressant exists—narcotics won't suppress a sneeze like it does a cough—there are a few old wives' tales that offer techniques to curtail sneezing, such as holding your breath and pinching your nose, for how long those ol' wives' tales don't specify, but I suspect until the first signals of air hunger suggest you take a breath. For the most part though, sneezing is treated by treating the underlying cause—that is, treating the cold or hay fever, or avoiding irritants like dust and animal dander—and just hunkering down and riding out the storm. Antihistamines which help dry up nasal secretions might aid in inhibiting histamine, the neurotransmitter of the sneeze reflex so antihistamines might work to suppress a sneeze as well as decrease the factors that trigger a sneeze. Which now brings us to how sneezing is wrapped up with the gods.

Ancient Greeks thought sneezing was a sign from the gods, usually a favorable one. Homer includes sneezing in his epic poem *The Odyssey*, which is the sequel to *The Iliad*. *The Iliad* tells the story of the ten-year Trojan War, and *The Odyssey* tells the story of one of the war's most important warriors, Odysseus (Ulysses in Latin), and his ten years of wayward travel after the war to get back to Ithaca. Desperately trying to reach home, Odysseus was thwarted by the gods, mostly Poseidon, who was seeking revenge for, among other things, Odysseus having blinded his son Polyphemus the Cyclops. Eventually, though, the gods finally let Odysseus return to Ithaca, an island off the west coast of mainland Greece in the Ionian Sea that separates the boot of Italy from Greece.

It is believed that Homer wrote the two epic poems around 700 BC, or about 500 years after the Trojan War supposedly took place. As we explored earlier, the Trojan War was fought between several Greek kingdoms led by King Agamemnon against the Turkish city of Troy, led by King Priam. The epic battle featured Odysseus, who was the king of Ithaca and the brains behind the trojan horse ruse of war, as well as the siege featured the finest warrior ever to live, Prince Achilles, son of Peleus, king of the Myrmidons, and the sea nymph Thetis, fighting for the Greeks against the finest Trojan warrior alive, Priam's son, Prince Hector.

While Homer's version holds that the Trojan War took place in order to rescue the kidnapped Helen, the real reason for this war around 1200 BC was simply that Agamemnon was a conqueror and had set his sights set on Troy. Across the Dardanelles is modern-day Hisarlik, Turkey, which is believed to have been the legendary Bronze Age city of Troy and its unconquerable walls. Agamemnon's war wasn't the last time the Turkish people, descendants of the Trojans, would suffer under the Greeks. Eight hundred years after the Trojan War, they had to deal with another Greek conqueror, Alexander the Great, who we ourselves met briefly at the start of this story. Of course, as Alexander was sweeping through Turkey, it wasn't the Turks he was interested in dispensing with—he was actually chasing down the Persians, who had previously crushed Greece and who were more or less still occupying Turkey, known as Anatolia in antiquity. Payback is a biatch.

Alexander was focused on decimating Persia mostly as retribution for Persia's conquering of Greece centuries earlier. The heart of Persia was on the other side of Turkey from Greece, and in order for Alexander and his invincible army to get there, he had to first conquer a bunch of other strongholds in the middle, some still under Persian control, some not. He ended up capturing Turkey in 334 BC, wresting control of it away from the Persians. Persia during its heyday had captured all of Mesopotamia, the Levant, the Nile Valley, the Indus Valley, Anatolia, and parts of Greece.

You'd think that once Alexander had Turkey, he would've marched right into the Achaemenid Empire of Persia and destroyed it. Persia was right next door to Turkey, after all. But interestingly, he didn't. Being the global conquer he was, and a clever military tactician, he decided to capture everything else surrounding Persia, even Egypt, to box the Persians in, giving them nowhere to flee, before he marched into Babylon, the heart of Persia. In a little over thirteen years, 336–23 BC, Alexander conquered nearly everything in the known world. That's what conquerors do.

According to the Greek historian Plutarch, when Alexander saw the breadth of his domain—Greece, Turkey, Persia, the entire Middle East, vast lands in India and Egypt—"he wept for there were no more worlds to conquer." If you were to think that such a supreme general as Alexander the Great died in battle or in a coup like Caesar or years later in his bed as an elder statesman, you'd be wrong. The great military genius Alexander the Great died at age thirty-two, from an infection, most likely typhoid fever—a tale to be covered in the next volume in this series.

The Trojan War, real or partially real, lasted for an entire decade, because Troy's walls were too mighty for any invading army to surmount—or so the Trojans thought. The Greek siege on Troy finally ended when the Greeks feigned defeat, pretending to depart the shores of Troy and leaving behind a gift from the gods, which was in fact a *ruse de guerre*—or "ruse of war" for us English speakers. That ruse, of course, was the Trojan Horse.

The massive wooden Trojan Horse sculpture, which was the brainchild of King Odysseus, had a small contingent of the Greek army's finest soldiers—Achilles and Odysseus among them— hidden inside. The Trojans believed the horse was a gift from the gods and happily dragged it through the impenetrable gates of Troy. When the Trojans had drunk enough in celebration of their supposed victory over the Greeks, when they were heavily fortified with alcohol, when they were stuporous beyond arousing, the *ruse de guerre* played out. The wooden belly of the horse opened and out climbed the hidden Greek soldiers, who simply opened the gate of Troy from the inside. The entire Greek battalion then poured into the city, unleashing all hell, destroying Troy.

Now that we've hit the main points of *The Iliad*, we're well on our way to the case of this "sneeze" in Homer's *Odyssey*.

After the Greeks defeated the Trojans, Odysseus began his journey back to Ithaca and back to his beautiful wife, Penelope, by setting sail west across the Aegean Sea into the Ionian Sea. It was anything but smooth sailing. Poseidon, god of the sea, had revenge on his mind. First one god, Apollo, kills Achilles in *The Iliad*, guiding that fateful arrow shot by Paris into his heel, and now another god has it out for the hero of *The Odyssey*. Apparently, Odysseus had taken credit for conceiving of the Trojan Horse when in actuality Poseidon gave him the idea. Poseidon was not happy, so he exacted his revenge by blowing Odysseus's ship of course all over the Mediterranean Sea for ten long years. The fact that Odysseus also blinded Poseidon's son Polyphemus the

Cyclops during his wayward voyage didn't help matters at all. This journey of Odysseus is where the word "odyssey" is derived from.

Add the ten-year siege of Troy—*The Iliad*—to a ten-year journey home—*The Odyssey*—and unsurprisingly folks back home in Ithaca were beginning to think Odysseus, having been gone twenty years, was dead and that their king would never return. His wife, Penelope, and son, Telemachus, however, never lost faith. But they were at the mercy of those who wished to seize the kingdom of Ithaca for themselves or wished to seize the beautiful Penelope for themselves or wished to seize both. Lustful, power-hungry suitors laid siege to the palace with riotous abandon, drinking Odysseus's wine, eating his food, and hoping to lure the queen into their bed. But their overtures were met with refusal. Penelope, despite her husband being gone now twenty years, kept faith, kept true.

Upon hearing one day from other seafaring travelers that Odysseus might be alive and nearing home, Queen Penelope vowed that she and Prince Telemachus would take reprisal on the suitors who had taken over the palace, eating and drinking their way through years of revelry at the expense of the crown. At the very moment Penelope was informed by travelers that Odysseus was alive and heading home, Telemachus sneezed loudly—the kind of sneeze that nearly scares the life out of a person. Penelope believed her son's sneeze to be a good omen from the gods and a sign that her revenge would soon be at hand. And revenge is precisely what Odysseus brought once he arrived home: he slayed every last suitor.

That, in a nutshell, other than the marvelous (mis)adventures—getting euphoric in the land of the Lotus Eaters, avoiding death on the island of the Cyclopes, dodging being a meal in the land of the cannibals, avoiding Sirens that lure ships to crash, and, my favorite, staying on Calypso's island, where she keeps Odysseus, for seven long years—is Homer's *Odyssey*.

The Iliad and *The Odyssey* are not just fun epic poems that Homer wrote so high school students in my day would be required to read 2,700 year old epic poems, and then be tortured with multiple choice exams. They also teach several important lessons, most of which were probably lost on us when we were teenagers as we attempted to explain their metaphors in fourth-period English.

From *The Iliad* we learn about good leadership. Good leaders must also be servants to those they lead, with Agamemnon being a poor leader, although he was victorious, and Priam being a good leader, although he lost. *The Iliad* instructs us about the evils of narcissism and self-absorption, such as when Achilles's narcissism leads to the death of many Greeks and many Trojans. *The Iliad* advises us that bad things happen to good people, like when Achilles kills Hector, who by all accounts was a noble, honorable warrior who didn't deserve to die nor to be paraded in death by being dragged behind Achilles's chariot. *The Iliad* advises us that man's master is force—not rules but force—such that civilized fighting will always lose out to ruthless fighting. When Hector wants rules for his duel with Achilles, Achilles declines them, and Hector dies; Achilles is a natural-born killer who follows no rules.

From *The Odyssey* we learn the importance of hospitality, showing kindness to strangers who more likely than not bid us no harm. Odysseus on his journey home was often the benefactor of many acts of reception. The more civil a person is, the more hospitable they are. *The Odyssey* instructs that loyalty and perseverance are vital—not just loyalty on the battlefield or Odysseus's perseverance in his wayward travels home, but especially the loyalty and perseverance of Penelope, who waits twenty long years for her husband to return. Other themes in *The Odyssey* include the dangers of vengeance, as demonstrated by Poseidon toward Odysseus. Vengeance is also exhibited by Odysseus, Penelope, and Telemachus toward the unwanted suitors during Odysseus's twenty-year absence. Homer's epic poem further discusses reality versus illusion, most notably in the episodes where the goddess Athena helps disguise Odysseus when in harm's way, such as cloaking Odysseus when he returns to Ithaca in order to protect Telemachus and the palace from the ire of Penelope's would-be suitors. Finally, *The Odyssey* teaches us about spiritual growth, especially that of Odysseus and his son Telemachus, both of whom come to appreciate the gods and the spiritual world.

Which is around about way to return where this chapter began, the various ways to parse infection, taking a cue from the surreptitious Trojan horses that invade our bodies, we'll begin parsing with mode of infection spread.

The most common infectious route, and the one that usually comes to mind first, is *droplet contact transmission*, which covers your basic coughing and sneezing so-I-can't-get-any-rest conveyance of infection, spewers of aerosolized pathogens that linger in the air for an unforgivable amount of time. It's estimated that microbes from a single cough or sneeze can linger in the air for up to forty-five minutes, but the usual is much, much less. Some studies dating back to the 1950s have attempted to parse how long droplets stay suspended, how far they travel after expulsion, number of scoundrel pathogens in each droplet, how many droplets per each expulsion, and on and on. The truth is, it is quite variable but for the most part, most droplets have dropped to the ground, or a surface that can be touched within a few minutes. Imagine that! You can walk into an elevator where floating viral particles in droplets have been riding up and down, up and down the elevator, like a kid pushing all the buttons, just waiting to land on someone's face. Or touching the handrail of an escalator where sneeze or cough droplets have just alighted.

Having said that though, most cough and sneeze droplets seldom travel more than a few yards and rarely stay suspended for more than ten minutes, usually only a few minutes. As I write these words, the world is dealing with the novel coronavirus, SARS-CoV-2, a portmanteau from <u>S</u>evere <u>A</u>cute <u>R</u>espiratory <u>S</u>yndrome-<u>C</u>oronavirus <u>V</u>irus-<u>2</u>, and which causes the infection disease COVID-19, also a portmanteau from <u>CO</u>rona<u>VI</u>rus <u>D</u>isease-20<u>19</u>), and so we find ourselves living in an era where there is no natural herd immunity, no vaccine to this novel scourge, necessitating with great interest the need for social distancing, washing hands and wearing face masks in an effort to dampen the pandemic. The only hoax when a pandemic hits the world is the deception and duplicity perpetrated by those who wish the pandemic not to be true: that is, the real hoax is not the pandemic but the falsehoods foistered upon those gullible Dunning-Krugers who wish to believe an alternative reality that fits their world. The corollary being is that those who claim facts to be "false news" are, in reality, the ones who easily swallow "false news." I believe the words you're looking for are "irony" and the consternation it causes those in possession of the facts are "flummoxed."

You don't even have to breathe-in cough (or sneeze) droplets to acquire most respiratory viral bad boys—it is enough to touch an elevator button or a stair bannister teeming with microbes, contaminated from aerosolized viral-laced droplets or from the snotty hands of the great unwashed. Whereas aerosolized droplets in the air might only survive several minutes before plummeting to the ground, infectious droplets on surfaces can last much longer. We are talking an hour, hours, even a few days. From there your own hands do the rest of the job. It's estimated that the average person touches their mouth, nose, and eyes about four times per hour, if not more. If you add that to the number of times you touch a contaminated surface, like, say, an airplane seat armrest or a cruise ship banister or a hotel elevator button, or any public door handle, it doesn't take much to imagine how we contribute to the pathogen's life cycle. Nor much to imagine how just a few people can launch an entire epidemic.

Nonliving items like elevator buttons and stair bannisters and the pull cord on buses that host infectious material chock-full of transmittable pathogens are called "fomites." While it sounds like "mites," which are tiny arthropod bugs, "fomites" have nothing to do with "mites." "Fomite" is from the Latin and means "tinder" or "wood," likely chosen for the contaminated infection surface to describe an inanimate object that provides a home for microbes that "fuels" infection.

Another common mode of infection spread, the one that gives me real heebie-jeebies, is *fecal-oral transmission*, which means exactly what it says. It's just so disgusting. A person with an intestinal infection like cholera or pinworms transmits that pathogen from their feces to their fingernails as they wipe their derrière, and if they fail to wash their hands, they can then transmit that microbe-laden fecal material on their hands to the next person's mouth, through foods eaten or water drunk. Disgusting, I know.

Such culprits that traverse from one person's rear end to another person's front end include the infamous norovirus, the bane of the cruise ship industry, which for the most part cruises go to great—and usually successful—lengths to keep their passengers free of infection. However, every so often we hear about a cruise ship so stricken with viral gastroenteritis that nearly everyone on board is ill, causing the cruise to be shortened so the passengers can scurry off that floating Petri dish. Truth is, most cruises are clean and uneventful. Like many things in life, we only hear about the bad, never the good. As the current outbreak of COVID-19 is all the rage, recall the infamous Diamond Princess cruise ship, which captivated the world, the first of nearly forty passenger conveyances hobbled by coronavirus, the Diamond Princess forced to anchor in Tokyo Bay before being allowed to dock. There were 712 confirmed cases of COVID-19, fourteen deaths, and forty-five who never fully recovered among her 3,500 passengers and crew.

Other butt-to-mouth diseases include cholera (like in the 1854 Broad Street pump fiasco), E. coli, and campylobacteriosis, both common causes of traveler's diarrhea, as well as these intestinal delights: giardiasis, hepatitis A, pinworms, and of course typhoid, which comes into a story about Typhoid Mary that we'll further discussed in the next volume in this series, *The Second History of Man*.

Pinworms are an especially nauseating infection to ponder. They are not microscopic pathogens like viruses and bacteria. There's at least something comforting about not being able to see the viral and bacterial critters using your body as their home; they're out of sight, out of mind. Not so for pinworms. They are tiny but still macroscopic worms, a little bit bigger than a kernel of rice, easily visible in poop. Their eggs easily lodge under fingernails, then swallowed from feces-contaminated food. These eggs take up residence in the rectum of their new victim where they morph into the adult worm. At night, pinworms crawl out of the rectum and onto the anus of the sleeping infected person, not only to produce an intense itching but to deposit eggs so that the victim's next act can successfully extend their parasitic life cycle. That act is the infected person, who is deeply asleep, unconsciously itching their butt, thus depositing pinworm eggs under their fingernails. Or fingernails become contaminated when using a poor technique to wipe after going boom-boom. Either way, it doesn't take a rocket scientist to figure out how those eggs then transfer from one person's unwashed fingernails to the next person's food. I don't care much for the onomatopoeic word "ew"—mostly favored by teenage girls—but when it comes to how pinworm eggs laced with feces are transferred from one's fingernails to another's mouth, well, "ew!"

Sexual transmission is yet another common route of infection spread, from the classic bacterial infections of gonorrhea, syphilis, and chlamydia, through the viral infections of hepatitis B, hepatitis C, HIV, genital herpes, and venereal warts, to the parasitic sexual disease trichomoniasis. All these microscopic pathogens are or can be transferred by sex, usually through coitus but also through oral and anal sex.

An ugly stepchild of sexually transmitted diseases is *transfusion transmission*, a phenomenon that we first became more aware of in the 1980s, through cases where hepatitis B entered the blood bank inventory, followed by hepatitis C and then HIV making their entrances into blood reserves. Fortunately, today accurate testing, improved blood processing and filtering out undesirable donors through screening allows blood banks to stay mostly clear of hepatitis- and HIV-tainted blood.

Sometimes transfusion transmission is referred to as a subset of *iatrogenic transmission*. "Iatrogenic" means that an illness is caused unintentionally by a doctor, the key syllable here being "un-" as in *un*intended, *un*planned. Iatrogenic transmission covers anything unintended that can happen through a medical examination, medical treatment, or medical prescription. The word "iatrogenic" these days carries a negative social connotation, but its meaning as derived from the Greek is simply "brought forth by the healer." In its purest form, an event being described as "iatrogenic" should not cast blame on the doctor for causing the disease; it merely describes the manner in which the illness occurred.

For instance, if someone who has no known allergy to penicillin has an allergic reaction, such as hives, to prescribed penicillin, then that allergic reaction is considered an iatrogenic event, even though the doctor nor the patient had no way of knowing the patient was allergic. That is an entirely different situation than a doctor prescribing penicillin to a person with a known allergy to penicillin who then suffers an allergic reaction. Both scenarios are termed iatrogenic events. However, the former could not have been anticipated and the latter should have been. In the end, the designation "iatrogenic" does not assign blame; it merely defines what mode of transmission led to the problem.

The medical lexicon has no word for where the mode of transmission is a shared intravenous drug needle, or at least not one I am aware of. Such infections are usually grouped under the heading of "infections related to IV drug use" and include the obvious, such as hepatitis B/C and HIV. This category also incorporates the not so obvious, such as the local or distant staph bacterial infections that can result from a junkie jabbing a dirty needle through even dirtier skin. While it is common knowledge that IV drug users can get hepatitis or HIV, the dangers of getting bacterial staph endocarditis of the heart or septic thrombophlebitis at the vein access site are less well known to the public, but very well known to physicians.

One day in another lifetime when I was a surgical intern, we admitted to our surgery service a person who had hit or nearly hit rock bottom with her IV drug use and had developed a septic thrombotic vein in one of her arms from a recent heroin dose. Staph bacteria no doubt. Our arms and legs have deep veins, but they also have superficial veins, and those superficial veins in the arms that are used to start an IV are the favorites of junkies. Over time a chronic IV drug user exhausts all their superficial veins, arms, wrists, hands, legs, ankles and feet. One of those superficial veins in her forearm had become contaminated with bacteria, resulting in bacterial thrombophlebitis, or a bacterial vein infection, that then clotted off the vein. It looked like a snake coursing up the skin of her arm.

As the surgical intern on the service that day, my scut list included starting an IV on her (we didn't have IV teams in the hospital in those days, no technicians whose specialty is slipping needles into veins; instead, it was us medical students and interns, at least in teaching hospitals). "Scut" is an interesting word. It might refer to the short tail of a hare or rabbit, but in medical school and residency training it means the most menial work to be done by the most menial member of the team, which—you guessed it—was either the menial medical student or the menial intern.

Starting an IV on an IV-drug-ravaged body to introduce antibiotics is not easy. The scarring caused from way too many needle sticks means it's hard to find a viable vein with which to start an IV, and I was unable to find one on this patient. I stuck her with the needle a few times, here and there, flailing and failing miserably, disastrously, much to her bemusement. I was embarrassed, but she didn't care at all that I kept missing; she knew I was an intern and needed the practice.

She wasn't much older than me, and perhaps she wasn't at all, but her long brown hair and dark brown eyes betrayed a face that had been worn down by hard years of heroin use, and god only knows whatever debasement she had had to endure to survive on the streets. Whatever youthfulness that should have been there, that once was there, had long since been pillaged by drug use. Somewhere deep down, I stupidly thought I could save her. I couldn't even start an IV on her, but somehow I thought I could save her. A fool's errand to be sure; addicts and alcoholics can only be saved by themselves.

After I had failed to start an IV—the frustration obvious on my furrowed brow, as I had other scut to check off my list—she quietly told me to let her give it a try, as she gently took the IV needle from my hand. Not surprisingly, she knew where her few remaining available veins were, and she had the experience of threading each and every one of them. She proceeded to easily thread a vein on the top of her foot, thus starting her own IV. The foot is never an ideal location for an IV, as the finnicky feet veins burst easily and tend to be persnickety—but if it is all you have, it is all you have. Later that day on afternoon rounds, the third-year surgery resident—the third-years run the service—chastised me for starting an IV in her foot (berating the intern was a daily amusement for some residents), and he wondered aloud how I could be so stupid. I simply responded with a cheerful, "I didn't start that IV—she did." He knew exactly what I meant, and he didn't say another word. We continued on with afternoon rounds.

I don't recall what happened to her. The IV antibiotics brought under control her infected bacterial thrombophlebitis vein within a few days as we followed her on twice-daily rounds, and she was discharged on oral antibiotics. Never heard about her again. And I did not save her soul.

A doctor will often speak about "making rounds," which means hitting the wards and managing your in-hospital patients and consults. The phrase "making rounds" apparently originated with William Osler, one of those four founders of the famous Johns Hopkins Hospital

in Baltimore previously mentioned. The other studs when Johns Hopkins Hospital opened in 1889 were William Halsted, professor of surgery, Atwood Kelly, professor of gynecology and William Welch, professor of pathology. But our attention is drawn to the architecture of the hospital itself.

Johns Hopkins Hospital's central tower, an historic landmark that still stands, is an open atrium from the ground floor all the way up to its Queen Anne–style dome. The atrium rotunda is graced with a statue of Jesus entitled *Christus Consolator*, a beautiful ten-foot marble statue that makes even the atheist feel welcome, Jesus' open arms conveying healing, compassion and hope. Somewhat unusually, the atrium tower is not circular but rather octagonal in shape, so when you traverse around it, regardless of what floor you are on (other than the first floor), you are basically walking in a stilted circular pattern.

Back in the day, off that tall central atrium tower stood two four-story medical wards, floors two through four, one tower for men and the other for women, built in mirror image. Every floor of the matched wards were connected to each other through that central octagonal atrium, such that, if you were to leave the third-floor "his" ward to head on over to the third-floor "hers" ward, you'd enter the atrium tower and walk "around" the octagon to the ward on the other side.

Imagine Dr. Osler, in his long white jacket, leading the bright but nervous medical students and residents from ward to ward, pausing for a few minutes in the atrium, the sunlight sparkling in, to discuss the clinical findings of the patient just visited, and then Osler and entourage proceeding around the atrium to the adjoining ward. Osler came to describe these twice daily ward visits as "making rounds"—around the octagonal atrium. Physicians worldwide still call our daily patient visits by this phrase: making rounds, the phrase likely originating with Osler. The original Johns Hopkins Hospital still stands, *Christus Consolator*, is still gracing the rotunda, but its once hallowed medical wards now tend to administrative duties, or dare I say, administrative "ills"?

Speaking of things that come out of our mouths, the *oral transmission* of infection spread is different from droplet contact transmission—cough and sneeze—and most certainly different from fecal-oral transmission. It involves direct kissing, such as with infectious mononucleosis, called "mono" for short and sometimes dubbed the "kissing disease." Mono is common among teenagers and college freshmen. Even sharing the same drinking glass within a certain period of time can transmit mouth-to-mouth pathogens. Other microbes in this group include the herpes cold sore virus and cytomegalovirus, which is a relative of the herpes family.

Direct contact transmission occurs where infected skin touches someone else's skin, either directly or through a secondary surface within a short period of time. Such skin-to-skin beasties include the classic fungus feet infection known as athlete's foot which can be spread through public shower floors we all avoid, or should avoid by wearing flip-flops, especially at those steamy, muggy, smelly fitness centers we all pay to rarely use. Stats suggest that only about 18% of members to fitness clubs actually use the center with any regularity, a not surprisingly low attendance fitness clubs bank on. Fungus is a microorganism that loves warm, wet surfaces, and locker rooms can be swarming with it. Some bacteria that cause the skin infection impetigo, a highly contagious strep or staph skin infection, can be transferred skin to skin. Pinkeye, especially bacterial, not as much viral, is easily transferred by contact. But it's not just bacteria and fungus that creep around damp floors: the wart virus likewise calls these dank surfaces home. Warts are caused by various varieties of human papillomavirus (HPV).

I can't speak to the condition of women's public locker rooms that adjoin fitness clubs, public swimming pools and public beaches, or, for that matter, the hotel swimming pool female locker room of what is not exactly a five-star hotel, but the men's equivalents are absolute war zones for skin-to-skin infections. And they stink of urine and mold, poop too. Between bacteria and fungus and wart viruses and little kids peeing on the floor and drunk men likewise missing the urinal and peeing on the floor, those wet, smelly, plague-infested, fomite-primed public locker room floors are to be approached with apprehension and caution. They are a hazmat condemnable zone.

The HPV warty virus can be transmitted skin to skin via locker room showers, but it can also be acquired from kissing someone with an HPV cold sore. An HPV cold sore is to be distinguished from its much worse but kindred spirit: the herpes cold sore. And both of these are distinguished from the noninfectious aphthous ulcer, also called a canker sore, which is usually a smallish shallow sore on the *inside* of the mouth—not on the outside on the lips, like a cold sore. An aphthous ulcer is usually the result of subtle often repeated trauma, such as run-ins with braces and errant toothbrushes, certain spicy foods, and even citrus juices are known to launch a canker sore.

As for the two types of lip cold sores, if an HPV cold sore is a bump in the road, the herpes type 1 oral cold sore is more like a crater in the road. Momentarily will review herpes genital sores, herpes type 2, which are an altogether different beast than the milder herpes type 1 oral cold sore. HPV and herpes type 1 are transmitted through similar routes—oral-to-oral kissing and through oral sex—and although the HPV version tends to be less symptomatic, it also tends to be more widespread across the lips and inside the mouth whereas the herpes cold sore mostly targets one specific area on the lips but with a vengeance, an embarrassingly conspicuous, frustratingly symptomatic, localized cluster of vesicles. Lip herpes results in painful blisters that ulcerate, crust, and leave quite the scarlet letter, as compared to HPV lesions on the lip which are indeed tender, but smaller, less conspicuous.

Herpes as mentioned comes in two flavors, the formal names are herpes simplex type 1 (HSV-1), often called oral herpes, which is milder and mostly targets the lips and mouth, and then there is its evil cousin herpes simplex type 2 (HSV-2), often termed genital herpes, which mostly but not always, targets the genitals. Genital herpes when it involves the mouth is a much worse character than when oral herpes targets the genitals; which is really just saying genital herpes, wherever it lands, is more the aggressor. I think you can imagine how type 2 genital herpes ends up on the lips, in the mouth or down the throat, and for that matter how type 1 oral herpes ends up on the genitals; no need to draw you a picture.

HPV warts and herpes of the fingers were both once common among dentists who put their gloveless hands inside the mouths of contaminated patients. Those dental patients—writhing in pain as cavities were drilled—would shower blood and saliva onto the ungloved fingertips of the dentist, a warty growth just moments away. Herpes of the fingers has a special name: herpetic whitlow, "whit" from the Dutch *vijt* for abscess and "low" from *flaw*. I guess, taken together, someone meant it to mean "an abscess flaw" of the fingernail.

These days, dental workers wear exam gloves, masks and protective eyewear—a standard of practice in the dental world that predates COVID-19—as they fill cavities and scrape plaque. But it's odd to think that not so many decades ago, they did not protect their fingers with latex or nitrile gloves. You would think that, given the preexisting knowledge of HPV and herpes cold

THE DENTISTS BY MUNCH

sores, the dental community would have wised-up sooner than it did. But it didn't. It appears wearing exam gloves and personal protective equipment (PPE) did not become a 100% mainstay of the dental world until HIV arrived on the scene in the 1980s.

Every time I go into a dental office, my teeth ache. Every time. The dental office is a good example of the adult version of the Pavlovian response, named for Ivan Pavlov's dogs, which salivated upon hearing a tone that the dog was trained to associate with being fed. Pavlov called their drooling "anticipatory salivation psychic secretion." But it's your basic "classical conditioning." Pretty much all dental offices have that same smell, you know what I mean, a combination of various chemicals like formocresol and meta-cresyl acetate, which are both topical bug-killing antiseptics. Dentists also use eugenol, which is a topical numbing gel derived from clove oil, alongside various acrylic monomers used to glue things together inside the mouth. When you add all those compounds together—formocresol, meta-cresyl acetate, eugenol, acrylic monomers—they produce a characteristic aroma unique to the dental office. As soon as I walk into any dental office, into that medicinally fragranced fog, it must subconsciously remind me of dental drilling, and my teeth begin to ache. I guess Pavlov might call that an "anticipatory drilling psychic pain."

Vertical transmission of infection is from mother to unborn child, that is, *in utero*, which is a term from the Latin meaning "in the uterus." In other words, an infection in the mother crosses the placenta barrier and infects the unborn child. Another mode of vertical transmission occurs when the newborn exits the safety and sanctity of the uterus into an infected birth canal, the vagina, a rigorous journey through an infected channel, that can be avoided with a C-section should the vaginal vault be deemed infected. C-section is more often selected over vaginal delivery due to failure of labor to progress, abnormal fetus position such as breach, fetal distress, a pelvis that does not lend itself to vaginal delivery, and cord prolapse. But, most certainly, an infected vaginal vault is another reason to take the C-section detour.

The most common *in utero* mother-to-fetus infections across the placenta, while the bun is still in the oven, are known as the acronym "TORCH" complex of infections: T̲oxoplasmosis, O̲ther (such as chickenpox, chlamydia, HIV, syphilis, and Zika virus), R̲ubella (also known as German measles), C̲ytomegalovirus, and H̲erpes simplex virus 2.

The more recent acronym for that set of vertically transmitted mother-to-fetus diseases—an acronym not available to me when I was coming up through the ranks of residency—is CHEAP TORCHES: C̲hicken pox, H̲epatitis, E̲*nterovirus*, A̲IDS, P̲arvovirus, T̲oxoplasmosis, O̲ther (including *Streptococcus*, *Listeria*, *Candida*, and Lyme disease), R̲ubella, C̲ytomegalovirus, H̲erpes simplex, E̲verything else sexually transmitted (such as gonorrhea, chlamydia, and HPV), and

Syphilis. Most of these CHEAP TORCHES infections transfer from mother to fetus across the placenta, but a few, especially venereal diseases, can or are only transferred to the infant while transiting through the birth canal, notably syphilis, gonorrhea, chlamydia, and cytomegalovirus. HIV can be transferred either way, across the placenta or during birthing.

Finally, there is *vector-borne transmission*—infections that travel from a source through an intermediary called a "vector," such as an insect, into a second unsuspecting victim. The vector is merely a host, and the most well-known example is malaria and its partner in crime, the mosquito vector. Other examples where a vector gives a parasite a free ride include African sleeping sickness (by the tsetse fly), Lyme disease (via deer ticks), and yellow fever (through mosquitos). To be clear, it is believed the vector does not become ill from the parasite as it transfers that pathogen from one point to the next; it is only an intermediary transmitting the infectious pathogen.

A bit of nomenclature is needed here first. When I say "insects," I also mean arachnids like ticks, which are technically not insects, they are in the class Arachnida. Both Insecta and Arachnida are in the phylum Arthropoda, but commoners like me often refer to spiders and ticks as insects, they are not.

Nearly all vector-borne infections—malaria, yellow fever, Dengue fever, West Nile virus, the plagues, Lyme disease, tick fever, Chagas disease, African sleeping sickness, and so on—are transmitted by members of the phylum Arthropoda, with a few exceptions. Schistosomiasis, a parasitic flatworm infection that targets the bladder, liver, and intestines, is transmitted to humans through snails, and snails are members of the phylum Mollusca.

Having said that the vector doesn't become ill, some researchers do think that the vector indeed gets a little ill, or more likely, it got a little ill in the remote past, thousands, if not millions, of years ago. In other words, there might have been a time when mosquitos became sick from malaria. But as time passed, the vector insect became immune to the parasite, or the parasite had bigger fish to fry, mosquitos no doubt taking great pleasure in not being ill and only serving as a vehicle for transmission.

How do we know this? How do we know the mosquito isn't queasy from malaria? It is not as though a mosquito can complain much. But from what we know about vector transmission, it only stands to reason that the vector insect should not get ill. By definition, a vector couldn't be good at its vector job *and* also get ill. It's one or the other. Sometime eons and eons ago, some type of symbiotic relationship arose between certain parasites and vector insects and arachnids for reasons that science is not clear on, foregoing a vector infection, and delighting in infecting us humans instead.

Whether they became ill or not, the likely reason some parasites chose vector insects to do their bidding, to be the transmitter of their lineage, is quite simple: the sheer numbers of insects and ticks available to parasites to act as transport vehicles. If you're a pathogen that needs a secondary species to continue your life cycle, to issue your progeny, you'd be a pretty dumb parasite to pick the dodo bird as your carrier—there was never very many of them, and now there are none. In other words, parasites and viruses first chose arthropods buzzing about because there were gosh darn so many of them…and not just so many, they buzzed about, they moved.

Especially viruses, but also parasites like malaria, chose the insect model as a vector perhaps 500 million years ago, since a gazillion of insects were flying every which way, and viruses, which are not terribly mobile creatures needing a large host population in which to survive, chose wisely. Perhaps in those days long since gone, the insect did become a little ill, but, over time, the virus must have decided to use the insect only as a vehicle for transmission from one mammal to the next mammal, for extending their life cycle; or, alternatively, the insect wised up enough to *not* allow itself to become ill. Insect and arachnid vectors are merely commuter trains. If you're a parasite choosing the vector mode of transmission, you choose insects and you don't make those insects very ill, if at all. Because if a parasite kills its vector, it kills itself. You don't bite the hand that sort of feeds you.

Not all vector insects transmit infections to humans by a bite, which is how the blood-sucking mosquito deposits its malaria-tainted saliva into humans. Some vectors transmit infection through a less elegant mechanical fashion. In Chagas disease, the *Trypanosoma* parasite lives within the poop of the triatomine bug. The vector triatomine bug climbs onto a sleeping person, usually their face—which, by the way, is why the triatomine bug is also called the "kissing bug"—and there on the skin of the face the bug nibbles about as if kissing the sleeping person, while also pooping, the slumberous victim unknowingly accommodates the whole process by scratching their face. A scratch is a break in skin integrity, which allows the Chagas larvae, having in the meantime deposited out of the poop the triatomine bug has casually been leaving behind on the facial skin, and burrow themselves into the skin, leading to Chagas disease.

African sleeping sickness, which is caused by a cousin of the Chagas parasite, enters the victim through the bite of the tsetse fly, much like malaria does with the bite of the mosquito. Somewhere in time two species of the *Trypanosoma* genus—the one that causes Chagas and the one behind African sleeping sickness—decided to go down different vector-borne transmission routes to extend their lineage. One *Trypanosoma* chose the mechanical poop mode—the "kissing bug" Chagas model—while its cousin chose the more elegant saliva mode of the biting tsetse fly, the African sleeping sickness model. Tsetse means "fly" in the Bantu language and is sometimes called the tik-tik fly. The origin of "tik-tik" is less clear but likely is an onomatopoeic word, that is, as the fly annoyingly buzzes around, it makes a "tik-tik" sound.

One more point to clarify is about the difference between vector-borne infection and zoonosis. Vector-borne diseases are infections transmitted to humans using the vector intermediary. Once the infection is in the human host, it is not directly passed human-to-human. Humans get malaria and yellow fever and tick fever and Lyme disease from a vector, not from other each other. Vector-borne infections are not contagious.

Zoonosis is different, it is an infection in an animal host that jumps into a human host, a sort of one-way leap, and once that infection enters the human arena, it can then be passed human-to-human. It is contagious.

The classic example of a vector-borne infection is the noncontagious malaria. The classic example of a zoonotic infection is pretty much all flu viruses, which originally came to us humans from the avian world, after our troglodyte ancestors made the bold move to domesticate wild birds and then live in close proximity to their prize flock.

A novel flu virus might emerge in a bird population or a swine population and then make the dastardly bold leap into the human population, where we do our bit to cough and sneeze that new flu virus to all corners of the globe. The zoonotic jump from animal to human can occur from handling or eating or being near the infected animal host. To make matters a tad murky, some zoonotic jumps involve an intermediary, a vector, but it is not the same as a true vector-borne infection, which by definition is not communicable between people.

To parse between true vector-borne infections and true zoonotic infections, ask yourself one question: Once the infection enters the human host, is it contagious human-to-human? If the answer is no, it is a true vector-borne infection (malaria); if yes, it is or once was a zoonotic infection (the flu).

Now that we've covered the modes of infection transmission, and before we deal with those clinical/subclinical and primary/secondary and acute/chronic/relapse/recurrent mumbo jumbo, our story is about to turn towards one of the more boring human infection pathogens: fungus. Hopefully I'll make it entertaining. But, before that, let us loop back to take a brief look at the sad story of the dodo bird, (in)famously driven to extinction by humans.

We're actually not even sure what the dodo bird looked like, because there is not even a single stuffed taxidermic specimen. The dodo was an ancient flightless bird that lived on the single island of Mauritius, east of Madagascar in the Indian Ocean, about 1,200 miles off the African coastline. Its closest surviving taxonomic family members are pigeons and doves. No natural predators lived on Mauritius, so it came as an unwelcome surprise to the dodo when, during the Age of Discovery in the 1500s, predators in the form of

We've lost the ability to fly Dad: To fly! How is that evolution?

Dutch, Portuguese, and English scallywags landed on the island for rest and provisions. It is believed the dodo was about three feet tall, but since there are few fossil records and only a few paintings of it, not much about the dodo is known for certain. Nothing is known of their diet, their mating rituals, the sound of their song. Nothing.

Because the dodo evolved in isolation from predators, when the explorers landed, the dodo was apparently fearless of humans, practically waddling up to salty swabs as if to greet them. And being dodos, when one squawked, whether it was being chased or even killed, the other dodos, instead of fleeing, came running to see what was up. With no fear of predators and no ability to fly, they were easy prey for that night's dinner. And what the water dogs who landed on Mauritius didn't eat, their real dogs (and cats) did. The first explorers heading towards the Malabar coast of India intermittently arrived on the island beginning in the early 1500s and before the end of the 1600s, the dodo was extinct. Two hundred years and poof! The dodo was gone.

This story for the dodo gets worse. Several partial and complete taxidermic specimens of the dodo found their way to various museums around the world, but most were lost over time. It is believed the last complete dodo taxidermy specimen was housed at the Oxford University Museum of Natural History in England. But in the 1700s, when the specimen became grossly dusty and moldy, a curator foolishly threw it on top of a trash heap, never to be seen again. It was a most inauspicious end to the dodo bird, moldy or not. And speaking of molds …

The Catholic priest and botanist Pier Antonio Micheli was the first, in 1729, to describe fungi, offering yet another example of men of the cloth contributing to the advancement of science rather than hindering it. After becoming ordained, Micheli, who was born in Florence, Italy, taught himself Latin and botany, eventually becoming a professor of botany at the University of Pisa through an appointment by Cosimo III de' Medici, Grand Duke of Tuscany.

In hindsight, Giovanni realised that it was probably a bad idea hiring a builder with a short leg!

The Leaning Tower of Pisa—famous not just because it leans but because, as we know, Galileo conducted gravity experiments from its highest reaches—was built in the twelfth century. It was constructed as a freestanding bell tower for the Cathedral of Pisa. The leaning—which is truly unintentional—began ever so slightly during construction, due in part to the ground underneath being unsuitable for such a massive structure. The leaning continued to increase over the centuries, eventually reaching 5.5 degrees from perpendicular.

In the late 1990s, over 600 years after construction was completed in 1372 AD, Italian authorities finally took steps to put an end to the tower's list, lest it fall over altogether. The unusual architectural angle was partially corrected and stabilized, brought to its current 3.9 degrees in 2007. The engineers did an excellent job: the Leaning Tower of Pisa currently does not show any

"No way is that going to fall down anytime soon."

penchant toward listing again, yet some lean is still preserved for posterity, nostalgia and tourism. And tourism. We can now assume the Leaning Tower of Pisa won't fall down.

And we can only hope the Italian government can similarly figure out how to keep the famed city of Venice from sinking into the sea. The issues threatening to send Venice into the realms of Poseidon—Neptune in Latin—are twofold: rising sea levels are an enemy as much as the compaction of the sediment upon which the city was built beginning in 400 AD. Venice was constructed upon approximately 100 interconnected small islands, none of which possessed durable earth. Basically, Venice sits atop a lagoon. Figuring out how to correct the Leaning Tower of Pisa was child's play compared to what it will take to save Venice. Without an engineering solution, Venice will go the way of the Lost City of Atlantis. Venice is a beautiful old lady, rich with history, truly a world treasure that needs preservation.

I could tell you a funny story about when I traveled to Venice with a girlfriend, arriving by train from Florence. She was, shall we say, unfamiliar with the fact that Venice, being a bunch of little islands, is a city held together by some 400 bridges that cross twenty-six miles of canals. No roads. No automobiles. No Mercedes sharing the road with the donkey. Just canals. As we emerged from the train station, she inquired where the taxis were to "drive" us to our hotel, a quip that was heard by several nearby travelers. I could mention how I rolled my eyes—how she saw and even *heard* my eyes roll. I could mention how I had to explain to her Venice had no roads, no cars, only canals with gondolas and pedestrian walkways and bridges. But I won't tell you that story. She'd kill me. Luckily, I was quick enough to spare her and me further embarrassment by mumbling that we would take a "water taxi" to our hotel, and on we went. Eventually it dawned on her that Venice was *the* city of canals. But, again, I won't tell you that story.

Atlantis is a fictional city of allegorical proportions that features in the writings of Plato, who describes it as a fake utopian society, a foil for an ideal city-state of Athens. In Plato's tale, Athens won the naval battle between the two city-states, meaning Atlantis lost and subsequently fell out of favor with the gods, who then doomed it to sink into the sea. Ouch. Plato's allegory tells us that Greece, the republic, is great, and certainly much better than other city-states and other forms of governance, which in his story are represented by Atlantis. And what forms of governance did Plato despise. Well he loathed timocracy which is power by wealth with little regard to social or civic responsibility, and oligarchy in which power rests within a small group of rulers was not appealing to Plato, nor was democracy or rule by the majority. Plato detested tyranny where there is a single absolute ruler (kings and queens fall into this latter heading). So what form of governance did Plato favor: aristocracy where power in vested into a small ruling class that is grounded in wisdom and reason. Socrates was forced to drink the hemlock when he too favored rule by philosophers over any form of democracy.

Despite Plato's story of Atlantis appearing to be fictional, some historians believe it must have some real-world inspiration. If there ever was a real Atlantis that was lost to the sea, such a city was likely an island centered in the Mediterranean or in the Atlantic Ocean just beyond the Strait of Gibraltar.

If it existed at all, the island Atlantis might have been consumed by a natural event such as an earthquake, volcanic eruption, rising sea levels after the last ice age, or some such seismic event.

There is historical precedence for this theory. The eruption of the Thera volcano around the seventeenth century BC apparently devoured the Minoan civilization on the island of Crete. And we know that 10,000 years ago, sea levels were lower, because more of Earth's water was locked up in the polar ice caps. Thereupon, as Earth came out of that ice age, sea levels rose, which certainly could have doomed low-lying city-state islands such as the ostensibly mythical Atlantis into the depths of the seas.

It is believed that before that time, more than 10,000 years ago, before the ice age ended and sea levels rose, our Neolithic cousins would have been able to wander from Ireland to England to France, and therefore possibly even to this utopian island of Atlantis. Then, as the ice sheets melted, as the oceans rose, Ireland separated out from England, which then separated out from Europe. The same situation is what allowed Neolithic peoples to cross what is now the Bering Strait—back then it would have been the Bering land bridge—and migrate into the Americas from Asia.

Back in Pisa, we return to the famed Italian polymath and astronomer Galileo Galilei standing at the top of the Leaning Tower, preparing to perform his gravity and motion experiments from its great height. As we saw earlier in this story, these experiments involved dropping things like cannonballs off the tower to prove they fell at the same speed even though they had different masses. That is how Galileo disproved Aristotle's near 1,800-year-old theory that objects fall at different speeds according to their masses. Then, 100 years after Galileo's discovery, Isaac Newton provided the math of gravity to support Galileo. In the physical world as humans experience it, Newton and Galileo were indeed correct. But in the world of Einstein, where mass bends space—those watch-wearing apples and time dilation discussed earlier—it turns out that in fact Aristotle is the one who was right (sort of), those cannon balls don't quite fall at the same speed, but for a reason he could not have possibly understood.

Both Galileo and his most important predecessor Nicolaus Copernicus learned their astronomy by studying the ancient writings of the great second century AD Greek astronomer Ptolemy. Up until Copernicus, astronomy was basically all Ptolemy, with a little Aristotle thrown in for good measure, just like medicine was all Galen. These second-century scientists, Galen and Ptolemy, created such convincing theories of medicine and astronomy, respectively, that they dominated the landscape and the skyscape for 1,500 years.

Some of Aristotle's fellow Greeks, dating from the third century BC onward—Eratosthenes, Pythagoras, Archimedes, Hipparchus—also dabbled in defining the cosmos and planetary motion, all placing Earth at the center of the universe, otherwise known as the geocentric model. To be fair, one Greek actually dilly-dallied with a sun-centric model—Aristarchus of Samos—in the second century BC but was likely dismissed as a lunatic. Four hundred years later, Ptolemy expounded upon the Aristotelian view—even though it was wrong—with convincing observations of the night sky. In fact, so convincing were his maps that they were used by every mariner, observing the night sky to sail the high seas.

The Earth-centric model also became the Christian Church's favored viewpoint, with the idea of Earth being the center of everything, and humans the center of that, and wives the center of that. Anything to the contrary of what the Church thought was considered heresy. Probably anything contrary to a wife's thinking is blasphemy, too. Ptolemy was an ancient polymath of Greek ancestry and a Roman citizen living in Alexandria, Egypt. As was said before: wars do that. Galen was also a Roman citizen of Greek ancestry, but he lived in Anatolia, modern day Turkey. Wars do that.

Ptolemy's model of the solar system—his night star charts were flawless—required a lot of fudging to explain the motions of the planets around Earth at the center. He was too smart to not have known better, but he didn't know better. Either he never thought to consider a sun-centric model, or he never allowed himself to consider a sun-centric model. Had the brilliant Ptolemy taken that moment of pause and thrown off the Earth-centric dogma that was forced upon him by the weight of history, by the weight of Aristotle, dogma that was not supported by the observations and motions of the celestial sky, Ptolemy rather than Copernicus, 1,400 years later, might have arrived at the heliocentric model of the solar system. But he didn't.

Oddly or actually not so oddly, even though Ptolemy's solar system was based on a woefully wrong premise, his star locations were spot on—so accurate, in fact, that seafarers continued to use them for well over 1,500 years. If you're a captain sailing the high seas, especially during the Age of Discovery from the 1400s onward— Christopher Columbus, Vasco da Gama, Pedro Álvares Cabral, John Cabot, Juan Ponce de León, Bartolomeu Dias, Ferdinand Magellan, and even Captain James Cook—whether the earth was at the center or the sun at the center of the solar system mattered not. All that mattered were your star charts, those distant stars not giving a wit what's at the center of our puny solar system. Age of Discovery captains navigated to the far reaches of the seas, daring to sail off the edge of the map, using Ptolemy's star charts, and never once, not once did they sail off the edge of the world.

Ptolemy arrived at his mathematical model of the solar system after compiling all knowledge then known about the planets and stars, beginning with the teachings of the Persian Babylonian astronomers from 1,000 years before his time and spanning to the writings of Aristotle and the treatises of the later Greek astronomer Hipparchus. Merging their written observations with his own celestial observations, Ptolemy produced the *Almagest*, the bible of the sky. The *Almagest* star maps were a compendium of all known planetary and star motions. It is an important point to digest. The stars are so distant and their progression across the night sky as the year unfolds so predictable, as well as the progression of the then known six planets, Ptolemy's approximate sixty-four detailed celestial maps were accurate viewing from Earth. Star locations week-to-week, month-to-month never changed, repeating the cycle every year, regardless if the Earth was or was not the at the center. In other words, Ptolemy's celestial night sky maps were independent of his solar system rendition. Imagine being the proud owner of a leather-bound folio of Ptolemaic celestial sky as you set sail from Spain looking for the Far East, navigating the high seas and plotting a course.

Although the second-century Ptolemy maps were improved in the tenth century by the Persian astronomer Abd al-Rahmān al-Sūfī, his Latinized name being Azophi, it wasn't until the sixteenth century that the maps of Tycho Brahe brought even more accurate star catalogues.

Tycho Brahe was a Danish nobleman and astronomer who attempted to merge the Ptolemaic Earth-centric planetary system of the second century known since antiquity with the Copernican sun-centric system of the sixteenth century. His collision of two vastly different planetary systems became known as the Tychonic system. Brahe ever so much desired to reconcile the religious doctrine of an earth-centric (geocentric) universe with the observed sun-centric (heliocentric) universe of Copernicus. Of course, Brahe was wrong, Copernicus was right. But Brahe's detailed observations of the night sky were nevertheless helpful to future astronomers and sailors. He was a clever astronomer and took great pains to detail the celestial sky, even if he spiritually conflated data to fit religious doctrine. You sort of wonder if, deep down inside, he might've known better

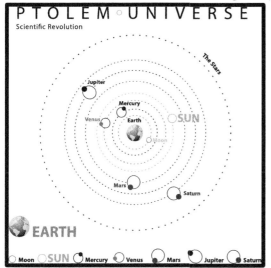

Ptolemy's earth-centric model lists the celestial objects circling a stationary Earth at the center in the following order, from closest to farthest: Earth is at the very, very center, then our Moon, followed by Mercury and Venus, followed by the Sun, then Mars, Jupiter, Saturn, and finally the stars (Uranus, Neptune and Pluto had yet to be discovered). Brahe's solar system also has Earth at the center, with Mercury and Venus orbiting the Sun as the three together orbit Earth.

In Ptolemaic solar system, the Sun circles earth somewhere between Venus and Mars.

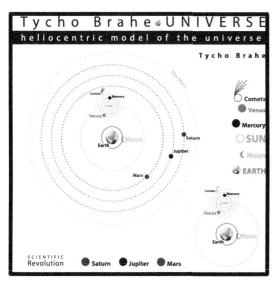

Although this became the accepted order, so many obvious aberrations existed as a result that it is just mind-boggling. The model just didn't make sense, didn't add up. To make the observed solar system sky make sense according to the theoretical Earth-centric model, Ptolemy had to add a bunch of fudge factors. The fudgiest fudge he added was that each sphere orbiting Earth—Moon, Mercury, Venus, Sun, Mars, Jupiter, Saturn—had its own local orbit around itself. Not just a spin about its axis but an orbit about itself. And the Tychonic map has that fudge factor of Mercury and Venus orbiting the Sun as the three orbited Earth as a group, sort of like Mercury and Venus were moons of the Sun.

Ptolemy's and Tycho's solar system motions had to be incredibly convoluted to make the math work, and the truth of it is, when the Sun was eventually placed at the center of our solar system—which is what a solar system is, a system of objects orbiting a star—which is what Copernicus precisely did; then, to quote Forest Gump: "and just like that" the math made sense.

The battle between the complicated, the-math-does-not-work-out, too-many-assumptions Ptolemaic-Tychonic Earth-centric universe and the simple, the-math-works-out, hey-look-no-assumptions Copernicus Sun-centric universe is perhaps one of the best examples of Occam's razor. William Occam was a fourteenth-century Franciscan friar who proposed the "law of parsimony," what we call today Occam's razor: "Entities should not be multiplied without necessity." What does that mean? When confronted with competing theories or hypotheses, Occam's razor states: "the simplest solution is most likely the right one." The word "razor," when used in philosophy, figuratively means using a sharp razor to "shave off" unlikely explanations, a philosophical razor, to arrive at a conclusion.

Ptolemy, despite his geocentric model, also demonstrated that Earth is a sphere, not a flat pancake, and suggested as much for the other heavenly bodies that transit across our night sky, especially our local neighborhood of planets. There are people today—not many, but a select very special few—who still believe Earth is flat. God help us. As Ptolemy's *Almagest* became the bible of astronomy, and then the bible of seafaring mariners, fortunately—despite being galactically wrong with its Earth-centricity—it saved more than one well-meaning but half-witted sailor from sailing off the edge of the world. As said above, whether or not Earth or Sun is at the center of our solar system, the stars are so very, very far away that their positions relative to Earth don't change. Or do they?

Well they do, only because the Universe is expanding, but given their distance from Earth those changes are practically immutable, at least in the short run, and when I say, "short run," I'm talking tens of thousands of years. Take Polaris, the North Star, the brightest star in the constellation Ursa Minor. It is directly above Earth's North Pole in the night sky (Polaris is actually a triple star system so its three stars), the North Star giving the appearance of a fixed, unwavering point. We were taught that in elementary science class, the North Star is a constant, find it and you'll be facing north. But the North Star hasn't always been so. According to ancient texts, 4,800 years ago during the early Egyptian dynasties a star named Thuban was the "fixed" North Star. In about 8,000 years from now the star Deneb will move into that exalted position and Polaris day in the sun, so to speak, will have been pushed off due north.

Why was an exact celestial map so important? Well, it wasn't one map but as I alluded to, a collection of maps depending upon the time of year. It was a *changing* celestial map: the positions of the constellations change as the year progresses due to the orbit of the earth around the sun. As earth orbits around the sun—the sun is irrelevant, it is the movement of the earth relative to the stars—every few days a couple new stars become visible in the east as others drop off in the west. After a week or two, an entire new constellation might rise in the east while one goes missing in the west. Ptolemy's *Almagest* didn't just contain a single map but those sixty-four maps, a bunch of star charts practically one for every week. Although the star map repeats itself annually, the path of the planets does not. Because the planets are in our local neighborhood and take such a ridiculous amounts of time to orbit the sun—Neptune takes 168 years, Uranus 84 years and the non-planet Pluto 248 years—you need an app downloaded to your iPhone to have an accurate local solar night sky map. Otherwise, Ptolemy's celestial star charts would still work for sailors.

It took those 1,500 years of darkness for humanism to arise like the phoenix, freeing from the Dark Ages, a construct we discussed earlier, that among many things, allowed Copernicus to finally establish the heliocentric model, with all visible planets—including Earth—as well as the moon and the planets whirling around the sun in what he assumed were perfectly circular orbits. Copernicus more accurately described than Ptolemy, and for that matter more accurately predicted, the observed

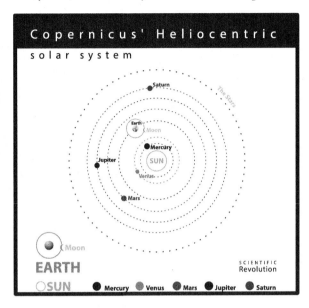

celestial planetary motion within our solar system with a much simplified celestial map. But as you probably have guessed, Ptolemy's navigational maps were still used after this revelation, unaffected by the sun's new position at the center of the universe. It's not the center of the universe, our Sun, but even Copernicus had no clue the size of our Milky Way, the size of the universe, let alone our solar system on some remote outpost.

Copernicus also introduced a vague concept of gravity to explain why all these heavenly bodies seemed to move in perpetuity around the sun. Just like the seventeenth-century Newton gave the math to the sixteenth-century Galileo's gravity experiment, Newton also gave the math to

the sixteenth-century Copernicus's understanding that a planet in motion tends to stay in motion, balanced by something pulling on it.

Whether it was Ptolemy with the Earth-centric model or Copernicus with the Sun-centric model, both believed that our solar system was the center of the universe. They believed that what has turned out to be our lonely little corner of the universe was the center of all things, and all the night stars, just like all our local planets, orbited around our sun. It wasn't until the 1700s, a few hundred years after Copernicus, through the works of Immanuel Kant, the great German philosopher and astronomer, that it dawned upon astronomers (pun intended) that our sun was not much the center of anything other than our own remote solar system, a lonely speck of cosmic dust on a far-flung spiraling arm of a massive galaxy now known

as the Milky Way. To Kant and others of the eighteenth century, neither our Earth nor our Sun had much to brag about. We Earthlings were on our way to finding out that not everything revolves around us. And it didn't end there. Even though Kant and pals taught us our solar system was not the center of our Milky Way, they still couldn't let go of the idea that the Milky Way was the *entire universe*. We call it the Milky Way because gazing up at the night sky on a really, really dark night, like from the top of a remote mountain, our galaxy looks rather milky smooth. It's not your eyes playing a blur game, the opalescence is real.

It took much less time for the belief that the Milky Way was the entire universe to come to a crashing halt, however. Around 1923, when Edwin Hubble was peering through that Hooker telescope we discussed earlier, the one at the Mount Wilson Observatory in California, he realized that some of what were believed to be bright distant stars or binary star systems, were in fact not stars at all, but rather whole other galaxies equipped with their own sets of billions and billions of stars. The universe was, a bit on the nose, galactic.

In 1929, Hubble published his thesis, demonstrating, among other things, that Andromeda, first identified in 964 AD by the Persian astronomer Azophi, was an entire galaxy in itself, rather than the "binary star" it was thought to have been. True binary stars are two stars that orbit around each other and many stars are precisely that. But, as it turned out, not just Andromeda but many other of these supposed binary stars were likewise entire galaxies of billions or trillions of stars. It's just that through a telescope from Earth they looked "binary" because they didn't look "solitary"—they were fuzzy and not perfectly spherical.

Hubble arrived at his conclusion based on Andromeda's observed Doppler redshift (describing, as you'll recall, how the wavelength of light moving away from the observer elongates to the red spectrum), which he used to calculate that Andromeda is at least ten times farther away from Earth than the farthest star in our Milky Way. The only way this could be true is if Andromeda was another galaxy. Also true was that its red shift meant it was moving away from our galaxy, an expanding universe.

When you consider the sobering fact that there are believed to be 10^{24} stars in the universe—that's a one followed by twenty-four zeros, a.k.a. a septillion—which are allocated across billions of galaxies, our own solar system comprising the sun and nine planets (go, Pluto!) is quite the lonely outpost indeed. If there is no other life in the vastness of the universe, it really is a lot of wasted space and a very lonesome place indeed.

In addition to presenting the heliocentric model, Copernicus further stated that Earth and all the planets are spinning on their axes, like spinning tops, a novel concept for the time. Spinning on an axis is different than orbiting the sun, and that spin is what gives us our day, our sunrises and sunsets.

Johannes Kepler, who we'll meet momentarily, observed that Earth's annual transit around the sun—as well as all the other planets' orbits, too—is not perfectly circular, as Copernicus had suggested, but rather elliptical, which is important for our four seasons on Earth.

"Sunrise, Sunset," a song written by Jerry Bock and Sheldon Harnick for the 1964 musical *Fiddler on the Roof* here becomes relevant to our story:

> Sunrise, sunset
> Sunrise, sunset
> Swiftly fly the years
> One season following another
> Laden with happiness and tears

"Sunrise, sunset /Sunrise, sunset" is Earth spinning once on its axis, which takes twenty-four hours. Sunrise is Earth spinning toward the sun, and sunset is Earth then spinning away from the sun. Even though we like to say the sun rises in the east and sets in the west, what is really happening is that Earth is twirling around its axis to bring the sun in and out of view from our own humancentric perspective.

"Swiftly fly the years" is Earth making one complete elliptical—not circular—revolution around the sun, which takes one year or 365 days (well, 365.2425 days, which, as we discussed earlier, is a timekeeping problem solved by leap years and leap seconds).

"One season following another" is partially based on Earth's asymmetrical, elliptical orbit around the sun, but, more importantly, our four seasons are actually largely influenced by the tilt of Earth toward or away from the sun, not as you would expect the elliptical orbit. Take winter and summer, for instance. One would think that in the winter earth is at its farthest distance from the sun and in summer at its closest. While such a thought might be understandable, it is wrong. Oddly, Earth is actually three million miles closer to the sun in winter, at ninety-one million miles away, than in summer, at ninety-four million miles away.

That Earth is three million miles closer in winter is of no consequence, obviously, when it comes to the amount of sunlight we receive or do not receive. Its negligible. Rather, it is the tilt of Earth on its axis combined with its elliptical journey around the sun that gives us our four seasons.

For the northern hemisphere, when Earth is tilted toward the sun, it is summer and the sun is higher in the sky, the days are longer, the weather is warmer because those northern latitudes are receiving more thermal energy—longer days with direct sunlight. This season begins with the summer solstice, the longest day of the year, which falls somewhere between June 20 and 22.

When tilted away from the sun in winter, the sun is lower in the sky, the days are shorter, the weather is colder because those northern latitudes are experiencing a decrease in thermal energy from the sun, with the winter solstice occurring between December 21 and 23.

For the Aussies and the Kiwis and the Pacific Islanders and pretty much all of South America and a big swath of Africa—the southern hemisphere—the opposite situation occurs: high summer (longest days) is in December and the depth of winter (shortest days) is in June.

As for equatorial Earth, it doesn't much matter what the tilt is, because the equator is tilted neither away nor toward the sun, the equator being the pivot point of the tilt, which is why equatorial earth experiences little change in seasons.

In fact, if Earth didn't tilt in relation to the sun, there'd be not much in the way of seasons for any of us. Interestingly, however, it is believed Earth was not always tilted. It is thought that soon after the cosmic schmutz formed Earth, a large asteroid struck it, not only kicking off enough material to make our moon but also tilting Earth on its axis 23.5 degrees, and even possibly messing with Earth's ecliptic orbit around the sun.

The spring equinox falls around March 20, which is about the same time we set our clocks forward for daylight saving time and lose one hour of sleep. The fall equinox occurs around September 23, and sometime around then we get back that extra hour of sleep lost in the spring. The distance Earth is from the sun during these two equinoxes is farther than in winter but closer than in summer, further proving distance from the sun plays little if any role in Earth's temperatures and seasons as we speed around our annual ellipse. A spread of three million miles, all things considered, is not enough greater or lesser sunlight to alter seasons. It's the tilt, baby, the tilt.

Curiously, what, if anything, does the asymmetrical, elliptical trip Earth takes around the sun do if not alter the seasons very much? Well, what it does alter is the climate over thousands of years. Although the hot topic of climate change is a subject better saved for another day, in a nutshell, three major celestial factors drive *normal* climate change—emphasis on "normal" for any climate change deniers out there.

First, that tilt of Earth is not stable but rather varies with a periodicity of about 41,000 years; Earth's tilt is said to be 23.5 degrees, but it does vary upwards of 24.5 degrees or downwards to 22.1 degrees. That might not seem like a big deal, but it is when it comes to climate. If tilt is what gives us our four seasons, small oscillations of that tilt would obviously alter climate.

Second, our elliptical orbit experiences what is called a "slight precession," meaning Earth doesn't follow the exact same path every year but one that varies slightly within defined parameters every 100,000 years. That varying elliptical journey, termed its eccentricity, does affect the thermal radiation reaching and absorbed by earth.

And third, the Earth is also wobbling about its axis, which is different than titling—sort of like a groom at his wedding both leaning and wobbling. Or a spinning top that as it loses its energy, as the spin slows down, a perceptible wobble appears. Earth's spin is not, at least not now, slowing down like a spinning top, but it does have a slight wobble with a periodicity of about

26,000 years.

These three periodicities—tilt, eccentric precession, wobble—plus a few other celestial influences, such as weather patterns on the sun (yes, the sun also has weather patterns) and the gravitational pull that the moon and other planets have on Earth, all conspire to induce Earth's climate to undergo periodic and normal climate change.

Added to these celestial influences are worldly forces such as volcanos and earthquakes and plate tectonics and oceans and mountain ranges, and you end up with further determinants of normal climate change. But, when you add to that the man-made contributions of atmospheric gases, mostly carbon dioxide which increases the heat retained by earth, a process termed "radiative forcing," Earth's entire atmosphere becomes a hothouse greenhouse—voilà, you have *abnormal* climate change!

The most classic example of the three periodicities plus celestial influences plus worldly influences—minus the human influences—is Earth going into and out of ice ages. Ice ages don't arise and don't go away because someone forgot to close the freezer. Ice ages are slow in the making and slow in the unmaking, the result of the doggedly unhurried celestial and worldly influences—tilt, precession, wobble, moon, volcanos, earthquakes, plate tectonics—that govern them.

As for the last line in the *Fiddler on the Roof* song— "Laden with happiness and tears"—that is decidedly not celestial in origin and isn't even earthly in origin. Rather, it is of the human domain, from the joys and the trials and the tribulations of being human. Unless, of course, you are a believer in astrology and are certain that somehow the positions of the sun plus some arbitrary selected celestial bodies at the time of your birth dictate the highs and lows of your life, in a nutshell, your happiness and tears.

The name "Ptolemy" comes from the Greek and means "son of war." The astronomer Ptolemy we've been visiting with is also known as Claudius Ptolemaeus, not to be confused with two other distinct but possibly blood-related, or quite possibly not-blood-related, Ptolemys. The first of these is Ptolemy, the ruler of Thebes, Greece (not Thebes, Egypt) who began the Ptolemaic dynasty of Greece. His bloodline ruled a large portion of Greece from 1200 BC up to around 325 BC.

The second Ptolemy is related to another ruling Ptolemaic dynasty—but this one was not in Greece, rather in Egypt, yet still of Greek bloodline. After Alexander the Great overran Egypt in 332 BC, Greece lorded over Egypt for the next 300 years, that is, until the Roman Empire took over. The inaugural Greek Hellenistic ruler in Egypt took the name Ptolemy I Soter, establishing a succession of Egyptian Ptolemaic rulers who were of Greek ancestry. The Hellenistic Ptolemaic dynasty in Egypt ended around 30 BC, when the Roman Empire conquered Egypt and kicked the Greeks out (leading also, of course, to the tragic romance of Antony and Cleopatra, but we already covered that). The Romans reinstated local rulers to lord over the now much diminished Egyptian nation, granting them limited powers. The Romans, you see, had learned from a few centuries of conquests the importance of maintaining local traditions and puppet governments in order to quell uprisings. When the Romans conquered the Holy Land, they installed Herod, a Jew, as the Roman client king of Judea. Those Roman conquerors were pretty clever.

Nicolaus Copernicus, the father of sun-centric solar system, died in 1543 from a stroke soon after his book on planetary motions was published, a book that was, for a time, kept from distribution within the realm of the Catholic Church, but not outside it. Since the Vatican wouldn't allow *On the Revolutions of the Celestial Spheres* to be published or distributed within its world, it was only available in the Protestant world, where it was printed in Nuremberg, Germany.

Nuremberg was also a sanctuary city for Martin Luther's Protestant Reformation for the same reason: it was out of the reach and beyond the influence of the pope and the Vatican. Luther, other than starting the Protestant Reformation in Germany is known for *Disputation on the Power and Efficacy of Indulgences*, better known as *The Ninety-five Theses* or *The 95 Theses*, a list of quotes from the Bible that laid the foundation for the academic disputations against the Catholic Church. Published in 1517, Luther supposedly nailed a copy of *The 95 Theses* to the wood door of the Wittenberg Castle church, formerly known as All Saints' Church, originally a Catholic consecrated chapel in 1346, today a Lutheran church.

Like his predecessor Copernicus, Galileo did not enjoy the blessings of the Church either. Although at one time a friend of Pope Paul V, Galileo fell out of favor with the publication of his supposedly heretical writings, which, in opposition to Church doctrine, supported and further validated Copernicus's heliocentric theories. Galileo's treatises were submitted as evidence to the Roman Inquisition that he had violated the Council of Trent. The Roman Inquisition was one of several Catholic Inquisitions, the other two being the Spanish Inquisition and the Portuguese Inquisition. These inquisition tribunals, in operation from the twelfth to the sixteenth century, were designed to snuff out heresy and affirm Catholic religious doctrine.

The Council of Trent was an ecumenical council of ecclesiastical theologians based in northern Italy's "religious fortress"—a fortress made not out of stone and mortar but out of Church doctrine. Its purpose was to thwart the religious heresy of the Protestant Reformation percolating in northern Europe in its attempt to invade the Catholic realm in Italy and southern reaches.

The Protestant Reformation can be described simply as a schism within the Catholic Church. Martin Luther in Germany first founded his version of Christianity, Lutheranism, in the early 1500s, then John Calvin in France brought about Calvinism in the early to mid 1500s, and eventually John Wesley in England fathered Methodism in the 1700s. The Church of England had already split from the Vatican in the 1530s, a move by Thomas Cromwell to provide a path for King Henry VIII to divorce his first wife Catherine of Aragon to marry his second wife Anne Boleyn. Wesley is noted not only for establishing Methodism but also for this poem:

> Do all the good you can,
> By all the means you can,
> In all the ways you can,
> In all the places you can,

At all the times you can,
To all the people you can,
As long as ever you can.

Inquisitions were Catholic church-sponsored institutions designed to squash heresy in whatever form it revealed itself, from religious heresy to scientific heresy. The first such Inquisition, prior to the big three—Roman, Spanish, Portuguese—was the Medieval Inquisition, which predates the rise of Protestantism, began in twelfth-century France. Trials of French heretics usually ended in punishment, some in executions. One of the more famous moments of the Medieval Inquisition is the trial of Joan of Arc in 1431.

Joan of Arc was a peasant girl born in 1412 in Domrémy, in the Bordeaux region of France, then under English control. Bordeaux is most famous for being home to some of the best wine vineyards this side of the cosmos, dating back to 60 BC when the occupying Romans began the cultivation and fermentation of grapes in the area. It's all rather odd, really: when you think of French wines, you think of Frenchmen. But the French wine industry indeed needs to thank the occupying Romans from Italy for laying its foundations.

And to make matters worse for the image of the French wine industry, in the mid nineteenth century an aphid blight destroyed a great deal of French vineyards. An aphid is a sap-sucking insect that in this instance attacked the roots of the vine. The aphid may have come from elsewhere, possibly even America, but what was apparent is that American vine roots were resistant to aphids. The French wine industry had two choices, watch their industry disappear or graft their upper plant vine fruit to the lower American vine roots. The horror of it all, French vineyards horse powered by American roots, French wine diluted by American plant genes. But that is exactly what many if not most French wine makers did, graft their vines to American roots.

Returning to Joan of Arc, in the 1400s, Europe was still reeling from the Black Death, also called the Plague, a rodent-driven bacterial infection that terrorized the continent (a plague upon man covered in *The Second History of Man*). Sixty years before Joan's story begins, the Black Death had wiped out 25 percent of Europe's population. This massive loss of life provides the backstory to the persecutions that sprang from the Medieval Inquisition that would follow. Why would that be so? Religious persecution sprang from the decay of the Black Death, finding someone to blame.

Around the tender age of nineteen, it is said that Joan received a "vision" that she should support the French monarch, Charles VII, in his effort to recover the Bordeaux territories from the English as part of the Hundred Years' War, a conflict between France and England over the line of succession to the throne of France. The Hundred Years' War began in 1337 and ended with a French victory in 1453—but not before the Joan of Arc story would unfold.

Since England was not as devastated by the Black Death as continental Europe—the English Channel saw to that—it had, for many, many decades of the Hundred Years' War, the tactical advantage. England's sheer numbers of foot soldiers meant it was able to maintain control over its conquered areas of France. But the British tactical advantage began to wane once the children of the French baby boom following the Black Death were of sufficient age to conscript into the ranks of the French military.

The situation of the foot soldier provides the perfect example of why Galen's miasma theory of infection spread stood for so long. The lower ranks of the military were mostly conscripted from the poor, and they stunk, and they were often ill, which all fell into line with what the Galenites believed about bad miasmas precipitating infection. That the infantry is populated by the poor is still true today, although we've thankfully unburdened ourselves of all that miasma hooey. The great American president Franklin D. Roosevelt once stated, "War is young men dying and old men talking," and to that I would add from history, not modern times as much, "War is often *poor* young men dying and old men talking."

There are a number of good, honorable reasons why young men, and now also young women, enter the military. Some serve out of a sense of duty to their nation. For others, the military is a way of life—their older brother did it, their father did, it's a family trade. And still others serve because quite frankly they need a job. When a shortage of soldiers arises during wartime, the military may conscript its warriors through a draft. It has often been true, less so now, these conscripted types of soldiers were primarily pulled from the ranks of the poor.

Where was the incorrect connection between bad miasmas, infection and soldiers made? Since antiquity, it has been the case that soldiers in trenches cannot afford the luxury of hygiene—and so they smell. These same soldiers historically also died in droves from infection, more so than from battle, especially from cholera, typhus, typhoid and like-minded pathogens. Galen connecting smells to infection was a sort of *post hoc ergo propter hoc* fallacy (from the Latin and means "after this, therefore because of this"). Because these soldiers smelled and then died in great numbers from infection, the smells must have caused the infection.

Returning to the Medieval Inquisition, during the Siege of Orléans southwest of Paris in 1428, Charles VII sent the young Joan of Arc as an emissary to Orléans on a relief mission. In 1430, the Burgundian faction—who were French but aligned with England—captured Joan at Compiègne and threw her in prison. A terribly biased council populated with English and Burgundian clergy under the ecclesiastical control of the French bishop Pierre Cauchon, a British puppet, placed Joan on trial for religious heresy. She was found guilty and burned at the stake on May 30, 1431. She was only nineteen years old. After the French victory over the English twenty-two years later, in 1453, ending the Hundred Years' War, the pope in Rome, Callixtus III, ordered the trial of Joan of Arc to be reviewed. The charges were debunked, and the pope declared her a martyr.

In the 1500s, Joan of Arc became the official symbol of the Catholic League, a religious political group and a major participant in the French Wars of Religion to stamp out the Protestant Reformation. Three hundred years later, in 1803, Napoleon declared Joan an official French symbol. She was beatified into heaven by the Catholic Church in 1909 and canonized as a saint in 1920. Her real name was Jeanne d'Arc, the English "of Arc" deriving from the surname of her father, Jacques d'Arc. "Arc" might have been derived from "*arcensis*" meaning "leader" or "superior."

After the Medieval Inquisition in France and elsewhere in the twelfth and thirteenth centuries came the three Catholic Church Inquisitions of the early modern period: the Spanish Inquisition (1478–1834), the Portuguese Inquisition (1536–1821), and the Roman Inquisition (latter half of the 1500s). The medieval period, also called the Middle Ages, is generally accepted to have spanned from the fifth century to about the fifteenth. Another term describes a good chunk of this period—the Dark Ages—precisely because as I've mentioned *ad nauseum* not much was happening in terms of humanism, science and industry. The light of humanity was turned off. As we know, the Church made sure of that after Rome fell. Between the Renaissance, Scientific Revolution and Enlightenment, the lights of humanity were turned back on.

The Spanish Inquisition involved not only Spain but also all of its colonies in the New World and was designed to maintain the strength of the Catholic Church. Its main focus was initially the conversion of Jews and Muslims to Catholicism, and when that failed, pogroms to drive them out of Spanish territories, and when that failed, the imprisonment of heretics, and when that failed, the torture of heretics, and when that failed, well, their elimination. When Martin Luther broke with the pope, the Roman Catholic, Spanish Catholic, and Portuguese Catholic Inquisitions had a whole new group of heretics to persecute and prosecute: Protestants.

But all bad things must (hopefully) come to an end. And the following is a key chapter in history, best described as the Enlightenment. Enlightened philosophers who developed theories of governance—John Locke in Britain, David Hume in Scotland, René Descartes, Jean-Jacques Rousseau and a mouthful of a name Charles-Louis de Secondat, Baron de La Brède et de Montesquieu in France, and Christian Wolff in Germany, the rise of governments with constitutions blossomed, or at least benevolent monarchies and emperors were reined in by the People, the influence of the Church started to diminish. Simply stated, governments emerged, religious influence declined, and humanism burst forth in a blaze. The last known supposed heretic executed during the Spanish Inquisition was a school teacher from Valencia named Cayetano Ripoll, who was executed on July 26, 1826, for teaching deism, a branch of Christianity that posits God exists, but that God does not interfere with daily life.

As pointed out earlier in the pages of this book, the Greek Empire and then the Roman Empire looked favorably upon science and humanism, or at least they didn't persecute heretics that much, with Socrates and Jesus making for a few notable exceptions. But as the Church filled the vacuum created by the fall of the Roman Empire, religious persecution became the name of the game, until the Renaissance arose to provide sanctuary. Interestingly, we still live in a world where some countries are totally religion based, and other countries are heavily influenced by religious doctrine. So much for the separation of church and state.

Like its Spanish counterpart, the Portuguese Inquisition was initially established to target Jews and Muslims, many of whom had fled Spain into Portugal. Once in Portugal, they claimed to have converted to Catholicism to avoid persecution, but the Inquisition accused them of secretly practicing their original religions—and so still persecuted them. Portugal, like Spain, included its colonies in the Americas in its investigations. And, again like its Spanish cousin, the Portuguese Inquisition's victims, besides Jews and Muslims, were Protestants, as well as supposed witches and bigamists. The Portuguese Inquisition ended in 1821, when the government of Portugal and its constitution began to wield more power than Portuguese religious leaders.

"The Inquisition is down the hall - this is Tax Audits."

Then there was the Roman Inquisition, the stepchild of the Medieval Inquisition, an interrogation that spanned the last half of the sixteenth century. It was a system of tribunals established by the Holy See in Rome, mostly directed at Protestantism, science, sorcery, immorality, witchcraft, and blasphemy, as well as literature deemed heretic. Some of the most famous targets of the Roman Inquisition include Copernicus, for heliocentrism; Galileo, for supporting heliocentrism; and Friar Giordano Bruno, for proposing that the stars are in fact distant suns like our own sun.

Friar Bruno further suggested that all those stars were surrounded by their own exoplanets, meaning our sun was not even the center of the universe. Copernicus declaring that our sun is the center of our solar system and the center of the universe was one thing. Friar Bruno declaring that our sun was not the center of much of anything in the universe was quite another altogether—and he was a man of the cloth. Imagine the horror felt within the Holy See? Probably a few tremors were felt within the old St. Peter's Basilica, a cathedral that stood from the fourth to the sixteenth centuries before the new St. Peter's Basilica was consecrated in 1626. For his heretic beliefs, Friar Bruno was burned at the stake in 1600 at the Campo de' Fiori in Rome. As morning follows night, as justice follows injustice, in 1889 a statue was erected of Friar Bruno in the very piazza where religious intolerance led to his being burned alive.

Galileo's book *Dialogue Concerning the Two Chief World Systems* was published in 1632 in Germany—because although he was Italian, Germany, as we know, was beyond the reach of the pope, a Protestant stronghold. While Galileo had his book secretly toted into Germany for publication, the great man himself did not undertake the same flight. Rather, with firmness of resolve, firmness in his beliefs, Galileo remained steadfast in his ancestral home of Florence and faced his accusers.

In his *Dialogue*, Galileo portrays a hypothetical person named Simplicio, who is a believer in the earth-centric world. You can see where this story is heading with a name like Simplicio. It appeared to the Church, and more specifically to the pope and the Council of Trent, that Galileo was not only favoring the Copernican sun-centric world, but simultaneously labeling the Church directly and the pope indirectly as "simpletons." That did not go over well.

The Council of Trent condemned Galileo to prison, a sentence commuted the next day to house arrest, under which he lived out the remaining decade of his life, at his villa in Florence under the patronage of Cosimo II de' Medici, Grand Duke of Tuscany. One of the greatest thinkers in all human history was confined to his home for the remainder of his life based on the fact that his theories—his *correct* theories—ran contrary to Church dogma. With instances like this, and so many, many more, it is no wonder that nations like the United States created constitutions that separate church from state, a concept that originated with those Enlightenment philosophers mentioned above. Sadly, too many Americans do not understand that part of the U.S. Constitution, do not understand that part of history, why forms of governance separate Church from State.

Several clerics and important figures, including the Grand Duke of Tuscany, Ferdinando II, who admired Galileo, wanted the scientist upon his death to be buried in the main body of the Basilica of Santa Croce in Florence, next to his father's body. Pope Urban VIII protested—after all, Galileo had implied the pope was a simpleton—so Galileo was instead entombed in some small room down a lonely, dusty corridor next to a minor chapel of the basilica. Nearly 100 years later, in 1737, upon irrefutable evidence of the heliocentric world,

GALILEO DESCRIBES HIS DISCOVERIES TO THE CHURCH

Galileo Galilei was reinterred in the main hall of the Basilica of Santa Croce, where a monument had already been erected in his honor. Galileo's tomb sits opposite another great Italian Renaissance thinker and artist: Michelangelo.

Galileo died on January 8, 1642, possibly from an infection that targeted his heart, as the symptoms preceding death were fever and heart arrhythmia. The most likely infection would have been a viral myocarditis targeting the heart muscle, although a bacterial endocarditis hammering a heart valve would be a near equal possibility in the differential diagnosis. Which brings us back to the differential diagnosis of Gurney #3.

When the initial lab results came back on Gurney #3, they showed nothing much out of the ordinary. Her hematocrit was a bit high and her urine a tad concentrated. She seemed dehydrated, which would result in concentrated urine. The kidneys know to balance the elimination of waste, mostly urea, which requires some water, against the need to preserve hydration. I went over the lab results of Gurney #3 with the attending physician, who, by this time, had his own set of ill and injured patients he was dialing. We decided to order a chest X-ray and run some IV fluids into her, in the hopes she'd feel better. When a person is dehydrated, there is nothing quite like a liter or two of IV fluids in the tank to perk them up. The initial chest X-ray was unremarkable—a status that would change later on.

I started an IV on Gurney #3. Being able to start IVs is an indispensable skill for a health care provider, especially for a soon-to-be resident navigating his scut list. Having IV access to a vein can mean a lot of things, and in some cases the difference between life and death. This was a point driven home during my plastic surgery residency.

Some years after I treated Gurney #3, when I was in my plastic training in Cleveland, we were in the operating room one fine morning getting ready to repair a cleft lip on a three-month-old infant. On much older children and all adults, the IV is started in the pre-op holding area, so when the surgical patient is brought into the operating suite on a gurney, the patient already has that lifeline IV securely in place. But in infants and young children, no way. No way are you going to place an IV on a screaming two-year-old before going into the operating room, much less a wiggling, irritable three month old infant. They're not going to hold still for that.

For infants and toddlers, you take them into the OR sans an IV, the anesthesiologist places their monitoring equipment, and then, using mask inhalation gas, masks-down the infant to where they're asleep enough to place an IV. So, whereas with an IV in place you first give the surgical patient an IV cocktail to put them under, and then place a breathing tube with oxygen and the anesthetic gas combo, without an IV in infants, you are forced to go straight to the mask inhalation anesthetic gas. As for then placing the IV, in the OR, it's usually the anesthesiologist or the operating room nurse who has the honor of starting the IV on newborns and infants. In the pre-op holding area, it's often a nurse who starts the IV on older children and adults. Out on the ward, back in my day, it was often the medical student or the intern who was tasked with starting IVs.

But on that day in that operating room, for reasons we were never sure of, as the anesthesiologist masked down our three-month-old patient, he lost the airway before the IV was started. A lost airway means the ability to oxygenate the patient is gone, often the result of muscle spasms in and around the vocal cords where the airway is at its narrowest, termed laryngeal spasm. Even with a mask you can't push oxygen into the lungs if that narrow opening is spasmed shut. With a lost airway and no IV to halt the laryngeal spasm with muscle relaxants, as the blood oxygen saturation drops, as the alarm bells and whistles on the anesthesia machine go alarmingly off, the heartbeat slows and slows and slows, ands life begins to ebb away. Stopping the mask inhalation anesthetic gas is not going to make the laryngeal spasm go away in a timely fashion, the spasm will continue until death.

What was needed in that frightful moment was an IV to shoot a muscle paralysis drug—succinylcholine (or "sux" in medical lingo)—into the infant's vein that would immediately halt the airway spasm, allowing mask anesthesia and then a breathing tube to be passed. But there was no IV. Just a panicked anesthesiologist trying desperately to mask the infant, alongside one plastic surgeon, two plastic residents, one nurse, and one scrub tech staring in disbelief at an infant turning blue, the beep, beep, beep of the heart broadcast through the anesthesia monitor frighteningly slowing down.

Everyone grabbed an extremity—nurse, attending surgeon, two residents—and began frantically scanning for a vein to cannulate an IV. The anesthesiologist continued attempts to mask the infant, to no avail. Little oxygen was getting passed that closed airway. The infant was turning darker shades of blue. I had grabbed the infant's right foot, scanned and palpated for a vein, took a moment of pause and asked God to guide my hands—which was odd because I'm not the religious type, but on that day, in that operating room, I asked for God's help.

It was for me a sort of Blaise Pascal moment, the seventeenth-century French mathematician who proposed Pascal's wager, positing that humans bet with their lives that God either exists or does not exist. Pascal argued that any rational person should live as if God exists, as you have nothing much to lose if he doesn't and everything to gain if he does. At that moment I surely was Blaise Pascal betting God existed. It should be pointed out that Pascal was also a Catholic theologian, he was not an agnostic.

Whether it was skill, whether it was luck, whether it was God, I managed to thread an IV into a foot vein, a vein I could barely see, much less barely feel. After yelling out "I got it," lickety-split the anesthesiologist leaped for my IV like a big cat hunting a gazelle, hooked up an IV bag, grabbed the syringe with "sux" and pushed it in full bore. Within moments paralysis set in, the laryngeal spasm ceased, the endotracheal breathing tube was placed, the reservoir bag of oxygen was squeezed a few times, and before our scared, frantic searching eyes the infant pinked up. All was well. I'm sure the anesthesiologist soiled himself.

As for Gurney #3, the middle-aged woman complaining of flu-like symptoms with an unremarkable exam, unremarkable labs, and an unremarkable chest X-ray, and an unremarkable medical student learning from her, we ran some IV fluids into her in the hope it would perk her up, burning daylight while chasing ghosts. We'll circle back to her after that IV fluid had had time to work its magic, or not work its magic.

Returning to eighteenth-century Florence, during that reinterment of Galileo in 1737 at the Basilica of Santa Croce: three fingers, one tooth, and one vertebra were removed from his corpse. Why those body parts and only those body parts were harvested from Galileo's 100-year-old corpse is anybody's guess. Where those body parts traveled, though, is a story in itself.

One finger—believed to be Galileo's right index finger—presumably preserved initially in embalming fluid (mostly composed of formaldehyde or formalin), is on display in a specially crafted glass jar in the Museo Galileo in Florence. The vertebra found its way into a museum at the University of Padua, Italy. The other two fingers—the thumb and middle finger—and the tooth were lost to history. Or so it was thought.

In 2009, the two fingers and tooth were found within the personal collection of an Italian family that wishes to remain anonymous. A glass-blown vase inside a carved wooden box adorned with a bust of Galileo had been passed down from generation to generation within that family. How that family came to possess those items from Galileo's remains unknown, largely because the family is unknown. But by 2009, the family, apparently having had enough of holding on to that rather bizarre set of macabre relics, bequeathed the glass vase containing Galileo's tooth, thumb, and middle finger to the Museo Galileo, where they joined their old friend the index finger.

In 1992, Pope John Paul II declared all things are new again when the Catholic Church rehabilitated Galileo, the pope poignantly stating the Church had erred in 1632. His sentence from 360 years earlier was vacated.

One of the problems with the Copernican heliocentric world was that, although Copernicus was correct in placing the sun at the center of our solar system, his perfectly circular orbits of the planets around the sun did not add up, that is, they were incompatible with what was observed. They were certainly better than the Ptolemaic system, and the Tycho Brahe rendition, but still there were inconsistencies. Perfectly circular orbits would create certain parallaxes, whereby viewing something from two different locations gives rise to certain alterations, which was not observed. Those parallaxes, which would be peculiar to circular motions, could only be explained away by positing that the planets instead have elliptical orbits around the sun.

You're probably wondering as I was, how and why did all the planets decide on an ellipse around the sun, why not circular, which would seem to make more sense. The answer apparently, lies during the early moments our solar system was being built from cosmic schmutz, a concept that would be true throughout the universe wherever there is a sun with orbiting planets. Without getting into Newtonian mechanics, and especially without getting into Einstein's world, the initial conditions of our solar system, the initial forces of a body wanting to centripetally jettison into the void countered by the sun's centrifugal gravity of wanting to spiral a body into its fiery depths, and add to those conditions all the planets tugging on each other, plus the fact that—and this is critical—neither the sun nor the planets are perfect spheres themselves, favored an oblong asymmetric ellipse around the sun in perpetuity. Moons about planets including our Moon about

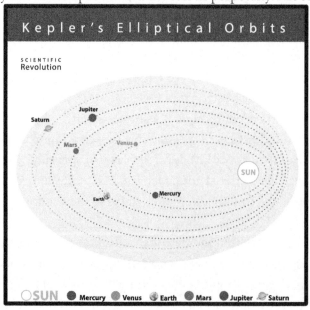

Earth come fairly close to circular orbits, but not quite. It is thought that throughout the universe perfect circular obits of planets around their star probably never occurs or occurs very rarely, only because perfectly circular stars do not exist. They're always fatter at their equator than pole-to-pole due to their spin on their axis.

This explanation came by way of Johannes Kepler, a contemporary of Galileo who I mentioned previously and as promised, we'd circle back to him. Kepler's argument for elliptical orbits and against Copernicus's circular ones might be stated as: if the sun were at the center of the "universe," and if the orbits of Earth and, let's say, Venus, were mapped out in perfect circles around the sun, the view of Venus from Earth wouldn't be the view of Venus that we actually see traveling across our night sky. However, if the orbits of Earth and Venus are instead understood as ellipses, the celestial planetary observations fall into place.

Kepler, born in 1571, was a German astronomer whose discoveries helped form the foundation of the theory of universal gravitation, which Isaac Newton would formulate in the late seventeenth century. Sir Newton—he was knighted by Queen Anne in 1705—was from England and is considered to be one of the most influential scientists of all time, a list that also includes such notable minds as Aristotle, Archimedes, Euclid, Ptolemy, Leonardo da Vinci, Galileo, Isaac Newton, Charles Darwin, Louis Pasteur, Marie Curie, Nikola Tesla, Albert Einstein, Niels Bohr, Max Planck, and Alan Turing.

Who the heck is Alan Turing, you ask? Turing is the father of modern computer science, that's who. It is just not many people outside the world of computer geeks knows much about him.

Alan Turing was born in London in 1912. When he enrolled in school at age six, his teachers quickly recognized his genius. As his education advanced, it became clear that his bent toward mathematics was novel, which is a nice way of saying that his thoughts on math ran contrary to a classical math education. This unique mathematical approach cast the young and brilliant Turing as an oddity. To make matters worse, during an era when being a homosexual was at odds with social norms and criminalized, Alan was gay. His first love, who he met at preparatory school, was Christopher Morcom, a fellow student at the Sherborne School in Dorset. Although only about age thirteen when they met, the relationship gave Turing wings with which to fly—a soaring cut short when Morcom died at age eighteen of bovine tuberculosis while at school. That's not human but bovine tuberculosis, an illness he had contracted from drinking unpasteurized milk years earlier.

There are three varieties of TB. The common one—the one you and I think of—is caused by *Mycobacterium tuberculosis*, which is mostly acquired by inhaling droplets, resulting in a lung respiratory infection that can spread to other parts of the body. Then there is *Mycobacterium bovis*, which percolates in cattle and is acquired by humans through unpasteurized milk as well as by inhaling droplets and eating contaminated meat. Bovine TB targets the lungs, liver, spleen, lymph nodes, and intestines. Avian TB, caused by *Mycobacterium avium*, is mostly an issue in the poultry world, with human infection occurring rarely.

But returning to Turing, beginning in 1931 he read his undergraduate work at King's College, Cambridge, where he achieved first-class honors in mathematics. It was at this time, around the mid 1930s, when Alan invented the basis of the Turing machine, a mathematical model of computation that replaced classical arithmetic-based formal computation with a computer algorithm, your basic first-of-its-kind logic board. For you and me and those of us in high school in the 1970s onward, it was the foundation of the "greater than," "less than," "equal to," and "if go to" computer languages that dotted the landscape. When I was in high school, I studied, or should

I say I *attempted* to study, BASIC (beginner's all-purpose symbolic instruction code), FORTRAN (formula translation), and COBOL (common business-oriented language). I was not very good at learning these computer languages, just like I failed miserably at learning Spanish and French.

After Cambridge, Turing crossed the pond around 1936 and arrived at Princeton University in New Jersey, where he did his doctoral work in mathematics and where he actually built the first Turing machine. It looked a bit like one of those old-style stock market ticker machines with its strip of tape pouring out Wall Street updates. No matter what it looked like, one thing is for sure: it was the foundation of what would become the logic board in all computers, even your iPhone. Many years later, in the 1960s, when the U.S. military tasked a new generation of computer-nerd geniuses with building the first iteration of the internet, the ARPANET (Advanced Research Projects Agency Network), as a telecommunications defense against a nuclear strike, several of those computer scientists went back and looked at Turing's theoretical computer designs. That is how advanced Turing was for his time. Funnily, they also found a "bug" in one of his designs. Imagine that: a theoretical computer bug in a theoretical computer.

In 1952, upon returning to Britain from New Jersey, Turing was convicted of homosexuality. When he was given a choice between imprisonment or probation that included chemical castration with synthetic estrogen, Turing chose the latter. The conviction also resulted in the loss of his British security clearance, which he needed to continue work on computer cryptography. On or around June 7, 1954, Alan Turing, the father of the computer, is believed to have committed suicide with cyanide. Some people claim that his death was accidental, since Turing regularly used cyanide to dissolve gold for his logic boards. In 2009, more than fifty years after his conviction for homosexuality and death by suicide, the British government issued an apology to Turing for his appalling treatment.

Now that we've reached the twenty-first century, let's do an about-face and head back to where we started this wild ride: Pier Antonio Micheli and the discovery of fungi in 1729. We have a bit more to explore with fungus, as fungi, as it turns out, are one of the archenemies of bacteria.

20 DR. LIVINGSTONE, I PRESUME?

"Out of damp and gloomy days, out of solitude, out of loveless words directed at us, conclusions grow up in us like fungus"—or so says Friedrich Nietzsche, who, when he penned that, was less concerned about how mold and fungus tend to appear anywhere, out of nowhere, and more concerned about the failure of humans to challenge the dogma that grows up around us. But that is exactly what fungus does: seemingly appear out of nowhere. And that is what dogma is: something that is definitely there, but often we have no real idea from whence it came. Recall those poor monkeys in that cage with its ladder, freezing ice showers, and a banana they were never allowed to eat.

Examples of macroscopic fungi—that is, fungus organisms that grow large enough to be visible to the naked eye—include molds and mushrooms. They seldom cause infections in humans but can cause intestinal illness if we ingest the wrong one, such as the lethal death cap mushroom. Other types of macroscopic fungi can cause a person to become all lit up—that would be the psychedelic psilocybin mushroom.

The death cap, known taxonomically as *Amanita phalloides*, is widely distributed across Europe. Its active poisons are at least eight types of amatoxins, and since these are heat resistant, even cooking the mushroom does little to diminish its lethality. The death cap looks a lot like another mushroom of the *Amanita* family: Caesar's mushroom. It doesn't take a rocket scientist to figure out where that fungus got its name, other than the Caesars, all of them, loved their mushrooms. Roman Emperor Claudius Caesar—not Julius, who was stabbed to death—is believed to have been assassinated by poisoning in 54 AD when his favorite mushroom, the delicious *Amanita caesarea*, was swapped out for its similar-looking cousin *Amanita phalloides*. The same fate befell Holy Roman Emperor Charles VI of Vienna in 1740, when he was likewise fooled, fed death cap rather than Caesar's shrooms.

As for psilocybin mushrooms, the active ingredient is a psychedelic compound that produces strong perceptual distortions, especially with the passage of time. A few minutes might seem like hours, almost as if time is standing still. Within the range of the psilocybin experience is a general pleasantness accompanied by a feeling of internal connection with others, even with animals, even with nature, heck, even with the universe. Occasionally, however, the psilocybin hallucination is unpleasant, termed a "bad trip," which is mostly characterized by fear. Good trips usually occur when the user is feeling happy going into the high, while bad trips happen when the user enters the trip sad or depressed.

Similar trips, good or bad, occur with other psychedelic drugs, such as lysergic acid diethylamide, better known as LSD, and mescaline. Mescaline is derived from the peyote cactus and similar cacti that are common throughout Mexico and the southwest United States. LSD, by contrast, is synthetic. It was first manufactured by the Swiss chemist Albert Hofmann in 1938 as a derivative from the ergot fungus. He apparently invented LSD to treat certain psychiatric illnesses, but it never caught on. It's not clear in the medical literature which psychiatric illnesses would have benefitted from tripping on LSD. Depression? Schizophrenia? No matter—LSD doesn't remedy psychosis anyway.

Ergot fungus is a member of the *Claviceps* genus, which occasionally infects rye and other grasses. Too much ergot ingestion can lead to ergotism, which is best characterized by body spasms, mania, psychosis, headaches, nausea & vomiting, and seizures. Ergot fungus can also kill body tissue, leading to gangrene.

During the Middle Ages, a combination of psychosis and dead tissue from the ingestion of ergot was termed Saint Anthony's fire—an unusual move to be sure, naming a hallucinogenic experience after a saint. Apparently, as the story goes, the good monks of the Order of Saint Anthony were especially adept at treating ergotism, so the disease was named after the cure, or more precisely, the curers. The order was founded in 1095 AD in Saint-Antoine-l'Abbaye, France, named after Saint Anthony the Great, who apparently lived to be 105 years old, born in 251 AD and dying in 356. It's not known if those monks who treated folks ill with ergot used actual medicinal techniques or merely relics and incantations to rid them of ergotism.

In the centuries that followed the 1692–93 witch trials in Salem, Massachusetts, it was speculated that the fourteen women and six men who were hung or burned at the stake, plus the five children including two infants who died in prison, were in fact not bewitched by sinister spiritual forces as was claimed. Rather it has been surmised all these hundreds of years later that the "evil force at work" in the lead up to the Salem Witch Trials was in fact ergot fungus poisoning, which the poor victims had ingested after it had contaminated a local farmer's rye field.

Derivatives of ergot, generally known as ergotamines, have been used medically to treat various ailments due to their unique ability to constrict blood vessels. Especially ergotamines have even been used to treat migraine headaches. The basic pathophysiology of migraines is the dilation of brain blood vessels, a true vascular headache, leading to the characteristic heinous and relentless pounding head—often described as the worst headache imaginable. Ergotamine theoretically constricts those brainy blood vessels, putting a nozzle on blood flow, lessening the excruciating migraine.

A lot of folks claim they get migraine headaches, but what they're likely experience is probably just bad tension headaches. Tension headaches involve the muscles of the head and neck, you know, *outside* the skull. And to be sure, they can be pretty bad. But true migraines are vascular headaches within the brain itself, and the resulting pain is dreadfully excruciating accompanied by nausea, vomiting, and photophobia (sensitivity to light). It's also true that tension headaches might *trigger* migraine headaches in those who are already prone to migraines. But, nevertheless, many

people who say they are having a migraine are in fact not. With true migraines, it's those brainy vessels that are at the center of the pain, and the ergot derivative drug can constrict them, reversing the pounding swoosh of blood dilation. Ergot wouldn't do much for a tension headache.

After Albert Hofmann figured out how to derive LSD from ergot, the CIA began messing around with the drug, too, believing it could be used as a form of mind control during interrogations, one type of "rendition protocol," which is a fancy phrase for torture. You gotta love the CIA and pals of similar ilk in foreign countries, coming up with the benign sounding "extraordinary rendition" or "rendition protocol" for what amounts to detention, interrogation, and torture.

The CIA experimented with LSD for rendition on God only knows who. The code name for that CIA program was Project MKUltra. The letters "MK" were a digraph, which is a pair of letters or characters that do not correspond to the normal value assigned to them. In other words, "MK" according to the CIA meant nothing. It wasn't an acronym for anything; they were just randomly selected letters, or so it was claimed. As if we're supposed to believe that—that someone in the CIA randomly selected "MK" out of the alphabet to name that particular LSD rendition program. Some have suggested "MK" stood for 'Mind Kontrol." After all, that was the purpose of MKUltra: interrogation and torture plus some hallucinogenic fun to "kontrol" minds. MKUltra began in the early 1950s, was officially sanctioned by the CIA in 1953, then reduced in scope in the 1960s, and officially closed down in 1973. The program was shuttered partly because it didn't work as advertised and partly because it was coming under investigation for unethical practices.

From Project MKUltra, LSD entered the counterculture world of the 1960s, alongside its cousins mescaline and psilocybin, in some small part thanks to Timothy Leary, a clinical psychologist at Harvard University. Leary began experiments with psychedelic drugs to study their therapeutic potential, especially in treating alcoholism, based on the idea that alcohol was addictive and psychedelic drugs were not, which is true. It was the old switcheroo, like when Freud tried to replace opiate addiction with cocaine, except, to Freud's chagrin, and the consternation of his patients, both are addictive.

Leary was eventually fired from Harvard, although he claimed he was intending to quit anyway. He wasn't canned for the experiments he was conducting on hallucinogens, at least not outwardly; rather he was fired for failing to show up for the classes he was assigned to teach. He was, after all, a professor and had teaching responsibilities. Maybe after all his experimenting, Leary was too lit up to even remember he had a course to teach, or maybe he just didn't care.

Psychedelic drugs are indeed technically not addictive. What makes a drug addictive is if it has one or two specific dependences: either the positive effects are so compelling that they are craved, or the withdrawal symptoms are so uncompelling that they are to be avoided, or both. Alcohol is sort of both—the buzz of a shot of Jack Daniels is desirable, and the withdrawal of alcohol is frightfully undesirable, especially for the alcoholic, leading to anxiety, irritability, labile emotions, bad dreams, and hallucinations.

Nicotine especially is both, likely more so than alcohol on either front. The stimulant is desirable—how it charges the brain to perform better while simultaneously relaxing the body—but the withdrawal of nicotine is terribly vexing. If it weren't for its addictive nature, nicotine would be a good drug. Consider this, it's perhaps the only drug that simultaneously relaxes the body while stimulating the brain to function at a higher level. Nothing I am aware of does that. But its delivery system—most often cigarettes—will eventually kill you. The inhaled, partially burnt hydrocarbons, that is, cigarette smoke, damage everything they touch, leading to such end-of-life diseases as emphysema and lung cancer. To make matters worse, smoking cigarettes adds a third addiction to the fray, not just the desired high, not just avoiding the withdrawal—there is also a behavioral addiction, the routine of smoking.

Psychedelic drugs are typically taken infrequently. The high is perhaps intriguing but not addictive; there are few if any withdrawal symptoms from discontinuing psychedelics. The same is not true of opioids, heroin being the most notorious among them. The high is euphoric, the withdrawal problematic.

Leary, along with his LSD experiments, is best known for the classic 1960s counterculture tagline "turn on, tune in, drop out"—a phrase that is often quoted and equally misquoted, and just as equally misunderstood. Despite what most folks think, the phrase "turn on" does not necessarily mean to take psychedelic drugs, or any type of mind-altering drug for that matter—unless those drugs help you to achieve the underlying meaning of the entire concept of "turn on, tune in, drop out."

According to Leary, "turn on" only means to activate and use your neural equipment to its fullest potential, to become sensitive to your environment and altered levels of consciousness. "Tune in" means to interact with nature and the world harmoniously. "Drop out" means to gracefully detach from those things in your life you wish to change or abandon. The phrase "turn on, tune in, drop out" does *not* describe getting high by "turning on" with drugs, does not mean "tuning in" to your favorite acid rock radio station, and nor does it describe "dropping out" of college.

In Leary's eyes, if you need drugs to turn on, tune in, and drop out, so be it. But if you can achieve that elevated level of consciousness without drugs, so be that, too. Timothy Leary died in 1996 at age seventy-five, not from a bad high as some might have predicted, not from a drug overdose as others may have thought, but from inoperable metastatic prostate cancer.

But let's get back to pure, un-laboratory-altered fungi. This time let's get to know the kind you *can't* see with the naked eye. Not the macroscopic mushrooms you buy at the grocery store for your salad, or the slimy black mold underneath that expensive shampoo bottle your hairdresser told you to buy and you never use—but rather those invisible microscopic fungus spores, such as some forms of pathogenic yeast and unicellular molds.

Fungi are neither animal nor plant, as they share characteristics found in both the animal kingdom and the plant kingdom. They likewise possess dissimilarities to both kingdoms, so they've been assigned their own taxonomic realm: Fungi kingdom.

I have fond memories of being a kid and enjoying the NBC show *Mutual of Omaha's Wild Kingdom* (1963–88), which was and is—revived 2002-2011—all about the world of animals. In my day it was hosted by the American zoologists Marlin Perkins and Jim Fowler. The show mostly featured African wildlife, but certainly wasn't limited to it, and even had episodes in the Florida Everglades and Australian Outback. I loved that show. But the rub is, I'm quite sure I would not have watched a single episode if it featured kingdoms other than the animal. "Mutual of Omaha's Wild Plants" or "Mutual of Omaha's Wild Fungi and especially Mutual of Omaha's Wild Bacteria!" No thanks. After the show went off the air in 1988, about fifteen years later the same insurance company, Mutual of Omaha, rekindled its unusual role in the animal TV show business: it launched a new *Wild Kingdom* segment from 2002 until 2011 through *Animal Planet*, a BBC Worldwide venture, which itself first aired in 1996 on the Discovery Channel, which itself was first launched 1985.

As you recall in relation to our earlier discussion about the kingdoms of life, when I was an elementary school knucklehead, the highest taxonomic classification were the kingdoms of animal and plant. That was it. All life was forced into one of those two spheres. Now, as we've also learned, the taxonomic system includes two superkingdoms or domains—prokaryote and eukaryote (or three, depending where you believe archaea should fall). The differentiation of domains is on the cellular level, not the macroscopic level.

Let's quickly review. Bacteria evolved on Earth first and have no nucleus; as such, they are prokaryotes—*pro* from the Greek meaning "before," and *karyote* also from the Greek, loosely meaning "nucleus." Archaea are bacteria too, and some say they are a side branch of the bacteria superkingdom domain, while others claim archaea constitute their own superkingdom domain altogether. Where the toad of truth lies does not concern us here.

But what does concern us here is that, to review from a much earlier discussion, after bacteria evolved a nucleus, they were no longer bacteria but became eukaryotes, the prefix *eu* from the Greek meaning "well," as in a "well-defined nucleus." The eukaryotic cell then gave us the four eukaryotic kingdoms: Protista, Plantae, Fungi, and Animalia. To keep it simple I prefer to place archaea with bacteria, which then makes the verbiage easier. Two superkingdom domains Prokaryote and Eukaryote, the former has one kingdom for bacteria that include archaea and the latter has four kingdoms for animal, plant, fungus and protists, giving us a total of five kingdoms across the board of two superkingdoms.

Recall that the image of the "hot soup" of early Earth was one put forth by the English scientist J. B. S. Haldane when he and the biologist Alexander Oparin hypothesized that the origin of cellular life on Earth likely began in the hot, soupy oceans 3.8 billion years ago with Stage 1 bacteria, with Stage 2 LUA arriving on the scene around 3.5 billion years ago and cyanobacteria a half billion years after that. Since it's estimated that Earth is 4.543 billion years old, that means for almost one billion years no cellular life existed on our planet—just a series of random chemical reactions within the oceans. But those chemical reactions were becoming ever more complex, heading down the road to proteins, fats, carbohydrates, and nucleic acids—on the road to organic cellular life, the stuff all life is made of.

The basis of organic life on Earth is the carbon (C) atom, an atom perfectly designed and positioned to take on its role of big-top ringleader due partly to its four electrons in the outer orbit that are available for pair bonding. Carbon is both literally and figuratively the ringleader of life—capable of building the organic rings of life, the carbon ring fundamental to proteins, fats and the RNA and DNA nucleic acid rings that carry the genetic code. To the carbon atom in your vintage primordial soup kitchen, throw in some hydrogen (H), sprinkle in some oxygen (O) and nitrogen (N), add a pinch of a few other elements like iron (Fe), sulfur (S), phosphorus (P), and fluorine (F) and, behold, you have most of the ingredients any growing boy or girl needs for life.

Silicon (Si) is a chemical element exactly one row below carbon on the periodic table, meaning carbon and silicon each has four electrons in its outer orbit, which to an organic chemist implies they form similar chemical bonds—generally two bonds for each two outer orbit electrons. Carbon and silicon, each having four outer valence electrons, pair with other elements similarly and they also both pair with themselves, carbon to carbon and silicon to silicon. This is important to appreciate because, due to this shared behavior, theoretical chemists have mused about the possibility of life somewhere in the universe based on silicon rather than carbon, a subject we broached early in this narrative.

Is it possible? Can there be silicon-based life out there in the vastness of the universe? Yes, it is possible, although not likely. One glaring issue with a silicon-based organic life theory is that, since silicon is a larger atom than carbon—carbon's atomic mass is 12 and silicon's is 28—silicon cannot as easily bind with itself as carbon can with itself. Why is this an important distinction? Ring structures, which are critical to proteins, fats, and nucleic acids, would not occur with silicon, or if they did occur, would be inherently unstable. Unstable ring structures don't and can't and wouldn't ever make life. Instability and life do not make for ideal bedfellows. So, if there is life out there in the expanse of the universe, it is almost certainly carbon based.

But all is not lost for silicon. It may not be qualified for the job of ringleader of organic life, but it plays prince of integrated circuits very well. It's, you know, the stuff our computers and iPhones and Androids are made of, silicon chips and circuits. The "silicon" in Silicon Valley. Why is silicon so invaluable to electronics and circuits while its blah in organic life? Silicon is a semiconductor, meaning it both conducts electricity and simultaneously insulates, as opposed to something like copper that just conducts but provides no insulation, and as opposed to elements like sulfur and chlorine which are good insulators but worthless conductors. Other than silicon, the only other semiconductor elements of the periodic table are of course members of related groups: germanium (named not after the flower geranium but rather the country Germany) and tellurium, *tellus* from the Latin meaning "earth." But there ain't much of either on the planet, so they won't be powering any circuit boards anytime soon, nor will we ever see a technology center called Germanium Valley or Tellurium Valley.

There is, luckily, a bunch of silicon on Earth, forming a whopping 28 percent of its crust. In addition to conducting electrons but not losing those electrons to the great void (insulation), silicon lattices of about eight atoms resists temperature changes, water, and steam, meaning you can bring your iPhone into the bathroom and take a hot steamy shower with nary an issue with your mobile phone short-circuiting. Dropping your iPhone in the toilet is usually, on the other hand, an irretrievable event.

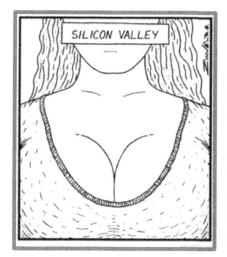

"Silicon" is also used to make "silicones," two words often confused; silicones are silicon compounds, and which find industrial applications as lubricants, sealants, cosmetics, and hair conditioners. And as for earth's crust being 28 percent silicon, silicates with the basic chemical structure SiO_4 are the most common element in granite rock. And perhaps importantly to some women and men—or at least as importantly as the circuit board thing—silicon is also used to make silicone breast implants, used both for breast reconstruction after a mastectomy for breast cancer and most frequently for cosmetic breast enhancement, a boob job, a different sort of "silicon valley."

The periodic table is not organized the way it is by accident. It's not as though its inventor, Dimitri Ivanovich Mendeleev in 1869 got to the end of a sheet of paper and placed the next element one row down, like hitting the return button on your keyboard. The table illustrates the periodicity or repetition of certain properties and at regular intervals. The horizontal rows are called "Periods" and every element in a given row has the same number of orbital shells. Think of those Russian Nesting Dolls, each new orbital shell is like adding another outer Russian Nesting doll. There are seven maximum orbital shell Periods.

The vertical columns are termed "Groups" because they share similar properties to each other, that is, their personality or chemistry are similar, mostly owing to available valence electrons in the outer orbital shell available to pair bond. There are eighteen Groups.

Let's take a closer look of vertical Groups and then horizontal Periods. Since we are talking about carbon being the ringleader of life, it is in vertical Group 14 which includes top to bottom carbon, silicon, germanium, tin, and lead. What do the elements in a vertical group have in common? Their outer orbital valence electrons are similar—that is, how many electrons are available to participate in the formation of a chemical bond—and therefore they exhibit related physical and chemical properties. Which is why it can be hypothesized that a silicon-based world in the universe could theoretically exist since silicon is right under carbon, sharing similar number of valence electrons in the outer orbital shell, regardless if it has two or three or four electron shells.

Germanium, tin and lead are in the same vertical Group as carbon (and silicon) but are just too gosh darn large to attempt the things carbon does in organic life. A silicon-based life is theoretically possible but germanium, tin or lead based life are not.

Then what about the horizontal rows, termed "Periods"? Every element in a Period has the same number of electron orbital shells. Take the second row, Period 2, which contains left to right lithium, beryllium, boron, carbon, nitrogen, oxygen, fluorine, and neon: all these elements have two electron orbital shells—two Russian Nesting Dolls—where electrons can live. As you move left to right along a Period, as one proton is added to the nucleus to form the next element, in example, going from carbon to nitrogen, so is one balancing neutron added to the nucleus and one balancing electron added to the orbital shell. Once you reach the last element in a Period, you have reached a point where adding one more proton and one more neutron forces the element to add the next outer layer orbital shell, where that one more balancing electron is forced to live, you add the next Russian nesting Doll.

How many electron orbital shell Periods are there? Seven. When you look at the far left Group 1, hydrogen has one orbital shell and the far right helium also has one orbital shell, you hit the return key on your typewriter, drop down to the next row, where lithium far left has two electron shells as does neon far right. After you reach neon, you again hit the return key, drop down one row where sodium has three electron shells and argon far right also three shells. After argon, hitting the return key the next row begins with potassium and four electron shells. After krypton five electron shells, after xenon six shells, and after radon seven shells. Then you are done.

How many Groups are there? Eighteen. What does that mean? It is a bit confusing. It is supposed to mean that there are eighteen flavors of electrons in the outer shell, eighteen personalities.

It's like going to the ice cream store, a small store called Period 1 Ice Cream Parlor has only two flavors, vanilla and chocolate, that would be hydrogen and helium. A store down the block Period 2 Ice Cream Parlor claims it has eight flavors— lithium, beryllium, boron, carbon, nitrogen, oxygen, fluorine, neon—but lithium far left, and neon far right taste an awful lot like hydrogen and helium above each. So this ice cream store has only added six new flavors. The next store Period 3 Ice Cream Parlor claims it also has eight new flavors— sodium, magnesium, aluminum, silicon, phosphorus, sulfur, chlorine, argon—but all of which taste exactly like Period 2 parlor's eight flavors.

Ice Cream Parlor 4, the Baskin Robbins store, beginning with potassium element 19 and ending with krypton element 36 claims it has eighteen new flavors, and indeed, ten flavors—scandium, titanium, vanadium, chromium, iron, cobalt, nickel, copper, zinc—are new, but eight are the same as the other parlors.

And that is where it ends. No matter how many more electrons you add as you drop down to Period 5, Period 6 and Period 7, the flavor, the chemistry, the personality of all new elements has been defined. There are eighteen flavors, no more. Even though Period 6 beginning with cesium and Period 7 francium have more than eighteen electrons in their outer shells—they each have 32 electrons in their outer shells—they still only have eighteen flavors. Why? The additional electrons in a series as the nucleus get bigger adds absolutely no change in personality. That is, the lanthanide series beginning with lanthanum at 57 electrons and ending with lutetium at 71 electrons, and the actinide series beginning with actinium at 89 electrons and ending with the theoretical element lawrencium at 103 electrons, all taste exactly, exactly the same. The universe gave us eighteen possible flavors or possible personalities of elements, no more.

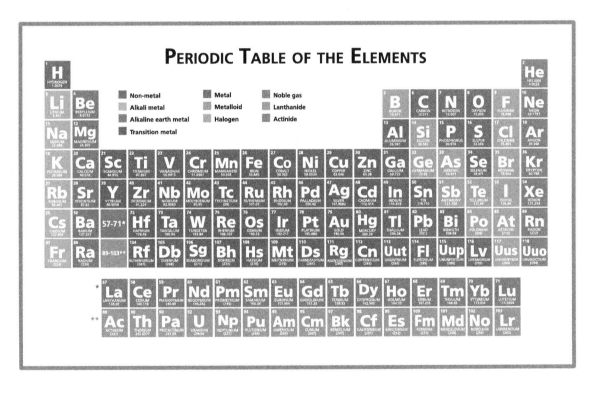

Just a few tidying-up points. The atomic number is the number of protons that defines the atom (or the number of electrons, since protons always, always equals the number of electrons). But don't confuse atomic number with the atomic weight, which is the number of protons plus the number of neutrons. For a stable element, these numbers are all equal. Carbon has an atomic number of 6 and an atomic weight of 12, for six protons and six neutrons and of course it has six electrons which add no atomic weight (supposedly). As you move toward the far end of periodic table and encounter the larger elements like uranium, the number of neutrons begins to vary—can you say isotope. This means that the number of protons and electrons in an element must always be equal, always, but the number of neutrons crammed into the nucleus can be variable, giving rise to radioactive decay qualities. Carbon-12 with six protons and six neutrons is the stuff of organic life; so is carbon-14 with two extra neutrons. It is stable enough to be incorporated into organic life, but it also decays, desiring to jettisoning its extra neutrons in a slow, rhythmic fashion. This is the basis of carbon dating.

We started this section with a quote from Nietzsche about fungus like conclusions grow out of damp and gloomy days, and it is there where we return: fungal infections. Human infections caused by fungus are called mycoses or mycotic infections. The two most common classifications of mycotic infections are *topical*, meaning they involve the skin and mucosal surfaces, and *invasive*, meaning they take up residence inside the body, especially in the lungs. The typical topical mycoses that target the skin and mucosal surfaces are your garden-variety athlete's foot, vaginal yeast infection, and oral thrush. Invasive fungal infections often target the lungs, termed pulmonary mycotic infections, and are acquired by breathing in fungal spores that then set up shop mostly in the bronchioles of the lungs. One classic example is San Joaquin fever, also called valley fever. It was first identified in the San Joaquin Valley in California, yet it is certainly not limited to the Golden State but prevalent throughout the Southwest United States. San Joaquin fever is an invasive fungal infection caused by several species of the *Coccidioides* fungi.

Fungi interact with humans in many ways, both good and bad, just like bacteria. We have beneficial bacteria in our gut, also found in cultured yogurt, which help digest roughage, and then there are the unbeneficial bacteria in our gut that give us *E. coli* gastroenteritis. Fungus may cause itchy rashes like athlete's foot and diaper rash, and it can grow deadly fungal balls in our lungs, but it also gives us penicillin to kill bacterial infections, delicious mushrooms to sauté and pair with our sirloin steaks, and psilocybin mushrooms to get high, if you're into that sort of thing. Bacteria and fungi span quite a spectrum, veritable Dr. Jekylls and Mr. Hydes.

Just like that particular fungus discovered by Alexander Fleming that produces bacteria-killing penicillin, some bacteria produce chemicals that not only kill other competing bacteria but even their archenemy: fungus. The classic example is the bacteria genus *Streptomyces*, responsible for a great array of both antibacterial and antifungal drugs, that is, once pharmacologists isolated the main killing compound and started tweaking it. These tweaks gave rise to streptomycin (historically for tuberculosis), neomycin (which has poor gut absorption and can be used orally to sterilize the bowel before intestinal surgery as well as topically on the skin), daptomycin (for leprosy), tetracycline (for acne as well as effective against some notable infections: cholera, Lyme disease, STDs), chloramphenicol (active towards some nasty anaerobic bacterial infections, as well as the Plague, rickettsial, and mycoplasma infections), and Fosfomycin (for urinary tract infections). And most importantly for our discussion, amphotericin B, which was isolated in 1955 from *Streptomyces* bacteria and has been and continues to be a powerful antifungal drug. Powerful. But the main point here is: *Streptomyces* bacteria produced a chemical designed to kill other bacteria and to kill fungus and Big Pharma ran with it giving the world some pretty nifty antibiotics and antifungals. Why did bacteria develop a chemical to kill other bacteria and kill fungus; and why did fungus do the same to kill bacteria? Probably because they were all competing for the same food source a billion years ago.

Once upon a time these antibacterials and antifungals had to be grown in the lab, or if synthesized, required a slick lab with a bunch of shiny beakers, pipettes, centrifuges and Bunsen burners. Once microbiologists appreciated that bacteria churn out natural antibiotics that kill other bacteria, and geneticists cracked codes and perfected genetic engineering, it allowed pharmacologists to turn bacteria into little drug-manufacturing machines by inserting a few codes into their DNA. These drug bioengineers insert the necessary genetic code for manufacturing a certain drug into a hijacked bacterium's DNA and keep it well fed and happy, and then that bacterium churns out the desired drug until the cows come home.

Today the field of hijacking bacteria to build microbiome drugs and perform other tasks is called synthetic biology. In addition to being coded to mass-produce antibiotics, hijacked bacteria can be programmed to manufacture vaccines faster than the old-fashioned way. The old-fashioned way is to grow viruses within embryonated chicken eggs and bacteria on an agar medium, then inactivate the infectivity of the resultant pathogen, which is termed the antigen with some chemical, but without wiping out that pathogen's signature. Often other things are added to the antigen broth to make it more stable for storage.

Then, when you then take that greatly weakened antigen, virus (or bacteria)—which is called a "vaccine"—and inject it into a person, the signature triggers the immune system to build immune antibodies and, what's more, remember that immunity. Then, when confronted with that pathogen that corresponds to that antigen vaccine in real life, the immune system theoretically responds and responds with much vigor. With microbiome drugs, you could or can theoretically skip all that Petri dish and chemical-bleaching nonsense and code bacteria to produce the signature antigen of the pathogen, already watered down and ready to go.

Faster production is especially important if, for example, we're faced with a flu pandemic with a rogue virus and no vaccine. And, as I write these words, the world is facing the COVID-19 pandemic, and the race for a quick vaccine is most certainly on. The problem with rushing to fast, such as with COVID-19 is the vaccine might not work, or the vaccine might cause too many side effects, or the vaccine might trigger such a weak immune response it isn't worth the trouble.

Outside of flu vaccines, other examples of microbiome drugs include the hepatitis B vaccine, insulin, human growth hormone, and interferons. And these sorts of drugs are just the *pharmacological* application of synthetic biology.

Bacteria can also be harnessed to produce a wide array of commercial products, such as enzymes in detergents that can eat away organic stains, like last night's pizza on your favorite white blouse, biosensors capable of detecting harmful toxins, synthetic hemoglobin to carry oxygen, and the one I like best, information storage.

In 2012, George Church, a geneticist and molecular engineer, encoded the entire book he had just published, *Regenesis*, into the DNA sequence of a bacteria. Why? Because he could. Another team of synthetic microbiologists successfully encoded the complete works of William Shakespeare into DNA, which sort of gives modern meaning to the words of the character Viola in *Twelfth Night* when she whispers, "For such as we are made of, such we be."

What is the advantage of DNA data storage, besides the fact it's pretty spectacularly cool? For starters, based on wooly mammoth archaeology digs in cold environments, it's known that if DNA is kept cool, dry, and dark, it can survive millions of years. By comparison, we have no idea how long the data on silicon-based memory chips, like that flash drive on your key chain, will last. Even the less complicated floppy disks of the 1970s might "forget" their memory. Encoded DNA properly stored might never forget its memory.

But, more crucially, the coding unit of a DNA base pair is extremely small, less than a single nanometer, whereas the current commonly used types of data storage out of Silicon Valley struggle to beat the ten nanometer threshold. DNA storage is not just ten times more compact than current storage drives—it's actually *1,000 times* more compact, because DNA is three-dimensional and twirls about itself, while silicon circuits are flat boards, that is, two dimensional.

For ease of storage and access, things like hard drives, optical data-storage discs like DVDs and CDs, and silicon flash-memory chips are the obvious choices to store that term paper due at eight tomorrow morning. In other words, data off a silicon storage device can be retrieved in milliseconds, whereas unzipping a DNA molecule to retrieve your 0s and 1s would be mind-numbingly slow. We're talking on the order of days and weeks. For long-term data storage over thousands of years, however, DNA might be just the ticket, or at least one of the tickets.

As an aside, grammarians cannot seem to come to a consensus as to when to use "disk" with a "k" and when to use "disc" with a "c." Even just in the computer world it is confusing. Optical things, like DVDs and CDs, use the "c" version disc, while the hard drive inside your computer goes with the "k" version disk. In medicine, the rule of thumb is to go with "disc," as in, "I slipped my disc picking up the heavy disk drive that contained images of a disk-shaped galaxy a gazillion light years away." Physicists, you see, prefer "disk" when describing celestial objects that are shaped like a plate. It is all so confusing. I think I'm getting a migraine …

Since we're on the subject of spelling and storage, what exactly is the language of DNA data storage, or more precisely, the alphabet? That is, how exactly were Shakespeare's works encoded into a nucleic acid sequence? To understand that, we must first understand the basic language of computers: the binary code of 0s and 1s, where each "bit," or "binary digit," has a value of either 0 or 1. Whereas a "bit" is either a 1 or a 0, a "byte" is a certain number of bits strung together, like 01001010, and once strung together, that byte can be a lowercase letter, an uppercase letter, a number, a punctuation mark, or something more complex.

For instance, my name, "John," using the binary 8-bit byte format is: 01001010 (J), 01101111 (o), 01101000 (h), and 01101110 (n). If I write my name as "John," that is called a text file; if I write my name as "01001010 01101111 01101000 01101110," that is called an 8-bit binary file of four bytes. The most common format for text files is the ASCII file, the American Standard Code for Information Interchange, which uses the 8-bit binary alphabet or strings of 0s and 1s. It was originally a 7-bit byte, but an eighth parity bit was added to the leading byte to allow for error detection or to be used when dealing with languages other than English.

Computer language, also called machine language, only understands binary code. Things like storage, such as central processing units (CPUs), are in low-level binary format. Software programs, like the Microsoft word processor I'm using to write these words, use higher-level languages like C++ and Java. Because machine language does not understand higher-level programming, an intermediary, an interpreter—which computer geeks call compilers, assemblers, and linkers—converts the program command into executable bits and bytes for the computer to understand. In other words, when I type out my name "John" on the computer screen in Microsoft Word, the higher-level language of the word processor converts those four letters into something the lower-level CPU understands: that long series of 0s and 1s in the above paragraph.

In digital data storage, everything, and I mean *everything*, from the storage drive on your Mac or PC to the flash drive on your key chain, stores data as binary bytes. Retrieving that data, whether it is a simple ASCII text file or a more complex JPEG file of your meal to post on Instagram (as if anyone cares) or an MP4 video of your puppy destroying the last roll of toilet paper on earth, uses the computer's operating system and installed software to retrieve and convert the bytes in storage into the desired format for the user interface. In the end, storage is just a bunch of 0s and 1s—which was a long set up to begin explaining how DNA can be used to store data.

As we know, DNA base pairs are always adenine (A) with thymine (T) and cytosine (C) with guanine (G), so, using DNA as "bits" is a straight conversion from a 0 or a 1 to a DNA nucleotide: adenine and cytosine can be assigned as 0, and thymine and guanine as 1. Then you unzip your DNA, and out unfolds in a series all the 0s and 1s to code that Great American Novel you've been working on for two decades but can't find the time to finish because you're too busy working and taking care of your parasitic adult children still living at home, children taking bits and bytes of your food and money.

Which brings us to parasite infections, also termed parasitism, an interesting group of infections. The word "parasite" comes from French and loosely means "one who eats at the table of another." A parasite infection is a nonmutual symbiotic relationship between two species, where the parasite lives off the host and the host gains nothing in return other than usually ill, but not ill enough to be killed. Parasite infections might make the host infirm, but not outright dead. That's because it would be decidedly disadvantageous for a parasite to kill its host—if a parasite killed its host it would kill itself. A parasite may eat at the table of another, but it doesn't destroy the table.

The only time a parasite kills its host is when it reaches a point where offing the host is required to continue its life cycle, to extend its lineage. That is, the adult parasite sacrifices itself to provide a way for its larvae to move on to the next host. Occasionally, an overzealous parasite overwhelms the host, the burden of parasitism becoming just too much, and the parasite accidentally kills the host and itself. But all things being equal, parasite infections rarely kill, but nor do they go away easily—instead, they smolder. They find that sweet spot of a nonmutual symbiotic relationship with the host.

Some of the earliest descriptions of parasite infections are found in ancient Egyptian medical texts dating back to 3000 BC and include the Ebers Papyrus writings of 1550 BC, discovered in Thebes, Egypt (not the Thebes, Greece mentioned earlier). When exactly the Ebers Papyrus was discovered is not clear, but we do know for sure that in 1873, Georg Ebers, a German Egyptologist, purchased the ancient document. For that reason, only, a commercial transaction, the document is named after him. The Ebers Papyrus is currently housed in the library of Leipzig University in Germany.

Written in Egyptian hieratic, the papyrus contains hundreds of disease descriptions as well

as potions to cure what ails you. Hieratic is a cursive writing system that was used primarily for government and administrative documents, while hieroglyphic script was a more symbolic written language used only for religious documents. Sort of like the difference between shorthand of court reporting, used for its brevity, versus normal longhand. The key that allowed Egyptologists to read Egyptian hieratic and hieroglyphic was literally a language primer: the Rosetta Stone.

The Rosetta Stone was discovered in 1799 by French soldiers, unearthed in an area called Rosetta along the Nile Delta. It was made in 196 BC on granite stone and contains a translation written in three scripts: hieroglyphic, hieratic, and Greek. The message itself is a decree on behalf of Ptolemy V issued in Memphis, Egypt, describing the divine cult of his monarchy, but that is hardly the point. What matters is that the person who inscribed the Rosetta Stone knew hieroglyphic, hieratic, and ancient Greek—making that hunk of rock, literally, a primer key. That single stone allowed for the translation of most Egyptian hieroglyphs and hieratic script into Greek, and from there into any other language.

To the ancient Egyptians, the city of Thebes—where the Ebers Papyrus was discovered—was originally known as Waset. Thebes is the city's Latin name, derived from the Greek version, Thebai, which was itself derived from the Egyptian Ta-opet, meaning "shrine." Waset, by contrast, means "city of the scepter." Why all these name changes for one city will be explained momentarily.

Thebes was situated on both sides of the Nile, the longest river on Earth, a river that played an integral role in the development of human civilization. The Nile as well as the Euphrates and Tigris rivers in Mesopotamia, modern day Iraq and southeast Syria are considered the cradles of all early human civilizations. Sadly, the Mystic River, which flows into Boston, did not play a part in early human civilization, despite some Bostonians believing they're the center of life. Although to be fair, Boston is believed to be the cradle of a good deal of modern medicine, which is at least partly true, so it's doing okay as far as the history books are concerned. And then there is that dynasty, the New England Patriots.

The Nile finds its headwaters in many African countries, including Ethiopia, Sudan, Kenya, and Uganda, and all those feeding headwaters eventually combine to form the mighty Nile as it flows north through Egypt and empties into the Mediterranean. The Scottish explorer David Livingstone spent the better part of his life looking for the source of the Nile, obviously not realizing there was more than one source.

David Livingstone was born in Blantyre, South Lanarkshire, in 1813. The hero of his own rags-to-riches story, Livingstone was born into poverty but rose to mythical status as a Scottish Congregationalist, African explorer, and medical missionary. His obsession with finding the source of the Nile river as well as his part in setting up Christian medical missionaries in Central Africa during the nineteenth century became the source of bedtime stories told throughout England and Scotland.

Livingstone was a living legend, so when the world had not heard from him for some time—pretty much none of his forty-four letters he dispatched after 1867 ever made it to Zanzibar, Tanzania, his connection to the outside world—with speculation that he had gone missing somewhere in the interior of Central Africa while looking for the Nile headwaters, the people of Britain especially became alarmed. The Welsh American explorer Henry Morton Stanley, while exploring the Congo Basin region, was dispatched in 1869 by the *New York Herald* to find Livingstone, because, quite simply, it would make for a great newspaper story. Stanley then received a communiqué passed down from the highest levels in Britain around March 1871, further adding to the pressure for him to find Livingstone. The Brits wanted their national treasure back safe and sound.

Eight months later, on November 10, 1871, Stanley wandered into a remote village near Lake Tanganyika, in modern-day Tanzania, where he encountered an ill and poorly provisioned European man. It was at that moment that Stanley uttered those famous words: "Dr. Livingstone, I presume?" After giving Livingstone some supplies, Stanley trekked to the eastern coast of Africa, reaching the Indian Ocean, where he arranged for further fresh provisions so that Livingstone could carry on with his missionary and medical work and carry on with his passion of chasing down the mysteries of the Nile.

Livingstone carried on as best he could, but a year and a half later, on May 1, 1873, his exploration came to a lonely end. As Livingstone lay on a cot in the remote village overseen by Chief Chitambo, located southeast of Lake Bangweulu in what is now Zambia, then known as North

"Well, I'm sorry Mr. Stanley, but if you don't have an appointment I can't let you see doctor Livingstone."

Rhodesia, he made one last journal entry. The journal closed, the ink pen carefully set aside, David Livingstone then took his last breath. Cause of death: malaria.

As mentioned above, the city of Waset (later Thebes) sat on both sides of the Nile. Its eastern bank was more dignified, solemn, and ceremonial, home to the temple of Karnak and monumental pylons. The metropolis on the west bank of the Nile was what you and I today would call a residential neighborhood, where folks lived and worked. Waset was the second seat of the Egyptian empire beginning around 2040 BC, located in the south, and the first seat of power was Memphis, established in 3100 BC, located in the north toward the Nile Delta. By the nineteenth dynasty, the capital was moved back north even further along the Nile, to the Nile Delta, which we know today as Cairo and the nearby port city of Alexandria.

When the Greeks conquered Egypt in 332 BC (that would be Alexander the Great's campaign) and installed their Ptolemy rulers (Cleopatra nearly the last among them), they changed the name of Waset to Thebes, probably because the original Ptolemies were from Thebes, Greece. Conquerors do that. Not much different than the American colonists moving in on the Traditional Territories of Native Americans and renaming them New England after England and New York after York. With the Muslim conquest of Egypt in 641 AD, establishing a caliphate, the name Thebes was replaced with Luxor, which is what it is still called today. Luxor is from the Arabic word *al-uqsur*, meaning "castle" or "fortified camp." In other words, over the course of about 3,000 years Waset (Egyptian) became Thebes (Greek), which became Luxor (Arabic).

Luxor is also the name of the MGM Resorts hotel in Las Vegas that boasts a grand pyramid entrance. Only one problem with that: the Egyptian pyramids were more of a Giza thing, outside of Cairo, not a Waset/Thebes/Luxor thing, further south up the Nile. (Although there were smaller ceremonial pyramids scattered about Egypt). It would be the equivalent of having the central feature of the New York-New York Hotel & Casino, also on the Las Vegas Strip, be the Washington Monument, located in DC, instead of the Statue of Liberty. Whoops.

The Thebes in Egypt as pointed out, should not be confused with the other famous city-state of the same name from antiquity: Thebes, Greece. The Greek Thebes besides home of the Ptolemies, is better known from Greek mythology than from real history, having played a role in the mythological stories of Oedipus, Cadmus, and Dionysus. Cadmus was the founder of this Thebes and its first king, as well as a great hero and slayer of monsters.

In a roundabout way, you already know the story of Oedipus—the namesake of the Oedipus complex, so named by Sigmund Freud. The Oedipus complex is the repressed desire of a child to engage in a sexual relationship with their parent of the opposite sex. Yuck, I know. According to Freud, this repressed desire occurs around ages three to six, and when healthily resolved through eventual identification with the same-sex parent, the child can settle the conflict and move on to the next Freudian crisis. In other words, eventually the child no longer desires sex with the opposite parent because the child has new issues to resolve. At least according to Freud.

We need to step through all of Freud's Oedipal analysis to find out why girls are better at relationships than boys—and I'm half-joking, but only half-joking.

As boys pass through the Oedipal phase—again, according to Freud—they form "castration anxiety" (ouch!) and as girls pass through, they form "penis envy" (ew!). Castration anxiety is the fear of being emasculated, either literally or figuratively, which is also a theme that runs throughout *The Big Bang Theory* sitcom through Leonard Hofstadter's neurotic relationship with his emasculating psychiatrist mother, Beverly.

For a girl, penis envy begins with the realization that she does not have a penis, followed by a transition from an attachment to her mother to competition with her mother for her father's attention and affection. Some modern psychoanalysts still believe it is precisely the resolution of this conflict that might make girls—whether you believe Freud's theories or not—better at relationships than boys. How's that?

The interesting thing about acquiring skills in relationships is that whether we are talking about little boys or little girls, both form their first parent attachment—and for that matter, their first human attachment—to their mother. Quite simply, mothers of course give birth to both boys and girls, often breastfeed those boys and girls, and typically provide the majority of nurturing in the early months, years for both boys and girls. Kissing boo-boos, making snacks, potty training—basically doing all the things stereotypical "traditional" dads, like most of the dads of Freud's time, and probably many today, would rather not do. Regardless of who the main nurturer ends up being later in life, the first attachment for most infants and toddlers is the mother, simply because she's the one that carries, births, and nurtures them.

Okay, so that's the first part of the equation. Both boys and girls are attached to mom first.

As time goes on, things change. Little boys in this scenario are first attached to their mothers, but then eventually identify with their fathers, and they do not need to develop much in the way of interpersonal skills to get along with dad. Boys get to be lazy about transitioning from mother to father, because it's easy to stay close to their mothers, who first nurtured him, while also hanging out with their fathers, throwing a baseball back and forth. According to Freud, boys develop few relationship skills in transitioning to identifying as a male and identifying with dad; they may learn how to tie fly-fishing knots but gain limited skills in relationships that might lead to tying the knot.

Boys *do* learn more important things from their fathers than outdoors skills—things they can also of course learn from their mothers, though Freud wouldn't have thought so. This includes such things as industry, financial responsibility, confidence, service, duty, strength, sacrifice, and, very importantly, the correct way to treat other people and especially women. Or at least they should.

Little girls, on the other hand, who are likewise initially nurtured by their mothers, see themselves as female but need to acquire feminine social skills in order to compete with their mothers for their fathers' attention. In the end, in Freud's Oedipal analysis, little girls learn more relationship skills attempting to win over dad: communication, expressing feelings, personal expression through clothes and coiffing—all skills that account for women being better at interpersonal relationships than men. By navigating from getting their mothers' attention to their fathers' attention, little girls become skilled at relationships while little boys become clumsy at them. Guys, as Freud would have it, just never stood a chance.

Leaving Freud's Oedipus and returning to the Greek tragic hero Oedipus, he was a mythical king of Thebes whose story is told in a classic Greek tragedy. The ancient Greeks loved their tragedies. In a twist of fate, Oedipus ends up unwittingly, unknowingly, killing his father, Laius, and marrying his mother, Jocasta, thus fulfilling a prophecy and bringing damnation and ruin upon his family. When the truth is revealed, Jocasta hangs herself and Oedipus gouges out his own eyes. Family, what can you do; a little knowledge can be a dangerous thing. The Greek writer Sophocles took the folklore story of Oedipus and immortalized it in his play *Oedipus Rex* from 429 BC, a play that only served to further the myth of Oedipus.

"I warned you, Oedipus, not everybody likes what they find when tracing their family tree."

As for Dionysus (Bacchus in Roman mythology), who happens to be the god of wine, fertility, and ritual madness, his mother was the mortal Semele, the daughter of King Cadmus, who founded Thebes, Greece, and his father was the mighty Zeus, king of the gods. The wife of Zeus, the goddess Hera, was terribly jealous and angry that her husband had fathered a child with Semele. As we have seen already, Zeus was a bit of a philanderer; among many other children, he also fathered Hercules with the mortal Alcmene. Anyways, Hera was pissed, so she disguised herself as an old crone, befriended the pregnant Semele, and convinced her that Zeus was not the father of her unborn child. Gods and mortals didn't have sex with each other in the conventional sense. Encounters between gods and mere mortals either happened while the god was in disguise as a human or were more like the Virgin Mary's immaculate conception of the baby Jesus—no touching needed. So, it was easy for Hera to convince Semele that Zeus was not the father of her unborn child, Dionysus.

Semele, tricked by Hera, demanded that Zeus reveal himself as proof that he was the father of her child. This demand, however, came with a hefty price, as us mere mortals cannot look upon the gods and goddesses in their real form, lest we wish to perish. Zeus protested Semele's request, reminding her of the price she would pay. But Semele had to know. Zeus relented and came to her in his real form. She, of course, died, which forced Zeus to take the unborn Dionysus from her womb and place him into his thigh. It was there in Zeus's thigh where Dionysus, like a parasite, lived until he was old enough to be born. Which is the turn our story is now taking, back to parasites.

Descriptions of parasitic diseases can be found in ancient Greek writings from 300 BC as well as ancient Arabic writings from around 800 AD. All these descriptions from antiquity chronicle two recurring symptoms that are not very specific but set out the most common symptoms of parasitism: recurring fever and recurring malaise. What specific parasitic diseases Hippocrates and others were describing will likely be forever unprovable, but the best guess is what we today call malaria and worms.

When I was at Antioch College in Yellow Springs, Ohio, for my undergraduate studies, one of the courses I took was Biology 101. In fact, it might have been the very first college class I ever walked into. I was quiet and always sat at the back of the classroom, a habit I carried into medical school. Antioch's science building was called just that, the Science Building, and despite the fact the money to build it mostly came from the inventor and businessman Charles F. Kettering, it was not named after him, likely because he was quite modest. Next door to the Science Building on the Antioch campus is the library Kettering's money also built, and in this case it is named after the great man.

Charles F. Kettering was a brilliant inventor from Dayton, Ohio, who had the brains and good fortune to invent in 1911 the first alternator for an electric starter, for the then bourgeoning Detroit automobile industry. It was a Cadillac, of course, that first incorporated Kettering's alternator. Prior to Kettering's invention, cars' engines had to be crank started, simply because there was no battery and no way to charge a battery if there had been one. Once the alternator-less motor was running, whatever electrical demand was needed by other components of the car, such as headlights, was generated by the running engine. But once the engine stopped, no stored electricity was available to start the car again—it had to be cranked. In stepped Kettering and his alternator.

The automotive alternator is a device that recharges a car's battery while the engine is running, thus storing some of the electricity while also directing electrical currents to other in-the-moment demands of the automobile, such as the lights, radio, heater, and air conditioner. I guess the alternator alternates between running the car's electrical demands and also charging the battery. Makes sense. After inventing the first useful alternator, Kettering founded one of several great Dayton, Ohio companies, the Dayton Electronics Corporation, which we know as Delco. Delco made the first ignition set for cars—basically your high-energy spark-ignition assembly. During World War I, Delco manufactured the only American-built plane to see battle action, the De Havilland bomber, a model built in conjunction with Wright Airplane Company, named after the Wright brothers, who flew the first airplane on December 17, 1903.

The most important electrical component of cars when I first learned to drive, the only thing me and my pals cared about, other than that our cheap cars kept running and had gas, was the car radio. We'd always dial in a station that played rock 'n' roll, and more times than not that radio was a Delco product. It wasn't long before Kettering was the head of research for General Motors, a position he held from 1920 to 1947.

Although Orville and Wilbur Wright were also from Dayton, which is where they designed and assembled their various flying contraptions, they flew that first powered plane, the Wright Flyer, near Kitty Hawk, North Carolina. They selected Kitty Hawk for two reasons. First, it's located on a beach with constant wind and updrafts, a desirable feature to help lift your nascent plane, and second, if you're going to crash from twenty feet in the air, a sandy beach or, better yet, seawater is a lot nicer landing than the rocky and forested terrain of the Miami Valley, where Dayton is.

WELL, WE JUST BECAME UNSPECTACULAR.

Kettering is also credited with inventing Freon, in collaboration with the Delaware chemical company DuPont. Freon is a coolant used to remove heat from refrigerators, freezers, and air conditioners to give us their useful coldness. Kettering additionally played a key role in the development of the Guardian Frigerator Company, out of Fort Wayne, Indiana, which invented the first self-contained refrigerator. Guardian Frigerator was bought out by General Motors, moved to Dayton, had its name changed to Frigidaire, and came under management by Delco and Charles Kettering. Sometime during the early to mid 1900s, the word "frigerator" was changed to "refrigerator," the "re-" added likely because it's easier to pronounce. It's just one of those quirks of language, where some type of sound sets up the word, "refrigerator" rolls off the tongue easier than "frigerator." Somehow, though, "Frigidaire" without any "re," is easy to pronounce. Go figure.

Overly compressed Freon, when allowed to then rapidly expand, converts into a much cooler gas, which is sent meandering through the coils of your refrigerator or freezer to draw off heat. It is a classic quality of a compressed gas, when it is allowed to rapidly expand, so few molecules of gas end up striking each other (as compared to when it is compressed) that the kinetic energy drops precipitously as does the temperature. That frigid volume of gas thermodynamically desires more than anything to warm up, so sending it through coils of a refrigerator, it will draw off heat. Freon and like-minded gases are pulling heat *out of* the refrigerator rather than the intuitively wrong thinking it is introducing cold *into* the appliance, keeping the milk from spoiling and the beer at a crisp 34°F—unless of course you live in Britain where serving cold beer is apparently a misdemeanor.

The lecture halls of the Antioch Science Building that Kettering built were designed in a classic theater style, with the rows set on a characteristic incline upward, meaning the professor stands down below at a dusty chalkboard (or at least it was a chalkboard in my day), and the students sit in immovable pedestal seats with wood tabletops. I still recall their reek of too many years and too many layers of Minwax wood polish.

You could always find me in the very top row in my science classes, all by my lonesome. I dubbed that row the "scholarship section," the implication being if you were at college on someone else's dime, you were of a lower social caste and, like the proverbial being instructed to sit at the back of the bus, you were required to sit in the last row of the lecture hall. While I was not at Antioch on an academic scholarship nor a merit scholarship nor a needs-based scholarship nor an athletic scholarship, I simply liked to sit in the back row and referred to it as the "scholarship section." I carried that nickname over into my medical school lecture halls, where I also sat in the last row. Even when I attend medical conferences these days, you'll still find me in the last row of the lecture hall.

Unlike the back row in college, which was nowhere near the exit door in the event the lecture was too uncomfortably boring to endure, the back row in medical school *was* near the exit, an attractive option should the occasion arise. The other benefit of the back row at medical school was that it gave a bird's-eye view of the entire class, providing great amusement as I watched the comings and goings of my fellow medical students and their various personalities.

We called the students in the front row "the gunners," as they were always the first to raise their hands to answer questions, always feverishly taking notes, and generally "gunning" for a better grade—type super-A. Those students who were bright enough to have been accepted to medical school but for whatever reason were struggling mightily also tended to sit near the front, lest they miss something. We called them the "cliffhangers," because they were barely hanging on in medical school.

Returning to Antioch and the first day of college, the very first question Professor Samuels asked my Biology 101 class was quite simple: "Are viruses alive?" The student discussion was lively—lively because we were all pretty stupid. Looking back, it was a delightfully grim exercise. We didn't know what the heck we were talking about but pretended we did. It was the classic "sometimes right, usually wrong, never in doubt" folly. Some in the class argued that we know bacteria are alive because they can reproduce—divide—on their own, and since viruses cannot divide on their own, they are not alive. Others argued that since viruses replicate their genetic code, regardless of how they reproduce, regardless if they're cellular or not, they must be alive, too.

What we learned that day, besides learning how alarmingly silly we all were, was that the answer to that question depends on your definition of "life." In fact, the answer to questions of all sorts depend on how you define the question in the first place. In biology, if you limit the definition of "life" to the ability to reproduce on your own, bacteria are alive, and viruses are SOL. If you define "life" as the ability to replicate your DNA or RNA through whatever means available, including by hijacking another cell's genetic machinery, then viruses are alive, too. It all comes down to how you parse the question. Although viruses are not enclosed cells, they do have the partial covering, called the capsid, that sort of cape-like type thing. It is the capsid that identifies the cell to invade and performs the docking procedure. Most folks who concern themselves with this dilemma consider cell division as the definition of "alive" meaning bacteria are alive and viruses are flukes of biochemistry.

A virus hijacking a cell is simply designated "viral entry," and it has three known mechanisms. Some viruses enter the cell with their capsid intact—termed *membrane endocytosis*—the preferred method of the flu viruses, poliovirus, and hepatitis C. The targeted cell is forced to essentially swallow the virus whole. Other viruses fuse their capsid with the cell membrane, like a mixture or a blend, injecting the genome minus the capsid—termed *membrane fusion*—which is how HIV does it, as well as the culprit behind the current SARS-CoV-2 COVID-19 pandemic. Finally, other viruses barely attach to the cell membrane and then inject their genome—termed *genetic injection*—a process most commonly seen when viruses invade bacteria. Yes, bacteria can become invaded with viral pathogens, and the resulting monster is called a "bacteriophage." The nature of bacteriophages in human disease is not well understood, so we'll skip it here.

One of the most common virus we are all familiar with is the seasonal flu. First the flu virus attaches to a cell membrane of the target host, such as a mucosal cell along the respiratory lining, and then the virus enters the cell by membrane endocytosis. The viral genome is now able to traffic within the host cell and hijack the genetic machinery, cranking out baby viruses.

The RNA viruses—most cold and flu viruses, gastrointestinal viruses, SARS, SARS-CoV-2, hepatitis C—perform most of their hijacking and replication inside the cytoplasm, where there's a bunch of goodies. Recall from a much earlier discussion that DNA inside the nucleus codes for the various normal RNAs that then leave the nucleus, especially messenger RNA and enter the cell cytoplasm in order to do what it is they do: assemble protein enzymes. It's the cell's normal RNA machinery inside cytoplasmic ribosomes that the RNA virus commandeers. RNA viruses, in other words, don't enter the cell nucleus in order to replicate.

DNA viruses—herpes, smallpox, HPV, and surprisingly adenovirus, the cold virus—after entering the cytoplasm must first traverse into the nucleus where there are enzymes that can unzip the DNA virus. They need to take this extra step precisely because they're DNA viruses and the enzyme they need to unzip is in the nucleus; there's nothing in the cytoplasm that can unzip their fly. Once unzipped, the unzipped DNA viral component then need to traverse back out into the cytoplasm where the strands of daughter DNA viruses are assembled and then the two halves reassembled in what are termed "replication compartments." Replications compartments are different cellular organelles than the ribosomes where normal RNA to protein manufacturing occurs.

Perhaps you're asking why DNA virus evolved—why not remain RNA and avoid the extra transit into and out of the cell's nucleus just to get unzipped? Simple really. It's for the same reason our DNA evolved to form a double helix in the first place: protection of the genome. At some point in remote history, those DNA viruses decided the double helix was worth the added effort; perhaps that is why at least the herpes virus can never be cleared once it's infected a person. It hunkers down in its double helix and the human immune system is helpless and hapless to extricate that bad boy. RNA viruses, by contrast, are easier to eventually extricate from the body, for the most part, since they're more vulnerable to our immune cells.

An odd poster child of RNA viruses is HIV, which takes a circuitous route to replicate unlike your bread-and-butter RNA viruses, and in so doing is better able to cloak itself from the immune system. HIV must first code itself into a complementary DNA sequence using the available cellular machinery, but that machinery is inside the nucleus. HIV does this by entering the cytoplasm, transiting into the nucleus, where it activates an enzyme coded by the RNA virus itself, termed reverse transcriptase. Reverse transcriptase then codes the DNA sequence of the RNA HIV—DNA normally codes for RNA, reverse transcriptase enzyme reverses that direction—and that HIV now in DNA costume, codes for messenger RNA, which then leaves the nucleus and codes the viral RNA HIV daughters out in replication compartments.

Whether RNA or DNA virus, often a point is reached where the host cell becomes exhausted from manufacturing too many viral daughters within its cytoplasm, and the process of "viral shedding" begins. Once a virus has overstayed its welcome, there are three ways it can leave a cell to either infect another cell within the same person or to be coughed and sneezed into a new host. *Viral budding*—sort of the reverse of viral endocytosis—involves the virus attaching to the inside lining of the cell and "budding" through the membrane, creating its capsid from cell membrane material as it buds. A second type of viral shedding is *apoptosis* of a cell—cell death or cell suicide—where the crippled cell basically bursts, termed cell lysis, releasing all those nasty daughter viruses who have already acquired their capsid. A third shedding technique termed *exocytosis* is when a fully formed virus replete with its capsid bulldozes its way through the cell membrane without killing the cell.

Before the existence of viruses was known, Louis Pasteur attempted to isolate the causative "bacteria" that caused rabies, a severe brain and nervous system disease that at the time was universally fatal—and still is fatal unless, treated promptly. It's caused by the bite of an infected animal, like a rabid racoon, skunk, coyote, fox, or bat (the most common carrier in the US). Pasteur could not have known at the time that rabies is caused not by bacteria but by a virus, and thus the causative contagion much too small for Pasteur to have been able to visualize using a conventional microscope, let alone to isolate or grow in some agar-filled Petri dish designed for bacteria growth, not viral growth.

In 1884, Pasteur in collaboration with his fellow French microbiologist Charles Chamberland, developed the Chamberland filter, which traps bacteria for experiments by allowing the unwanted fluid to pass through the filter and be discarded. In filtration, the stuff that cannot pass through the filter is called the "residue" and the stuff that can pass through is called the "filtrate." Unknown to Chamberland and Pasteur, there were creepy things—viral things—much smaller than bacteria that were able to pass through the filter and skulk around in the filtrate. The Chamberland filter would in time become key to isolating viruses, because one day someone decided to investigate what was in the usually discarded filtrate that was smaller than bacteria and that could cause infection.

Chamberland is also known for developing in that same year, 1884, an early prototype of the autoclave, a device that would eventually lead to the first fully functioning autoclave for sterilizing equipment. The inside of an autoclave's steel tank reaches high enough temperatures and pressures to kill all microbes. Where did the idea for such a contraption arise? From a French chef, no less.

A few hundred years before the autoclave's invention, in 1679, another Frenchman, Denis Papin, introduced what he called a "steam digester" not for purposes of microbiology but to achieve culinary delight. It's what you and I would today call a pressure cooker and make beef stew in, or, since it was a French invention, make *ragoût de boeuf.* Chamberland used Papin's pressure cooker concept to design the first autoclave for sterilization in medicine, based on the concept that high temperatures achieved through heat and pressure will kill everything alive. Imagine that— from *ragoût de boeuf* to dead microbes, all in a little over 200 years. Bon appétit.

In 1892, the Russian biologist Dmitri Ivanovsky was the scientist who decided to take a closer look at the normally discarded filtrate from Chamberland's filter system. Eventually, Ivanovsky was able to isolate the first virus ever discovered, which was tiny enough to pass through the filter and was lurking there in the filtrate, known today as the tobacco mosaic virus. He didn't actually visualize the virus nor was he able to grow it, but he knew something infectious was sneaking around, something that did not follow Koch's postulates for bacterial infections. Yet he was able to demonstrate that this "whatever" caused infection.

The tobacco mosaic virus is an RNA virus that infects plants, especially tobacco plants. At the time of Ivanovsky's discovery, it was causing quite the blight in tobacco crops. "Blight," in one of its definitions, describes the various infections of plants that devastate crops, usually due to fungi but also bacteria and viruses. "Blight" can also mean "spoilage," as in "the crack-house people across the street are a blight on the neighborhood." Another of its meanings is "deteriorated condition," as in "the volcanic ash blighted the sun for several years, ushering in a mini ice age."

A few years after Ivanovsky's virus discovery, in 1898, the Dutch microbiologist Martinus Beijerinck repeated the French scientist's experiment and showed that the agent passing through the Chamberland filter could only survive in other cells that were alive. Outside of a living cell, the agents could survive for only a short period. He called the new infectious agent *contagium vivum fluidum*, from the Latin, which translates to "contagious living fluid." Eventually, from "*vivum*," or "living," the name "virus" was derived.

Over the next thirty years, researchers isolated and cultured many different viruses, as they now appreciated that to grow viruses you need a host cell for the virus to attack, hijack the genetic machinery of, and then replicate inside. You see, bacteria just need food to grow and divide, so plating bacteria on a delicious algae Petri dish is just the ticket. But viruses, they don't "eat" in the conventional sense. Viruses need genetic machinery to replicate, which means they must go through viral entry and viral shedding gymnastics. Viruses are needy creatures that require live host cells to offer up themselves and their genetic machinery for the viruses to latch onto and replicate within. Bacteria grow on agar whereas viruses replicate in cells.

The scientist credited with inventing the first viable growth medium for viruses is Ernest William Goodpasture, who in 1931 successfully and repeatedly was able to grow viruses in embryonated chicken eggs. "Embryonated" means the chicken egg had been fertilized by a male rooster. Why viruses prefer fertilized chicken eggs to unfertilized chicken eggs was and continues to be quite the mystery—you'd have to ask the virus. Goodpasture's student Thomas Francis was the first person to grow the most common flu viruses—influenza A in 1934 and influenza B 1940—in embryonated chicken eggs. In a classic passing of the torch in medicine, Francis's own star student Jonas Salk, using the same chick embryo model, gave to the world the first polio vaccine in 1955. Science doesn't get much better than this.

The isolation of viruses, despite them remaining shadowy, unobservable, evanescent creatures during that period of history, allowed for some very important early viral vaccines to be created. When you think about it, it was a rather elegant situation: early viral researchers, later to be dubbed vaccinologists, couldn't actually see a virus under a microscope like they could a bacterium, protozoan, or worm, but they certainly knew viruses existed, certainly knew how to isolate the little beasties, and certainly knew how to make a vaccine against them—all from an invisible pathogen. These early vaccines included ones for the vaccinia poxvirus family, which includes the deadly smallpox virus and the equally disturbing poliovirus.

Your classic microscope uses regular light—photons—to illuminate the microscope slide from underneath, and that glass microscope slide contains the item one wishes to observe using the magnifying optics positioned above. That's your basic setup. Most modern microscopes have two glass magnifying lenses, known as compound microscopes. The eyepiece usually magnifies 10x and then three rotating objective lenses below the eyepiece, right above the glass microscope slide, magnify 4x, 10x, and 40x, which, coupled with the 10x eyepiece, deliver magnifications of 40 times, 100 times, and 400 times.

Older model microscopes needed a mirror below the slide to redirect light from a peripheral source up through the slide, since they didn't come equipped with their own built-in light source. The source was often a light bulb, and before electricity was discovered and harnessed, the source was likely a candle or lantern of some sort. Despite the light source limitations, the early microscopes of Antonie van Leeuwenhoek still possessed great magnification, believed to have been 200 times. After all, pure glass lenses in the 1600s bent light the same as pure glass lenses do in the twenty-first century. Bent light is bent light. Modern microscopes have a light bulb in the base that not only turns on but has a rheostat to deliver bright light or dim light, depending on what is needed to visualize the specimen on the slide.

You remember from high school biology preparing a slide, placing it under the optics, and viewing illuminated and magnified strange and wondrous things, right? No? Well, high school biology wasn't for everyone.

To understand the limits of light microscopy and why viruses couldn't, and can't, be seen while bacteria, red blood cells, and Van Leeuwenhoek's sperm could, we need to turn to the visible light spectrum. Although a visible light wavelength is a pretty short wavelength, measuring between about 390 to 700 nanometers, it is not short enough to visualize a virus. The average bacteria size is 500 to 1,000 nanometers, which pairs nicely with that visible light wavelength spectrum. Simply put, light photon wavelengths are of an appropriate size to etch bacteria and similarly sized objects.

Viruses, on the other hand, are much smaller, perhaps 10 to a 100 times smaller than bacteria, on the order of 50 nanometers, if that. Even with the best magnifying optics, a light wavelength longer than the size of a virus will result in no visualization. None. The visible light wavelengths just can't etch something shorter than themselves. However, modern-day electron microscopes, where the wavelength is 100,000 times shorter than light wavelengths, on the order of 0.0025 nanometers, or 2.5 picometers, can catch a peek of those most elusive viruses.

The units of diminishing metric length most of us are familiar with are $1/1000^{th}$ shorter or 10^{-3} for each descending unit, or in other words: meter 10^0, millimeter 10^{-3}, micrometer 10^{-6}, and nanometer 10^{-9}. Beyond nanometer, there is picometer 10^{-12}, femtometer 10^{-15}, attometer 10^{-18}, zeptometer 10^{-21}, yoctometer 10^{-24}, and finally the Planck length.

Metrically challenged Americans know what a meter is, about a yard—the height of a kitchen counter or the length of a guitar. After that, many Americans struggle with centimeters and millimeters or even something long like the kilometer, much preferring feet and inches and miles, the imperial system. But in the microscopic world, the imperial system is not used. In other words, we Americans and similar imperial devotees hold tight to our inches, feet, yards, miles, and pounds. But when it comes to microscopic thingamajigs, we are forced to accept the metric system because, quite simply, the imperial system has been abandoned.

One centimeter is about the width of a fingernail, one millimeter the thickness of a credit card, one micrometer the size of a small bacteria, one nanometer the size of a small virus or the wavelength of an X-ray, one picometer about the wavelength of your average electron zipping along, one femtometer about the diameter of that electron zipping along, one attometer the diameter of your garden-variety proton, one zeptometer the size of a neutrino flying out of the sun or any other basic sub-subatomic particle, your quarks and leptons, which were pretty much all consumed during the Big Bang, and, finally, one yoctometer—well, nothing that small has been measured, but it could be a more basic sub-subatomic particle, like a really, really, really small neutrino. As for the Planck length, it is a theoretical length; for all intents and purposes, any two things separated by a Planck length are likely occupying the same exact space.

In 1931, Ernst Ruska from Heidelberg, Germany, and Max Knoll from just up north in Wiesbaden, both electrical engineers, together invented the electron microscope. Because the electron wavelength is so much shorter than the photon, or light, wavelength, the electron microscope possesses vastly superior resolving power. That is, traveling electrons can visualize, can etch sizes much, much smaller than bacteria, even things smaller than viruses. In fact, the most powerful electron microscope available today can image down to the level of an atom. Some recent real-time electron-microscope images actually caught atoms forming a bond. Now that is impressive. If only Linus Pauling, who first elucidated the nature of the chemical bond, could have been around to see atoms forming a molecule within an electron microscope.

Ruska and Knoll obtained the first images of a virus around 1931. But others were involved in the development of the electron microscope who also deserve credit. It is yet another example of how science is built on many people's shoulders. For instance, in 1926, Hans Busch built the first electromagnetic lens that could grab passing electrons and deliver their image onto an appropriate medium for resolving. The back of our eyes, our retinas, wouldn't work for electron microscopes, as our eyes cannot image electrons, only photons. Negative film—for what is now, in the digital world, ancient—also couldn't be used to illustrate an electron wavelength, because it too

lacks the resolution capabilities.

Busch's electromagnetic lens works by focusing electrons passing through an electric field depending upon how they enter that field, meaning according to what the electrons do or do not pass through, what they etch. Those electrons are then delivered to either a piece of electron micrograph paper, which develops according to what it receives, or delivered to a monitor screen, like a TV, which reproduces the image based on how the electrons are received.

Visualizing viruses—which are on average about 10 nanometers in size or so—is child's play for an electron microscope. But how does it capture an image of the much tinier atom? The average atom is about 0.5 nanometers, but the wavelength of an electron is even shorter, at 0.001 nanometers, or 1 picometer, so an electron wavelength can image anything about its size and larger. That is the key, really—the imaging wavelength needs to be shorter than the image to be imaged. Imagine that.

A modern standard light microscope is like the kind you had in biology class, the size of a blender used to make frozen margaritas and protein shakes. The professional version of a light microscope that microbiologists and pathologists use is about twice as big. But an electron microscope, that is a different order of magnitude altogether. It is much bigger; the original electron microscopes were perhaps the size of a CT scanner or MRI machine or a Volkswagen Beetle. Today they, like most things in technology, are smaller, perhaps the size of the classic 6-foot high floor-model Dairy Queen machine that makes the most delicious soft serve ice cream this side of heaven.

I'm not exactly sure why Volkswagen named their iconic car the Beetle, but likely because it looks like one. Interestingly, in the early 1930s, the design of a simple "car for the people" was requested by villainous Adolf Hitler, who wanted an inexpensive car that could be mass-produced. The lead designer on that team was none other than Ferdinand Porsche, who of course gave us the Porsche car company and a few of the early racing-style Mercedes-Benz automobiles.

We previously listed the more or less common bacteria that cause human infection. A similar list before we move on for viruses include common cold (rhinovirus, coronavirus, adenovirus), various forms of the flu (influenza A, influenza B), mononucleosis, viral hepatitis (hepatitis A, B, and C), chickenpox, smallpox (extinct), measles, German measles, mumps, herpes (type 1 oral, type 2 genital), and HIV (responsible for AIDS).

"HIV" is the acronym for "human immunodeficiency virus" and "AIDS" is the acronym for "acquired immune deficiency syndrome." HIV, the virus, is generally accepted as causing the initial infection, and AIDS, the syndrome, results when the part of the immune system called the T cell—actually the T helper cell—takes a nosedive.

Much earlier we began the process of parsing infections, as to type of pathogen and modes of transmission. And it is parsing infections other ways that we now turn to.

21 THE OPPORTUNITY TO BE OPPORTUNE

We know there are four main types of infections, classified according to the offending pathogen: bacterium, fungus, parasite, and virus. We've also reviewed the nine modes in which infections can be transmitted—airborne droplet, fecal-oral, sexual, blood transfusion, vertical across the placenta, oral, direct contact, iatrogenic and vector like via an insect. By what other means are infections classified?

One designation places infections into three "presentations": clinical, latent, and subclinical. A *clinical* infection implies just that: the symptoms are apparent. Your basic sniffling, sneezing, coughing, aching so you get no sleep infection. These symptoms, or any defining symptoms of any clinical infection—or for that matter most any disease—are the most common thing we think of when we think of infection, if we wish to think of infection at all. A clinical infection is for the most part obvious: obvious to the patient, obvious to the physician, and obvious to nearly everyone else a person might be sneezing on.

A *subclinical*, also termed occult, hidden, or inapparent, infection is not that common. Perhaps the best known is the quiet smoldering of viral hepatitis B, which slowly, ploddingly can destroy liver cells. Subclinical infection means a pathogen is living inside a person, but the symptoms of the infection are so imperceptible that the pathogen flies under the radar of medical attention, even under the radar of the patient, who's unaware there's a nefarious creature skulking about their body.

Other examples of subclinical infection besides hep B are some of the sexually transmitted diseases (STDs). For instance, the period of time between an initial HIV infection (which is not only an STD) and the onset of AIDS can span months or years, this period of time is classified as a subclinical state where the HIV is replicating but the person has no symptoms. The category of subclinical STDs also includes gonorrhea and chlamydia, two VDs that can be so inapparent as to be unnoticed by an infected person, who nonetheless remains quite infectious. Because of the subclinical nature of STDs, the state of Nevada, where prostitution is legal, requires sex workers to be tested for gonorrhea and chlamydia weekly and for syphilis, HIV, and hepatitis B and C monthly, even if they're completely asymptomatic.

Still further examples of subclinical infections can include the flu, where some people can acquire that year's flu and not even know it. The current COVID-19 pandemic, which is technically a flu virus, was obviously characterized by some really ill patients, but apparently others have or had the virus and are flying under the radar; they don't develop coronavirus symptoms, but they certainly can pass it on to someone who very well might die from it. Which is one reason why wearing masks and social distancing have been mandated in many communities, not just to protect the individual, but to protect the community.

There is a long list of pathogens—a list that includes bacteria, parasites, and viruses—that can result in no-infection infection in some people, a subclinical state where it is possible to dig up some symptoms, some sign, some laboratory abnormality, but by and large, *nada*.

Subclinical infection is distinguished from a carrier state, which is a distinctly different condition altogether. A carrier state describes where a person has no infection but nevertheless has a bug quietly hiding in some dank, dark corner of their body. The carrier state also implies the pathogen can be transmitted through whatever approach is its *modus operandi*. If hepatitis B is the poster child for subclinical infection, then the mascot of the carrier state is Typhoid Mary, whose story will unfold in *The Second History of Man*.

Other infections are designated as *latent* infections, such as a recurrence of a herpes cold sore months or years after the initial herpes infection. Latent infection means there is a time lag, often a long one, between the initial infection and either the recurrence of the infection or the next version of the infection. Although more common with viruses—such as hepatitis B, herpes type 1 (oral), and herpes type 2 (genital)—latent infections can also occur with bacteria, the most widely known example being the gap between the second and third stages of syphilis.

Primary syphilis, the first stage that occurs about twenty-one days after exposure, is characterized by painless sores called chancres, usually genital but sometimes oral (if you're reading between the lines). Stage one is soon followed by stage two syphilis, or secondary syphilis, one to several months later, which is a more diffuse version with a rash on the hands and feet, swollen lymph nodes in the groin, neck, and axilla (armpits), a sore throat, headaches, patchy hair loss, fever, and fatigue. Stage two lasts three to six weeks as the syphilitic microbe navigates through the body.

Then comes the latent phase after secondary syphilis, a period of time that can last years and years, where the syphilis bacteria are alive but not causing any clinical symptoms. This latent phase of syphilis is also called subclinical syphilis, because even though the patient has no apparent symptoms, the syphilis spirochete is quietly, subtly, imperceptibly whittling away at the heart, brain, peripheral nerves, and bone. Some people don't enter the latent phase but instead have the unfortunate pleasure of going straight to the third stage, known as tertiary syphilis. Others wait years for this last phase to arrive on their doorstep.

Tertiary syphilis is not a pretty sight. It's most noticeable features include the onset of incoordination and gummatous lesions throughout the body, including the face, making hiding the condition nearly impossible. When syphilis starts to affect the brain and nerves, several other notable presentations also occur, including an overall general numbness throughout the body, dementia with a penchant for drooling, seizures, meningitis, and pupil abnormalities.

The generalized numbness and the spinal cord nerve involvement produce what is called "tabes dorsalis," a condition that would have been most evident in the ambling of an infected person wandering the streets of seventeenth-century London, Paris, or Rome during the epidemic rise of syphilis. Tabes dorsalis is the loss of lower leg coordination coupled with a loss of sensation on the bottom of the feet. Betwixt the two problems, a person could not, cannot, feel themselves walking, and so they would need to "slap" their feet upon the cobble stones in order to *hear* themselves walking rather than *feel* themselves walking. *Slap, slap, slap*—the dreaded, languid shuffle of a shell of a human being slowly walking toward their demented doom.

Another pathognomonic sign of tertiary syphilis is Argyll Robertson pupil, a syphilitic pupil that does not constrict when exposed to light, as described by the Scottish ophthalmologist Douglas Argyll Robertson in the late 1800s.

The underlying pathophysiology of the third stage of syphilis—the bedrock of the irreversible injuries to cells of the heart, brain, motor nerves, sensory nerves, skin, and bone—is the spirochete-shaped syphilis bacteria's relentless and irreversible destruction of blood vessels. This is the basis of what tertiary syphilis does: knocks off blood flow, ending in the ischemic death of whatever body part has lost its blood.

Stage one, stage two, and early latent syphilis are still infectious, but late latent and tertiary syphilis are not as much infections as they are the aftermath of infection. It's odd really: the tertiary stage of syphilis is after the syphilitic bugger has achieved most of its damage, yet it's not very contagious. The same holds true for treatment. Antibiotics, especially penicillin, will kill off the syphilis spirochetes in stage one, stage two, and early latent syphilis. Technically late latent and tertiary syphilis *can* be treated with penicillin, to annihilate the syphilitic pathogen, if still present, but it won't reverse the damage done to the organs, and so won't save the patient from the carnage. Such a person cleared of the syphilis bacteria in the late tertiary stage of the disease might live a tad longer, drooling dementedly, or the penicillin might not make a bit of difference. Either way, approaching death is all there is.

When syphilis was all the rage across Europe, from the early 1500s into the early 1900s—that is, before penicillin could cure it—poor souls with gummas on their faces, with their ischemic dead noses sinking into and vanishing from the face, couldn't hide from the world that they had syphilis. And it did not take a rocket scientist (or back then, a microscope or a light bulb to switch on) to know it was a sexually transmitted disease. Especially for married men afflicted with the great pox, signs of tertiary syphilis carried added shame, as it meant they had spilled their semen outside the marriage bed.

So that's it for clinical, subclinical, and latent human infections. But we're not done with the classifications of infections yet. Any type of infection can be classified in many different ways beyond what has already been discussed.

However, before we move on to other designations of infection, we should take a detour to look at an interesting condition in medicine—a real diagnosis—that dovetails with the concept of subclinical infection just discussed. It is termed "fever of unknown origin," or FUO, not to be confused with UFO. While this is as good a time as any to discuss FUO, I think we'll skip UFOs and Area 51.

A fever is normally a symptom of an underlying infection. It is a clinical finding. But what about a person who has a persistent fever with no known infection? If a person has a fever of at least 101°F continuously or repeatedly for a two-week period without any apparent underlying cause, it is designated a "fever of unknown origin." Sometimes the source of the fever is later discovered, sometimes it is not. Most cases of FUO turn out to be caused by a subclinical infection. However, others might not be from an infection at all but rather from inflammation associated with an autoimmune disease or even a reaction to certain drugs.

I will not bore you with a comprehensive list of infections that can be so subtle that the only symptom is a fever, rendering the infection an FUO. But here are a few of the more common ones: many parasite infections, but especially malaria and giardiasis and those nasty worms; sinus infection; bone infection, or osteomyelitis; gallbladder infection, called cholecystitis; early appendicitis; infectious diverticulitis of the large intestines; low-grade pelvic inflammatory disease of the ovary apparatus; early or undiagnosed tuberculosis; mono; hepatitis; and toxoplasmosis. All of these can be so subclinical as to go unnoticed—unnoticed, that is, except for a fever.

Noninfection causes of FUO includes the gamut of autoimmune diseases and connective tissue diseases, before a firm diagnosis has been made. These include systemic lupus, early rheumatoid arthritis, sarcoidosis, non-infection diverticulitis, and inflammatory bowel diseases such as Crohn's and ulcerative colitis.

Other FUOs can be caused by things that are downright undesirable, such as undiagnosed lymphoma, leukemia, lung cancer, colon cancer, and pancreatic cancer.

Some processes associated with inflammation might only show up initially with a fever. These include a clot in the leg, either superficial or deep vein thrombosis; a clot in the lung, termed pulmonary embolus; and, of course, an overactive thyroid gland, called hyperthyroidism, can present with a fever.

Surprisingly, or maybe not surprisingly at all, some drugs are known to cause fever of unknown origin, and these include—to make it a double whammy—many antibiotics such as the penicillins, cephalosporins, sulfonamides, and amphotericin B. It's always a frustrating juncture when you treat a patient for a clinical infection with an antibiotic and all the signs of infection disappear, yet the patient still has a fever. What's with that? Is the infection smoldering or is it a fever of unknown origin due to the antibiotic? Other drugs associated with fever are phenytoin (Dilantin) used in epilepsy; the barbiturates such as pentothal, which you and I know as "truth serum"; allopurinol for gout; methyldopa for Parkinson's; procainamide and quinidine for heart arrhythmias; and propylthiouracil, which is sometimes used to treat hyperthyroidism.

Now let us return to our regularly scheduled programmed classification of infections. Another categorizing method for infection includes primary, secondary, and opportunistic human infection, which are different designations than clinical, subclinical, and latent. *Primary infection* is the initial infection, like a cold, like the flu, like strep throat, like whatever. Occasionally, a primary infection can develop secondary issues or other stages of the infection, which are *not* secondary infections. In example, strep throat evolving into rheumatic fever or scarlet fever, or primary syphilis progressing through from stage two through latent to stage three syphilis. A *secondary infection*, by contrast, is a different pathogen that either occurred incidentally after the primary infection or, more likely, occurred because of the primary infection.

One of the most common examples of secondary infection is a viral cold followed a few weeks later by a bacterial sinus infection. The viral cold is primary; the bacterial sinusitis is secondary. Another example is a yeast vaginitis as a secondary infection after a woman has taken antibiotics for something entirely different. In Deborah's situation, she had a primary viral cold or flu, and then developed what sounds like a secondary lower respiratory bacterial infection, a pneumonia, a few weeks later.

And that's where our story now returns to.

That phone ringing.

When Deborah was ill several weeks earlier, she likely had your garden-variety viral cold or flu. We don't know for sure, but it sounded like a head cold, which would have been your basic primary viral cold infection. She likely recovered from that head cold and enjoyed several days of respite before becoming ill again, although the history as presented through the intermediary to me was frustratingly patchy on that subject. What was clearly conveyed to me, though, was that the earlier cold had perhaps relapsed. But this could not have been the case; it had to be, according to medical science, most likely a secondary infection.

Colds don't relapse. Colds just don't do that. Our immune system, unless frightfully compromised, makes sure of that. A person can't have the same cold virus back to back within weeks, can't get bulldozed by the same virus in such close succession. The immune system builds some memory of a recent cold virus, its telltale markers, its signature, enabling it to block a second attack. It is possible, however, to have two different cold viruses, one after the other. But the thing is this: the viruses have to be different enough serotypes, and it is not common to have cold or flu viruses of different serotypes steamrolling victims in the same community. For travelers, well, it might be a different story.

What was clear to me was that some type of secondary infection had reared its ugly head in Deborah's world. Her initial infection had been in the upper reaches of the respiratory tract—the nose and sinuses—making it your basic rhinitis cold or flu. Sometimes the cold virus can crawl into the sinuses, giving rise to a viral sinusitis, which is still just a cold, but a geographically wider one. If the sinuses develop a *bacterial* infection on top of the cold virus, that would be a secondary infection. A secondary infection by definition involves a different pathogen.

What is clear is that Deborah a few weeks previously had an upper respiratory infection, likely a viral cold. We'll never know. What is known is that, for a few days anyway, she appeared to have been on the mend, had recovered. Then something went awry. Bacteria somehow seeded her lower respiratory reaches, either from something like a secondary bacterial sinusitis that arose on the heels of a viral cold, or just straightaway, bacteria entering her pulmonary passages, bypassing an infection up above, and settling down in her lungs.

Relapse is different. Relapse is when you have something, and it goes away (or mostly goes away) but then comes back. You can suffer from an ulcer, treat it, and it will seem to have gone away, only to rear its head again. Cancer, unfortunately, is one of the most common examples of something that relapses, such as when chemotherapy eradicates the vile cancer cells of leukemia, followed by a period of remission, only to have the leukemia cells recur, relapse. But can infections "relapse" in this same way? Is there a difference between relapse and reinfection?

Infections that relapse are usually from infections that were never eradicated in the first place. Take, for example, tuberculosis (TB). Because most antibiotics primarily work during cell division, which is when the bacteria cell is most vulnerable, and TB bacteria grow so slowly, undertreating tuberculosis can result in relapse. Bacterial infections hidden in areas with poor blood flow—sinuses, the middle ear, bones—where the immune cells and prescribed antibiotics attack in a more impotent way. And, so, the bacteria might beat a retreat, hibernate but not be wiped out. When the antibiotics are stopped and the immune system has been tricked into believing the enemy has been vanquished, the same pathogen charges back into the field of battle. This would be a relapse with the same pathogen. Relapse equals same pathogen.

Reinfection is precisely that: it is another infection, but not necessarily with the same pathogen; this is what distinguishes it from relapse. The first pathogen is completely wiped out, and once the combat zone has quieted down, reinfection—more formally termed secondary infection, as we've just covered—in the now worn-out and weary battlefield can occur with a different bug. When a pathogen has truly been wiped out and its encampments in the body completely destroyed, relapse is not possible, but reinfection is.

Having said reinfection has to be with a different pathogen, that is not quite true. Reinfection *can* occur with the same pathogen, but it needs to be that same type of pathogen re-entering the body, not a relapse from a hiding pathogen. I suppose the classic example would be acquiring gonorrhea from a sexual partner, successfully treating yourself but not the partner, and then becoming reinfected through a hookup with that same partner.

Having said that, though, the immune system is built not only to wipe out an invading pathogen but also to be primed to prevent that same pathogen from relapsing or reinfecting anytime soon. It has to do with both immune memory and residual sentries—T cells and antibodies—still circulating about. Consider this, you have strep throat circulating within a family and once the first person, little Billy, completely recovers and all strep bacteria are wiped out, little Billy is unlikely to get another strep throat from his older sister Nancy's current strep infection. This holds true with viruses as it does with bacteria. If relapse occurs in a short period of time with the same pathogen, it most likely means the infection was never truly wiped out. And if reinfection occurs it means the immune system perhaps did not do a very good job remembering the pathogen's signature.

But—there are always exceptions to the rule, like the gonorrhea example above. Some pathogens even if completely wiped out can cause another infection in a short period of time; that is, a reinfection. It has to do with some microbes that do not elicit enough of an immune memory and immune surveillance, making reinfection possible. In addition to some of the STDs, intestinal worms can do that. You can treat pinworms and rid your body of every last vestige, and once all the pinworms are gone and all the residual antiworm drug gone, all you'd need to do is consume another contaminated food source and *wham!*, your reinfected again.

The current COVID-19 pandemic, among all the things that are troubling about it, is that entire immune memory thingamajig. The seasonal flu triggers enough of an immune response—circulating antibodies—that it is unlikely a person will get the flu again that season. But with this coronavirus, it appears some people who get ill and recover and have their blood tested afterwards, are not producing a profound immune response, at least not in terms of immune memory. The implication being two-fold: first the person can get COVID-19 again and second, a poor immune response might explain why this novel virus is so devastatingly morbid and even fatal.

After primary and secondary infection, the next characterization is *opportunistic infection*, which like relapse and recurrent, needs to be distinguished from a secondary infection. An opportunistic infection indeed can also be a primary or a secondary infection, but special circumstances distinguish it from your garden-variety secondary infection. Unlike the typical secondary infection just discussed, like a secondary bacterial sinus infection in a battlefield of a viral sinus cold, an opportunistic infection by definition takes the "opportunity" to violate a host who is compromised for underlying reasons. The compromise can be immune system related or combat zone related or whatever. The other qualifying characteristic is that, in opportunistic infections, the pathogen is one that ordinarily does not commonly cause infection in humans. This is key.

The poster child for opportunistic infections is pneumocystis pneumonia in patients with AIDS. Pneumocystis pneumonia is not a common infection in healthy bodies, but a patient immunocompromised by AIDS—meaning a drop in the function of CD4+ helper T cells, in which HIV has taken up residence—cannot fight off the yeast-like fungus *Pneumocystis jirovecii*, the pneumonia's source pathogen. Mycobacterium tuberculosis can target an AIDS patient, but that would not be an opportunistic infection since TB targets pretty much anyone. But its cousin *Mycobacterium avium* does not target just anyone—it targets the immunocompromised, and as such occurs in AIDS patients as an opportunistic infection with much zeal. I'd be remiss to leave you with the impression opportunistic infections only target AIDS. To the contrary, they take the opportunity to target other immunocompromised folks, cancer patients, those on chemotherapy, uncontrolled diabetics, long-term steroids for another ailment—all these scenarios weaken the immune system war footing.

The most common setup that leads to an opportunistic infection is indeed a compromised immune system as just outlined. Occasionally, however, a compromised battleground will be the culprit, such as a viral pneumonia laying the injured foundation for a bacterial pneumonia to grab hold. In burn patients, the burnt skin is more susceptible to odd bacterial infections that don't normally infect the skin.

There are a gazillion bacteria in our gut, and most are terribly friendly lads, chomping away at food we cannot otherwise digest. Other intestinal bacteria friends synthesize vitamin K which we absorb and use for making blood clotting factors. Some of our gut flora bacteria actually act as sentries keeping in check sinister bacteria hell-bent on causing trouble. But occasionally, the friendly gut flora gets all messed up, like after a long course of antibiotics. And when that ensues, it is an opportunity for an otherwise held-in-check friendly bacteria to be opportunistic, and none perhaps more notorious than *Clostridium difficile* bacterial infection of the colon. Often called *C. diff*, which appears on the heels of an altered gut flora due to antibiotics, it can if not treated promptly and aggressively sap the life out of a person. *C. diff* is most often caused by certain antibiotics altering gut flora balance yet paradoxically it is treated with other antibiotics—sort of fighting fire with fire.

Is does bear mentioning that other previously discussed secondary infections can also be classified as opportunistic infections, such as yeast vaginitis in a woman on antibiotics for a bladder infection; this is both a secondary infection and an opportunistic infection, they're not mutually exclusive. But the real key to understanding opportunistic infection is that the pathogen involved infrequently causes human infection. Not to put too fine a point on it, but an opportunistic infection is the result of an opportunistic pathogen taking the opportunity to be opportune.

Which brings us to the last, somewhat obvious, set of classifications for infections: acute, chronic, and carrier state. An *active infection* really doesn't need much explanation. It is just that: active. It might be primary, it might be secondary, it might be opportunistic, but it is also active.

Chronic infection is also just that: an infection that won't go away. It lingers, but that lingering is or can be by design, it is what the pathogen does, like malaria. Being designated a chronic infection can be either an infection that should have gone away but didn't, like chronic sinusitis or an infection that by its very nature tends to smolder, like TB. And some chronic infections are really acute infections that are taking their sweet patooties to resolve, like a lingering flu.

Still other chronic infections result when the pathogen finds a place to hide, a place under the radar of the immune system and out of reach of antibiotics, where it can continue to smolder and is not dormant. If it instead went dormant and then woke up, that would be a relapsed infection.

Then there are those chronic infections that, as mentioned above, by their very nature, by design, are chronic—they come to roost and have no plans to leave anytime soon. Classic examples of chronic infections include parasites such as malaria and some viruses like HIV, the avatar for chronicity, as well as hepatitis C and genital herpes (remember: sex lasts but for a moment; herpes lasts forever). And then of course there's those wormy miscreants—tapeworms, flukes, roundworms, and hookworms—which are all basically parasites, and most parasite infections are chronic.

A *carrier state infection*, which we briefly touched on earlier, is where the patient is not ill from the pathogen but can transmit the pathogen. It doesn't meet the definition of a chronic infection because there is no infection. A famous example of this carrier state is Typhoid Mary, who lived in Long Island, New York, in the early 1900s. Although I keep referring to Typhoid Mary, I am sorry to say the exploits of Mary won't be fleshed out until *The Second History of Man*.

Perhaps it's now time to recap the several classifications of infection, appreciating that one infection might fit into several categories. I forget them too, so it is worth rehashing here.

The most obvious classification is according to the type of heartless pathogen that plague humans: *bacteria*, *virus*, *protist* (amoeba), and *fungus*.

Next up is the mode of transmission, which we discussed many pages ago, and which includes *droplet contact*, as in inhaling coughed viral droplets floating in the elevator you just stepped into; *fecal-oral* (ew) like cholera, typhoid, and the cruise ship gastrointestinal norovirus; *oral*, as in swallowing a wormy parasite or the kissing infection mono; *sexually transmitted disease*, as if that needs further explanation; *contact*, like with athlete's foot from locker room floors; *transfusion*, as in from a contaminated blood bank; *iatrogenic*, which means it came from the healer to the patient; *vertical*, an infection passed mother to fetus; and, finally, *vector*, those infections like malaria and yellow fever delivered to our doorstep by insects and their ilk.

Then there are the various presentations of infections. One category parses *clinical*, *subclinical*, and *latent* based on symptoms. Another category parses *primary*, *secondary*, and *opportunistic*. And yet a third form of presentation interprets *acute*, *chronic*, *relapse*, and *recurrent*.

That about does it. Or does it? We have journeyed this far, and I've really not given a good explanation of the difference between contagious infection, also termed communicable infection, and noncontagious infection. A *contagious infection* means it spreads directly, or sort of directly, from person to person. The cold and the flu are the poster children for contagious infection. HIV can be contagious (person-to-person transfer) and noncontagious (use of a contaminated needle or a transfusion of tainted blood).

Noncontagious infections include such examples as the Plague and rabies. You cannot "catch" it from someone who has it. You have to acquire it from a nonhuman source: the flea for the Plague and a rabid animal for rabies or breathing in fungal spores. Of course, the Hollywood film *World War Z* (2013) starring Brad Pitt turned the viral scourge rabies into a human-to-human transmission, where infected rabid zombies bite a person, who then turns into a rabid zombie and goes ahead and bites another person. It's a fun film, though it takes quite a few liberties with infectious diseases, especially with the speed with which a victim turns into a rabid zombie. In *World War Z*, this process takes about ten seconds which is of course ludicrous, even most poisons don't work in ten seconds, let alone the rabies virus! Such a virus would need several days to morph a bitten human into a foamy-mouthed zombie creature.

"Let me guess...it's contagious!"

When you think of contagious infections, just think of the mode of transmission and you'll likely have your answer, usually. Contagious infections are transferred person-to-person within reasonable proximity of time and distance, to include droplet (sneeze/cough), fecal-oral, STD, vertical placenta, contact (touching a fomite surface), and oral (mono) transmission modes. The noncontagious infections are transmitted by transfusion, dirty needles, vectors, oral (swallowing a parasite), contact (as in athlete's foot off that smarmy locker room floor) and of course breathing in a noncontagious microbe, like fungal spores or anthrax.

To give an examples of how an infection might be parsed into these various categories, let's take the common cold. It is an acute, contagious, primary, clinical, viral infection, hitting five categories. If a bacterial sinus infection develops during a cold, that infection is classed as an acute, noncontagious, secondary, clinical, bacterial infection.

But we have one more set of classifications to cover. Infections can be further defined as outbreak, epidemic, pandemic, and endemic. Oddly, as we will find out, those same four categories can apply to noninfectious, noncontagious diseases as well.

An *outbreak infection* is the sudden appearance of an infection at a specific time and location. Its occurrence is not anticipated, making it sporadic. Outbreaks are generally confined to a small population and are usually self-limiting. A classic outbreak infection is intestinal food poisoning, which will hammer one group of people, such as when bacterial *E. coli* gastroenteritis ravaged members of a recent church picnic who ate Aunt Hilda's famous—and now infamous—egg salad, or a hepatitis A outbreak originating from the busboy of a single restaurant.

An outbreak infection can develop into an epidemic infection—well, Aunt Hilda's egg salad can't, but little Johnny's preschool cold outbreak certainly can. The term "outbreak" can also apply to noninfectious medical illnesses, such as from poisons. In 2001, a neurotoxin outbreak caused by perfluoroisobutylene—used to make other industrial chemicals—was traced back to contaminated dialysis machine filters.

Perhaps one of the most famous outbreak infections was the July 1976 Legionnaires' disease, which first appeared as a result of the American Legion convention in Philadelphia. It's so famous because, prior to that convention, Legionnaire's infection had not been identified—it was a novel outbreak. Of the 2,000 attendees of the convention that hot muggy July week at the Bellevue-Stratford Hotel—where air-conditioning units were running wild due to the humid heat—182 people became ill and a staggering twenty-nine died.

Legionnaires' disease, as it has become known since its maiden appearance, is a bacterial pneumonia characterized by shortness of breath, cough, high fever, and rather impressive muscle aches and pains.

The initial illnesses from that 1976 American Legion convention, and the subsequent deaths, actually did not begin to present until well after the convention was over, after all the legionnaires had returned to their scattered homes, setting up for what could have become an epidemiologic nightmare. Fortunately, one doctor, Ernest Campbell, an astute physician located in Bloomsburg, Pennsylvania, realized that three of his patients, who had all died from some novel pneumonia within a few days of each other, had been at the convention the previous week. He connected the dots and contacted the Pennsylvania Department of Public Health, which then contacted the national sleuths at the Centers for Disease Control and Prevention, who then went to work. Cases started popping up along the Eastern Seaboard, clustered in the Philadelphia area, and all were Legion conventioneers. A previously unknown bacterium was subsequently isolated, as was the source for that outbreak: contaminated filters in the hotel's air-conditioning system. The bacterium was formally named *Legionella pneumophila* and the infection Legionnaires' disease.

A rather bizarre example of a noninfectious outbreak is the Jamaica ginger paralysis outbreak of the 1930s. Jamaica ginger, also called jake, was a nineteenth-century medicinal brew that contained 70 percent alcohol, and in the 1930s was sold in the US as a way to bypass Prohibition laws. Jake was marketed as a medicinal concoction—not an adult beverage, which is what it really was—to treat upset stomach, headaches, motion sickness, morning sickness, general nausea and vomiting, arthritis, and a few other maladies. Ginger does, after all, possess natural anti-nausea properties. And at 70 percent alcohol, I imagine after sipping jake a person couldn't have cared less about *any* symptom they were experiencing.

The Jamaica ginger imported for sale contained the compound tri-o-tolyl phosphate, intended to hide the concoction's bitter taste. This additive turned out to be the not-so-brilliant idea of two American bootleggers. When the bootleggers ran out of natural ginger oleoresin flavorant, they substituted a synthetic version—tri-o-tolyl phosphate tastes similar to ginger, you see, but it is also a neurotoxin.

The neurotoxin adulterant improved the taste of jake but resulted in an abnormal stride in those who drank it. This bizarre ambling became known as Jamaica ginger paralysis, characterized by an accompanying *tap-click*, *tap-click* gait sound, also known out in the streets as Jake leg or Jake foot. Within a few months, the chemical cause of the Jamaica ginger paralysis outbreak was identified, and the offending additive removed from the concoction—but not before affecting some 50,000 Americans, many of whom never recovered. Not sure what happened to those two enterprising bootleggers who added the neurotoxin. They likely didn't drink their own product.

An *epidemic infection* is the rapid spread of a microbe to a larger population in an adjoining or nearby geographic location over a short period of time. One of the first identified infection epidemics occurred in 430 BC, when a sweeping infection, likely typhus, burned across Greece, killing 100,000 people. In 1713, a measles epidemic spread throughout all thirteen British colonies in America, but not the French territories to the north, nor the Spanish territories of Florida, nor west beyond the Appalachians.

I illustrate this last epidemic—there were many before it, many since—so as to gain an appreciation of how an epidemic contagious infection might and might not cross borders. In other words, why were the French Canadians and the Spanish Floridians spared? It's simple really: folks didn't move around like we do today, jet-setting across the country, across borders, across international waters. In 1713, pretty much only merchants, and maybe a few vacationers, moved about the colonies. Bostonians might have ventured to New York and New Yorkers to Philadelphia and Philadelphians to Charleston, but not much outside British America. The measles like the colonials who had measles kept within assigned geographic region.

In 1910, a plague epidemic took the lives of 40,000 people in a specific district in China, but it didn't spread outside that province. In 1972, a flu epidemic swept the United States, but it did not venture into Canada or Mexico, which is a tad odd, because by the 1970s Americans were certainly vacationing north and south of the border.

A *pandemic infection* is an epidemic infection that has crossed borders, has crossed oceans, and is spreading globally. As I sit here, editing these words, the world is in the throes of the COVID-19 pandemic. Lucky us.

The most notable pandemic infection in history is, of course, the Black Death, an eight-year scourge, from 1346 through 1353, caused by the *Yersinia* bacterium, which was transmitted to humans from the bite of fleas living on rats. All told, the Black Death reduced the population of Europe by 25%, some areas in Europe by 50 percent and worldwide killed an estimated 200 million people, reducing the global population by nearly 25 percent. All in just eight years. Of course, the Black Death involved an intermediary, the flea, so it was not person-to-person. A whopper of a person-to-person pandemic infection was the 1918 Swine Flu that swept across the world, killing between 50 million and 100 million over a 2 ½ year span.

Returning to the fourteenth century plague, without understanding how disease was spread—other than according to Galen's miasma theory—the finger-pointing blame game during the Black Death grew more frenzied as the corpses piled up ever higher. Between good old-fashioned superstition, a stunningly incorrect understanding of infection, bigoted religious intolerance, and breathtakingly treacherous ignorance, the supposed causes of the fourteenth-century Black Death pandemic covered the gamut of human fears: punishment by God for who knows what, the alignment of celestial bodies (notably Mars, Jupiter, and Saturn), and of course the Jews and Gypsies, easy targets back then to blame anything and everything on. Pogroms against Jewish settlements during the Black Death were swift and vicious. Entire communities were wiped out. And I don't just mean homes were burned to the ground—I mean entire populations were murdered.

Why were the Jews targeted by other Europeans as the cause of the Plague? At the top of the list was age-old anti-Semitism, which was all the rage during medieval times. You know the story: the Jews were blamed for the death of Christ, those Christians not appreciating the salient fact Jesus had to die in order to redeem humanity and in order to be resurrected. A point lost on some Christians. Never did understand that duplicitous dichotomy. Second, despite living in the Jewish ghettos of Eastern Europe, which you think would be overrun and foul by the definition of "ghetto," it was instead the case that Jewish laws about cleanliness and bathing—the kosher stuff—kept the Plague-carrying fleas at bay. No one ventured into those ghettos carrying the plague, Jews did not venture out of their ghettos, and kosher law kept things relatively clean. It seemed to many gentiles that the Jews had been spared by the pandemic, but suspiciously so, not for the real reasons.

Finally, the Black Death was brought into Europe along the Silk Road, that merchant trade route out of China and into Europe. The Plague bacteria did not originate in China or the Far East, but within the steppes of Central Asia. As merchants traversed the steppes, the fleas harboring the *Yersinia* bacteria joined their merchandise, crossing the Bosporus into Europe. Jews, however, did not covet, did not buy or desire, the silk and spices from the Far East, so Plague-infested merchants and merchandise simply did not enter their ghettos. Jewish communities dodged that Black Death–bullet.

So, when the Black Death was steamrolling its way across Europe, Jews did not die in droves. This gave rise to what in logic theory is called "an affirmative conclusion from a negative premise": since Jews were not dying, they must have caused the Plague. Like all good epidemic vermin, the *Yersinia* bacteria needed a population in which to extend their lineage. Jewish ghettos in Eastern Europe were precisely not that: they were isolated communities where few ventured in, and still fewer ventured out. And where laws of kosher were strictly followed. The Black Death, like the angel of death before it, most often skipped over Jewish homes. For that, Jews were singled out and murdered.

Pope Clement VI, a Benedictine monk from France, did his best to protect the Jews, or so they say, issuing two papal bulls stating, basically, that the Black Death was the devil's work, intended to turn Christians against the Jews. Pope Clément's decrees were partially undone by Holy Roman Emperor Charles IV of Czechia, who decreed Jews' property title forfeit should they die from disease or be killed in a riot, giving incentive for people to run roughshod over Jews just to claim their property. As the Plague ran its course and began to subside in 1351, so did the Jewish pogroms and persecution.

In more recent times, relatively speaking, that 1918 Swine flu pandemic (sometimes called the Spanish flu) was also profoundly devastating—one of the granddaddy pandemics of all times. A viral infection, the Swine flu ungraciously emerged in the world toward the end of World War I. Although also called the Spanish flu, it's not proven it actually began in Spain. In the first twenty-five weeks of the pandemic (a mere six months), some twenty-five million people died worldwide. Digest that arithmetic for a moment: one million people died per week during the first six of the 1918 Spanish flu pandemic. No corner of the earth was spared, and when the scourge had run its wretched three-year course, it had taken the lives of between 50 and 100 million people worldwide.

In comparison, World War I, sometimes called the Great War, a conflict that ran four years, from 1914 to 1918, witnessed the death of seventeen million humans. Tragic to be sure, but it's still no comparison to the damage wreaked by the Spanish flu that followed on its heels. When comparing number dead over period of time, the 1918 Swine flu was about six times more lethal than WWI.

An *endemic infection* is an infection that sort of smolders in a certain geographic region, and the numbers of infected go up and down slightly over the course of years, but the microbe persists more or less unperturbed. Malaria is a classic endemic infection in Africa. Caused by a parasite transmitted to humans by the bite of a contaminated mosquito, malaria has been endemic in Africa for perhaps 100,000 years. Another endemic disease in Africa is African sleeping sickness, a parasite infection caused by a *Trypanosoma* species and transmitted to humans by the bite of the tsetse fly. Then there is yellow fever, a viral infection also transmitted to humans by the mosquito and is endemic in Africa. Those annoying mosquito vectors transmit quite a few infections to humans, and the infections they transmit tend to remain endemic.

Just as the term "outbreak" can be applied to noninfectious diseases, such as an accidental industrial or environmental exposure or Jake foot, "epidemic" and "endemic" can likewise be applied to noninfectious diseases. Black lung disease, also known as coal workers' pneumoconiosis or anthracosis, is the result of coal miners inhaling anthracite into their lungs. Anthracite is the hard-as-rock version of coal. Black lung disease reached epidemic proportions in the United States in the mid-1900s, and since it was confined to certain coal mining regions, it was both endemic as well as epidemic. With the Federal Coal Mine Health and Safety Act of 1969, the incidence of black lung disease steadily declined, as miners were now protected from anthracite inhalation.

Some epidemiological reports suggest diseases like diabetes have reached epidemic proportions in the US due in no small part to the wide availability and high consumption of sugary foods and drinks, which we humans find hard to resist. Foods that are bad for us but are so enjoyable. My personal Achilles' heel is chilled chocolate.

However, not all diabetes is a result of lifestyle choices. Type 1, or insulin-dependent, diabetes is distinguished from type 2 diabetes, as type 1 is a congenital, or inherited, illness and it appears to be an autoimmune disease; it is not the result of a person indulging in sugar, although that will certainly make type 1 diabetes worse. While its exact cause isn't known, the current theory is that insulin-dependent diabetes type 1 results from the person's immune system attacking the cells in their pancreas that make insulin. Insulin is a hormone needed to regulate serum sugar levels as well as few other things, such as the synthesis of lipids and fat storage.

Type 2 diabetes, by contrast, is often developed from years of dietary indiscretions, such as overconsumption of sugar and fatty foods coupled with a sedentary lifestyle and obesity. However, type 2 diabetes also runs in certain families, so there appears to be a genetic component there that cannot be ignored. Regardless of whether it is a result of lifestyle choices or an inherited disease, those with a predisposition to type 2 diabetes should be able to stave off the disease, or at least manage it, through healthy lifestyle choices.

And speaking of lifestyle choices, ultimately adults are responsible for their own health and healthcare decisions. Interestingly, when it comes to health-related issues women reign supreme. Not only do females take better care of themselves than their male caveman husbands or boyfriends, but women often make the healthcare decisions for their husbands, children, parents, even in-laws. Women make the medical decisions and guys control the TV remote. Seems fair.

Frustrated by Jim's reluctance to take his high blood pressure medicine, Tracy took things into her own hands.

Although it is comparing apples and oranges, Gurney #3 presented to the Dog House ER not quite as ill-looking like Deborah must have appeared in those few days prior to New Year's Eve. Gurney #3 came to the ER by personal transport, her husband drove her since she really was not presenting that sick. She walked into the check-in waiting room on her own energy, checked-in with few signs of discomfort, was evaluated and prioritized by the triage nurse, and then escorted to gurney #3, which would become her station for the next several hours, and which is where her and my paths crossed.

Yet despite a relative paucity of symptoms, something deep inside her must have told her to go to the ER. Despite her feeling she only had the flu, which would not necessitate a visit to the ER under ordinary circumstances, there must have been a voice deep inside telling her something was not quite right. Something intuitive told her to pay a visit to the ER. It was a lifestyle choice, and as it turned out, an important choice. Was something lurking.

When we think of lifestyle choices, what comes to mind are the basic tenets of a healthy life, commonalities such as eat healthy food and don't depart radically from human nature, by overeating, smoking, drinking and sedentary living. But taking care of oneself also includes other healthcare choices, like going to the physician as needed, dealing with emergent diseases and illnesses that crop up, having a willingness to go the distance managing chronic, long-term ailments, and taking prescribed medicines and therapy. It also includes listening to that little voice inside that might be telling you, screaming at you something is amiss.

As for Deborah, in the days leading up to her becoming more ill, she likely had the faculties about her to perhaps make a different lifestyle choice, to go to her doctor, but perhaps not. We do not know; we will never know over such a distance of time if she had a little voice sitting on her shoulder, nagging at her to get checked out, to make a visit to her primary care physician, her PCP, or even make a stop at an urgent care clinic for an evaluation. Sometimes in life, we are so busy with work, with family, with children that we often do not take the time to tend to ourselves. It is like the classic war scene where a soldier is shot and his pal says "you're bleeding" and the soldier retorts "I don't have time to bleed." Patients usually know, usually sense intuitively when they've fallen off the curve. Whether that internal "sense of falling off the curve" actually rises to the level of consciousness, that is, a patient self-aware they have fallen off the curve, or, a subconscious feeling, a bad vibe that drives one to the doctor, makes no difference so long as a visit to the ER is on the horizon.

Unfortunately, sometimes issues with our personality or our beliefs, including and probably especially our false believes snuff out both conscious and subconscious acts of self-determination. Consider this, in the US as the COVID-19 pandemic rages, as I write these words, even from the highest levels of government, cult followers believe, were made to believe, *wanted* to believe that the coronavirus was and is fake. It fit nicely into their false paradigms. I read recently of a nurse from a state in the US, state that paid little attention to CDC recommendations, describing the near final words of a cultish American dying from COVID-19. Paraphrasing, the nurse related this man's near final words, words before he was too ill to even talk: "This can't be happening, the virus is not real." Imagine that, imagine being so deluded that you're denying a virus exists as that very virus saps the last breaths from your life.

In *The Second History of Man* we'll discover how religion might have played a role in the death of Jim Henson, the creator of the Muppets. Like Deborah, Henson presented to an ER rather late in the course of an infection that eventually ended up claiming his brilliant life. Would the same fate await Deborah?

The NHS wants patients to make more decisions about their treatment...so here's your blood tests and a prescription pad, I'll be back later.

As a physician I often ask my patients what they think, what is their gut feeling. Most are glad to offer an assessment, some ramble, some stay on point, but either way it is a great way for a physician to get a sense of which way the wind is blowing, especially for climactic events. Patient's usually know intuitively if they're off the mark. A few patients however often answer politely but innocently or circumspectively: "you're the doc, you tell me." And then I remind those few reticent souls that patients usually harbor a visceral hunch, and it becomes my job to coax that inkling out. If I can.

With Deborah's storyline, whether or not a PCP or urgent care doc would have cued into an early pneumonia during the days leading up to her New Year's Eve demise cannot be assured over such a remote event, over such a distance of time. My best guess is they would have. Consider this, if I gained was able to gain a sense that something was amiss, an ER doc would have been hit over the head like a ton of bricks. Such a clinical presentation that Deborah was churning, like a mid-Atlantic hurricane, would have been difficult to miss, even by that doc-in-the-box physician previously discussed. Just looking at her probably would have been enough to surmise she was tottering. To that you add an examination which would have revealed a fever, junky lungs on stethoscope exam, a blood count with a bump in crime fighting white blood immune cells and a chest X-ray imaging patchy infiltrates in the lungs, characteristic of pneumonia. Collectively with those signs and symptoms a diagnosis of pneumonia could have and can be easily made. Of course the physician, armed with that work-up would then need to look at the overall picture, the forest, not the trees—infirm, ailing—and ascertain the next path to follow: is this an outpatient "walking pneumonia" oral antibiotic scenario or are we dealing with an inpatient need those Avada Kedavra IV antibiotic scenario?

And speaking of that earlier doc-in-the-box story where I took over the shift of that physician mired eyes-deep in a world of hurt, recollect he was poorly dialing a patient on the edge. That doctor was working-up a patient for respiratory and intestinal red herrings, all the while a heart attack was unfolding right before his eyes. He had a bunch of trees—physical exam, lab tests—but was clueless to the forest on fire. He couldn't even smell the smoke. Sometimes you don't even need to ask the patient a single question, don't need to perform a single physical exam nor order a single test. You gaze at the patient, stare at the patient for several moments, you look at the forest from the trees, and the path to follow becomes self-evident, it bubbles to the surface that you are dealing with a patient who has fallen off the curve. Gurney #3 did not present with such an obvious forest fire; I'd like to think that Deborah, had she presented days earlier would have presented with a galactically apparent forest on fire.

Which is where we will turn in *The Second History of Man*, the next book in this series. We will continue with the saga of Deborah and that of Gurney #3, to their eventual conclusions, all the while introducing new characters into our overarching narrative of patients I have encountered. The next volume of this narrative will explore what happens if we *can't* stave off an infection on our own, when our immune system is unable to thwart an invasion, our shields are down—for you Star Trek trekkies—and we need more firepower to eviscerate an invader. These medicines are collectively known as anti-infectives—a catch-all term for the Big Four: antibacterial (more commonly called antibiotic), antiparasite, antifungal, and antiviral.

The Second History of Man will also explore preventive measures for infection, discuss such issues as is hygiene genetically hardwired into us sapiens, cover some specific common bacterial infections of historical significance, throw in several plagues of biblical proportion, and of course, my favorite part, we'll travel into all those jumping off points of human history, those darlings, from Darwin and the Silk Road through Florence Nightingale and the Charge of the Light Brigade to the Rat Pack and Oscar Wilde.

I hope you have enjoined *The First History of Man*. These stories, like a soap opera, are continued—without soap commercial interruption—in *The Second History of Man*.

Printed in Great Britain
by Amazon